"Winchester could probably write circles around most writers on the planet."
—*San Francisco Chronicle Book Review*

"A historical thriller."
—*BusinessWeek*

"A cautionary tale about the kind of disasters we tend not to think about until it's too late."
—*USA Today*

"A riveting account."
—*New York Daily News*

"*A Crack in the Edge of the World* may be the most gripping account of a temblor ever written."
—*Outside* magazine

"IF THERE'S ANYONE QUALIFIED TO WRITE A BOOK ABOUT THE 1906 SAN FRANCISCO EARTHQUAKE AND FIRES, IT'S SIMON WINCHESTER."
—*Entertainment Weekly*

"Winchester's **exhilarating** new book describes what happened that fateful spring morning and, just as crucially, why. It is the **dramatic** story not just of a city's destruction but also of a planet's construction. . . . **Hugely entertaining.**" —Bernard Cornwell, *Mail on Sunday*

"Winchester turns his **formidable storytelling** talents to the earthquake that destroyed San Francisco on April 18, 1906 . . . [but] doesn't lose track of the human toll caused by natural disasters."
—*People* (**four stars**)

"A **richly companionable** volume. . . . **The joys of Winchester's book are many.**" —*San Francisco Chronicle* (a Best Book of the Year selection)

"Winchester's **timely** new book teaches—reminds, really—that we should have quite precise worries about the incurably unstable ground on which scores of millions of Americans live."
—George Will, *Washington Post*

"**[An] important contribution**. . . . Winchester brings his formidable storytelling talents and his passion for history, but also solid earth-science credentials. . . . **A grand and accessible adventure**." —*Science*

"**Brims with colorful detail**—which makes for lively reading . . . **a richly textured, informative, compelling account**."
—*Times Literary Supplement*

"**Spellbind[ing]**. . . . This is nonfiction, but Winchester, also a novelist, utilizes the strategies of fiction: playing with the reader's growing anticipation of the event itself . . . Winchester [is] our geological Cassandra."
—*Miami Herald*

"A virtual guidebook to the world of tremors and earthquakes and the geological conditions that produce them. . . . A **thoughtful and illuminating** book by an author in **impressive** command of his subject."
—*Houston Chronicle*

"A **grand** tour of the world's tectonic plates and the major seismic events through the ages." —*Forbes*

"**Winchester is at his best** in *A Crack in the Edge of the World* . . . **dramatic** . . . fascinating . . . [written in] lovely prose."
—*New Orleans Times-Picayune*

"Without slighting the human suffering of the victims of earthquakes, tsunamis, and other natural disasters, Winchester places their tragedies in an almost **cosmic** context." —*Boston Globe*

"A **wonderful, entertaining** history of the Great San Francisco Earthquake and a history of the development of the field of geography and plate tectonics. **Winchester is a master**."
—*Salt Lake City Tribune* (a Best Book of the Year selection)

"Highly readable, **brilliantly researched**." —*Hartford Courant*

"**[A] near-novelistic narrative** . . . puts the tremendous earthquake into a planet-wide context, while still illuminating the characters at the center of the story." —*San Francisco Weekly*

"**Superb**." —*Sunday Times* (London)

"A **magnificent** testament to the power of planet Earth and the efforts of humankind to understand her. A **master storyteller** and Oxford-trained geologist, Winchester effortlessly weaves together countless threads of interest, making **a powerfully compelling narrative**."
—*Publishers Weekly* (starred review)

"**A must-read**. . . . Winchester's tale excels at unfolding a complicated scientific story as part of a narrative." —*Minneapolis Star Tribune*

"Engagingly, **captivatingly readable** . . . written with a passion and intelligence that makes it compelling as Winchester describes the fragility of our world." —*Independent* (London)

"A **panoramic** blend of history and science that captures the event and its destructive aftermath with **gripping** immediacy." —*Tucson Citizen*

"**Eloquent** . . . a very good book." —*San Jose Mercury News*

"**Fascinating**. . . . Winchester combines a gift for storytelling and a passion for geology to compelling effect." —*Time Out* (London)

"**A story that still resonates**. . . . Winchester puts these past natural disasters in their proper place: in the context of geological as well as human history." —*Denver Post*

"**Truly compelling**." —*Contra Costa Times*

"A **gripping** account of how and why the earth's largest moving parts slip, slide, strike, and subduct as they do . . . fascinatingly described." —*Sunday Telegraph*

"**A great story** in its own right, but juxtaposed against current events, it is **prescient**." —CNN

"A lyrical vision. . . . **Read it and shiver**." —*Herald* (Glasgow)

"**Required reading**. . . . Winchester details the destruction of San Francisco—not just from the quake, but the fires that followed—and the geology behind it." —*New York Post*

"**An illuminating analysis** of the way Western cities cope (or not) with natural disasters. . . . **Fascinating** . . . a seminal story." —*Times Educational Supplement*

"A good writer can make complicated topics easy to grasp, and **a great writer** can make them **entertaining** as well. Winchester personifies this ability. He brings on his journey a storyteller's excitement, making the events and facts, which often include his own experiences, read **more like a novel than a history or science book**." —*Tokyo Daily Yomiuri*

"**A memorable tale**." —*BusinessWeek*

"Arguably **Winchester's best**. In it, Winchester's talents as a writer and journalist are yoked to his training as a geologist in his Oxford days. The effect is **spellbinding**. Thorough, witty, and written with an assurance that comes from the author's knowledge of his subject inside-out, this is popular history at its best. . . . **This book has it all**." —*Edmonton Journal*

A Crack *in the* Edge *of the* World

Also by Simon Winchester

The Meaning of Everything

Krakatoa

The Map That Changed the World

The Fracture Zone

The Professor and the Madman

In Holy Terror

American Heartbeat

Their Noble Lordships

Stones of Empire

Outposts

Prison Diary: Argentina

Hong Kong: Here Be Dragons

Korea: A Walk Through the Land of Miracles

Pacific Rising

Small World

Pacific Nightmare

The River at the Center of the World

A Crack *in the* Edge *of the* World

America and the Great California Earthquake of 1906

Simon Winchester

Harper Perennial

New York • London • Toronto • Sydney

HARPER PERENNIAL

Grateful acknowledgment is made for permission to reprint the following copyrighted materials: Page ix: Excerpt from "Carmel Point," copyright 1954 by Robinson Jeffers, from *Selected Poetry of Robinson Jeffers,* by Robinson Jeffers; used by permission of Random House, Inc. Page 157: Lyrics from "San Andreas Fault," by Natalie Merchant, copyright © 1995 by Natalie Merchant/Indian Love Bride. Page 231: Excerpt from "Once by the Pacific," in *The Poetry of Robert Frost,* edited by Connery Lathem, copyright 1928, 1969 by Henry Holt and Company; copyright 1956 by Robert Frost.

An extension of this copyright page appears on page 463.

A hardcover edition of this book was published in 2005 by HarperCollins Publishers.

FIRST HARPER PERENNIAL EDITION PUBLISHED 2006.

Designed by Elliott Beard
Maps on pages x–xiii and 276–77 by Laura Hartman Maestro
Maps on pages 60–61, 75, 80, 168, 183, 190, and 260 by Nick Springer

The Library of Congress has catalogued the hardcover edition as follows:
Winchester, Simon.

A crack in the edge of the world: America and the great California earthquake of 1906 / Simon Winchester.—1st ed.

p. cm.

Includes bibliographical references and index.

ISBN-10: 0-06-057199-3 (alk. paper)

ISBN-13: 978-0-06-057199-3 (alk. paper)

1. San Francisco Earthquake, Calif., 1906. 2. Earthquakes—California—San Francisco—History—20th century. 3. San Francisco (Calif.)—History—20th century I. Title.

F869.S357W56 2005

979.4'61051—dc22 2005046009

ISBN-10: 0-06-057200-0 (pbk.)
ISBN-13: 978-0-06-057200-6 (pbk.)

06 07 08 09 10 ❖/RRD 10 9 8 7 6 5 4 3 2 1

With this book I both welcome into the world
my first grandchild,

Coco

and offer an admiring farewell to

Iris Chang

whose nobility, passion, and courage
should serve as a model for all,
writers and newborn alike.

Now the spoiler has come: does it care?
Not faintly. It has all time. It knows the people are a tide
That swells and in time will ebb, and all
Their works dissolve.

ROBINSON JEFFERS, *"Carmel Point,"* 1954

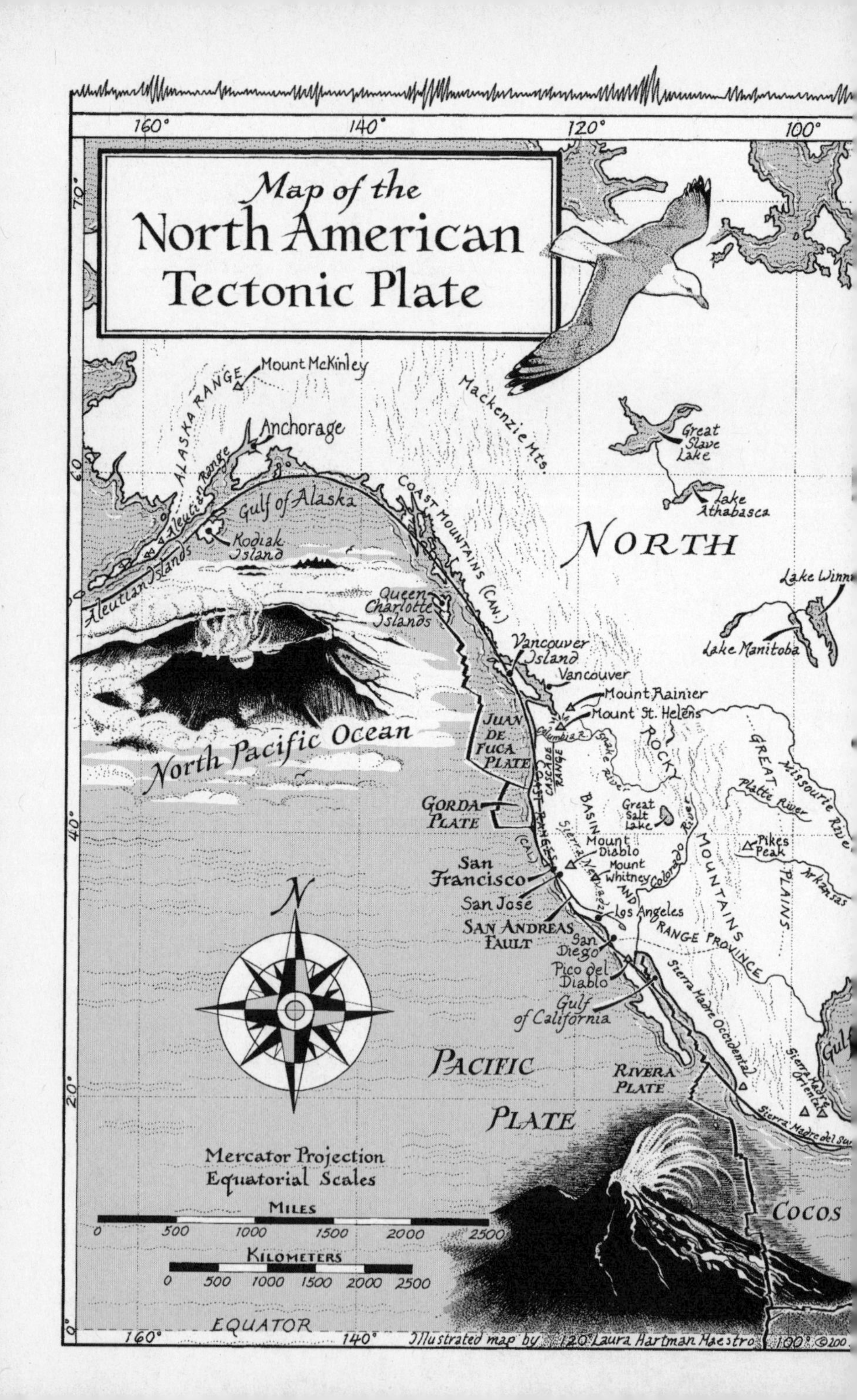
Map of the
North American
Tectonic Plate
160°
140°
120°
100°
70°
60°
40°
20°
0°
Mount McKinley
ALASKA RANGE
Anchorage
Aleutian Range
Gulf of Alaska
Kodiak Island
Aleutian Islands
Mackenzie Mts.
Great Slave Lake
Lake Athabasca
NORTH
Lake Winn
Lake Manitoba
COAST MOUNTAINS (CAN.)
Queen Charlotte Islands
Vancouver Island
Vancouver
Mount Rainier
Mount St. Helens
Columbia R.
Snake River
ROCKY
GREAT
Missourie River
Platte River
JUAN DE FUCA PLATE
CASCADE RANGE
COAST RANGES (CAL)
North Pacific Ocean
GORDA PLATE
BASIN
Great Salt Lake
Sierra Nevada
Colorado River
Mount Diablo
Mount Whitney
Pikes Peak
Arkansas
PLAINS
San Francisco
San José
SAN ANDREAS FAULT
Los Angeles
AND RANGE PROVINCE
MOUNTAINS
San Diego
Pico del Diablo
Gulf of California
Sierra Madre Occidental
Sierra Madre Oriental
Sierra Madre del Sur
N
PACIFIC PLATE
RIVERA PLATE
COCOS
Mercator Projection
Equatorial Scales
MILES
0 500 1000 1500 2000 2500
KILOMETERS
0 500 1000 1500 2000 2500
EQUATOR
Illustrated map by Laura Hartman Maestro ©200

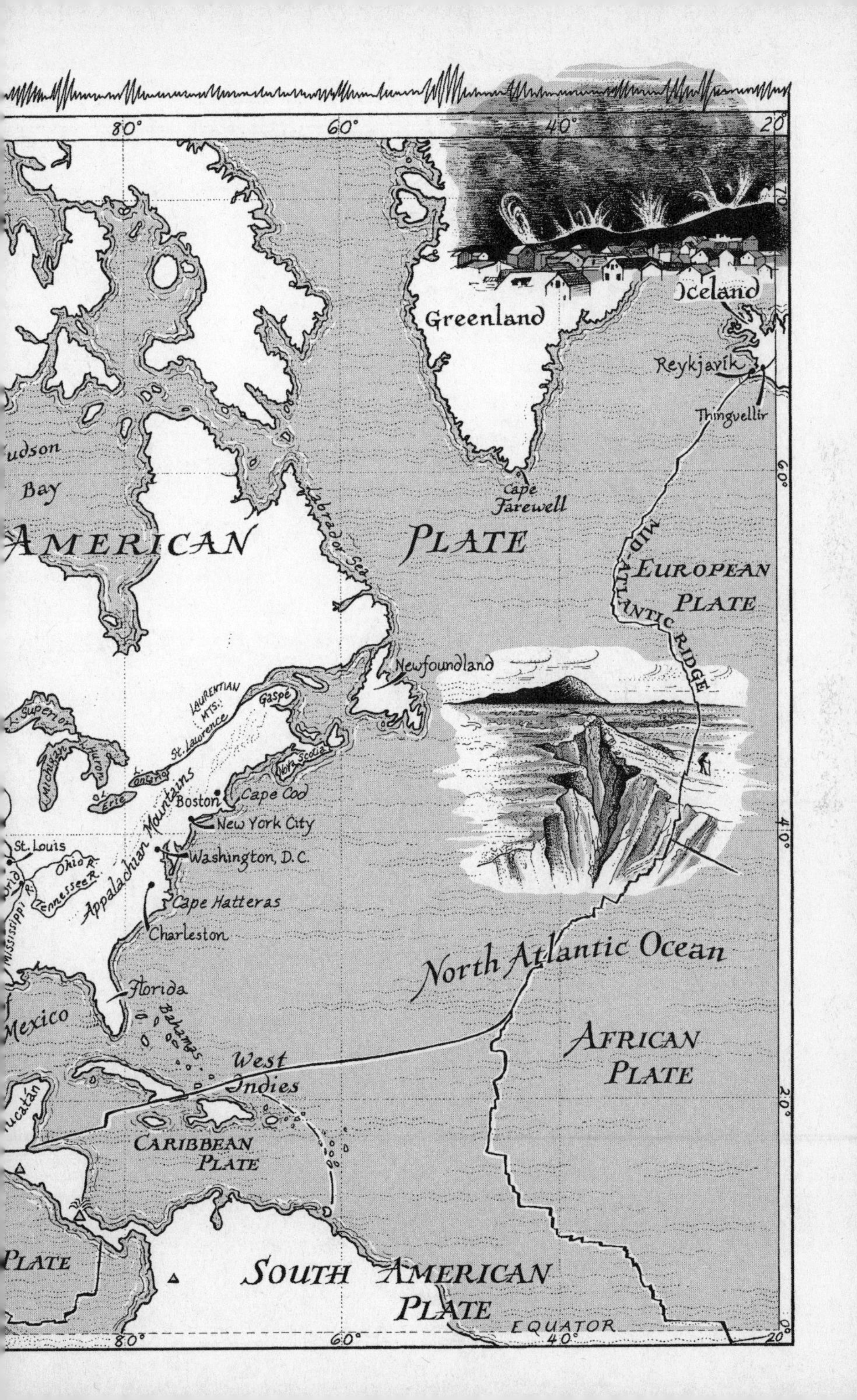

80°
60°
40°
20°
70°
60°
40°
20°
0°
Greenland
Iceland
Reykjavík
Thingvellir
Cape Farewell
udson Bay
AMERICAN
PLATE
Labrador Sea
MID-ATLANTIC RIDGE
EUROPEAN PLATE
Newfoundland
LAURENTIAN MTS.
Gaspé
St. Lawrence
L. Superior
Michigan
Huron
Ontario
Erie
Nova Scotia
Appalachian Mountains
Boston
Cape Cod
New York City
St. Louis
Ohio R.
Tennessee R.
Mississippi R.
Washington, D.C.
Cape Hatteras
Charleston
North Atlantic Ocean
Florida
Bahamas
Mexico
AFRICAN PLATE
West Indies
ucatán
CARIBBEAN PLATE
PLATE
SOUTH AMERICAN PLATE
EQUATOR

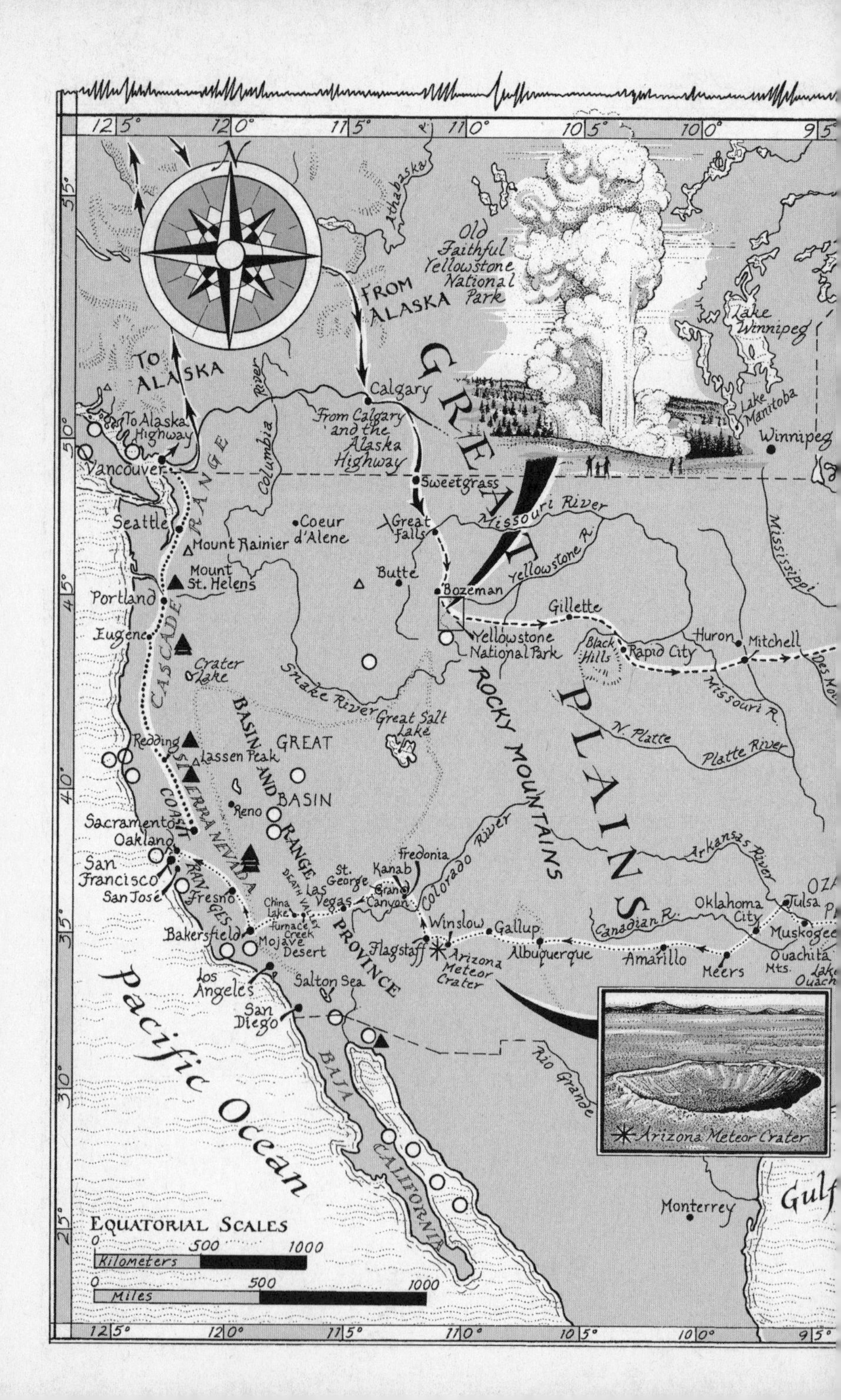
Old Faithful Yellowstone National Park
FROM ALASKA
TO ALASKA
GREAT PLAINS
ROCKY MOUNTAINS
Athabaska
Lake Winnipeg
Lake Manitoba
Winnipeg
Calgary
From Calgary and the Alaska Highway
To Alaska Highway
Vancouver
Columbia River
CASCADE RANGE
Sweetgrass
Missouri River
Mississippi
Seattle
Coeur d'Alene
Great Falls
Mount Rainier
Mount St. Helens
Butte
Bozeman
Yellowstone R.
Portland
Gillette
Eugene
Yellowstone National Park
Black Hills
Rapid City
Huron
Mitchell
Crater Lake
Snake River
Great Salt Lake
Missouri R.
N. Platte
Platte River
BASIN AND RANGE
GREAT BASIN
Redding
Lassen Peak
SIERRA NEVADA
COAST RANGES
Reno
Sacramento
Oakland
San Francisco
San José
Fresno
Colorado River
Arkansas River
Fredonia
Kanab
St. George
Las Vegas
DEATH VALLEY
Grand Canyon
China Lake
Furnace Creek
Bakersfield
Mojave Desert
PROVINCE
Winslow
Gallup
Flagstaff
Arizona Meteor Crater
Albuquerque
Canadian R.
Oklahoma City
Tulsa
Muskogee
Amarillo
Meers
Ouachita Mts.
Los Angeles
Salton Sea
San Diego
Pacific Ocean
BAJA CALIFORNIA
Rio Grande
Arizona Meteor Crater
Monterrey
Gulf
EQUATORIAL SCALES
Kilometers 0 500 1000
Miles 0 500 1000
125° 120° 115° 110° 105° 100° 95°
55° 50° 45° 40° 35° 30° 25°

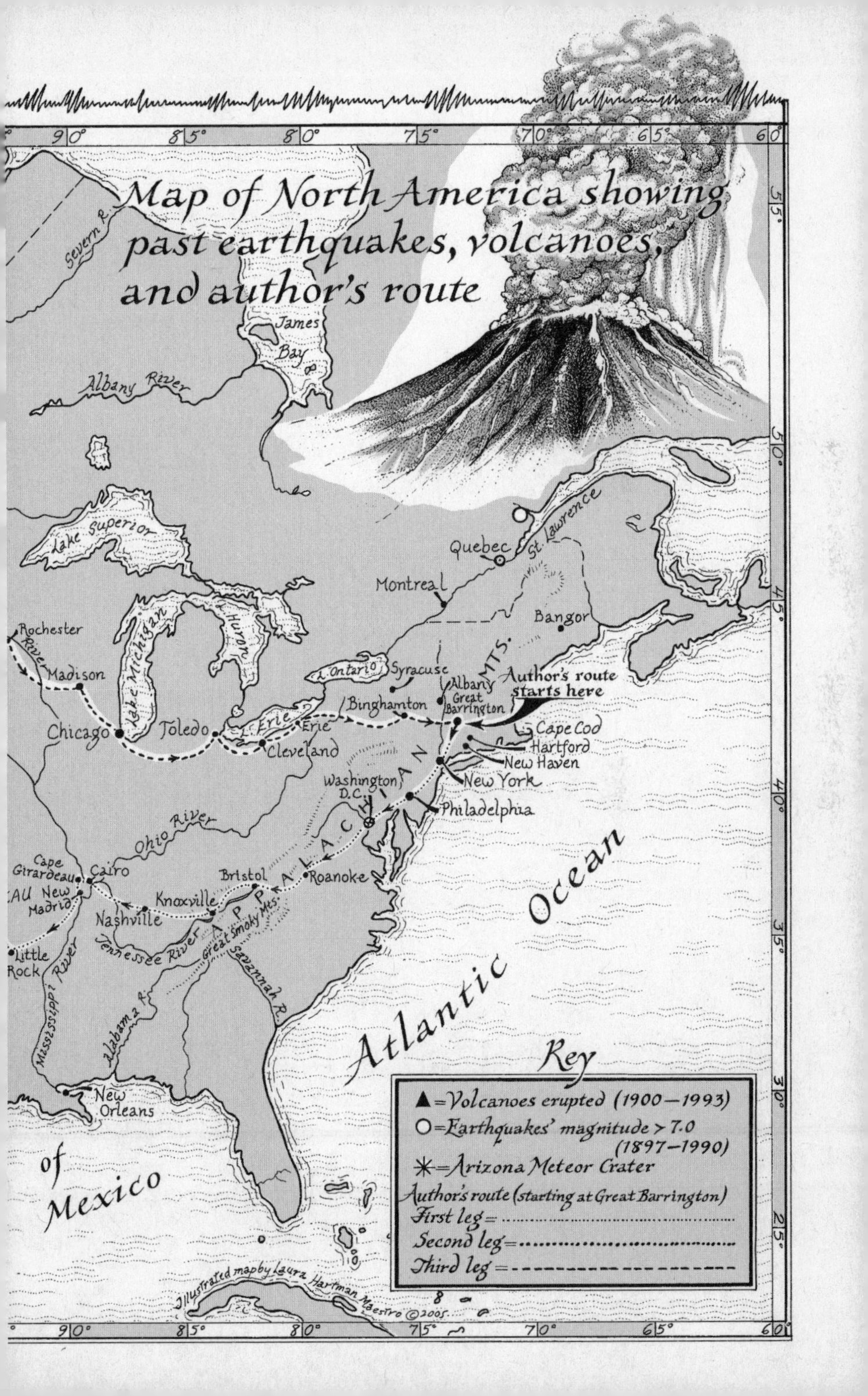

Map of North America showing past earthquakes, volcanoes, and author's route
James Bay
Albany River
Severn R.
Lake Superior
L. Michigan
L. Huron
L. Ontario
L. Erie
St. Lawrence
Quebec
Montreal
Bangor
Rochester
Madison
Chicago
Toledo
Erie
Cleveland
Syracuse
Albany
Great Barrington
Binghamton
Author's route starts here
Cape Cod
Hartford
New Haven
New York
Philadelphia
Washington, D.C.
APPALACHIAN MTS.
Ohio River
Cape Girardeau
Cairo
New Madrid
Bristol
Roanoke
Knoxville
Nashville
Great Smoky Mts.
Tennessee River
Little Rock
Mississippi River
Alabama R.
Savannah R.
New Orleans
Atlantic Ocean
of Mexico
Key
▲=Volcanoes erupted (1900–1993)
○=Earthquakes' magnitude > 7.0 (1897–1990)
✳=Arizona Meteor Crater
Author's route (starting at Great Barrington)
First leg=
Second leg=
Third leg=
Illustrated map by Laura Hartman Maestro ©2005.

CONTENTS

MAPS

ILLUSTRATIONS

A Crack *in the* Edge *of the* World

PROLOGUE

The created World is but a small
parenthesis in Eternity.

Sir Thomas Browne, 1716

O wad some Pow'r the giftie gie us
To see oursels as others see us!

Robert Burns, *"To a Louse," circa 1785*

The Well-Illumined Earth

Some while ago, when I was half-idly browsing my way around the Internet, I stumbled across the home page of an obscure small town in western Ohio with the arresting name of Wapakoneta. It rang a distant bell. Once, very much longer ago, I had passed by the town on what I seem to recall was a driving expedition from Detroit down to Nashville. But, so far as I remember, I didn't stop there, not even for a cup of coffee. It only struck me at the time as being a rather attractive name for a town—a name that was (I subsequently read) a settler adaptation from a word in the language of the local Shawnee Indians.

The town these days is nothing too exciting—which is what one might expect of a place that lies just off that part of the Eisenhower Interstate Highway System known as the I-75, not very far from the rather better-known and quintessentially midwestern Ohio city of Lima. It has some 10,000 inhabitants, and the way in which it was built, ordered, and settled a century or so ago makes it very similar to uncountable other cities found between the bookends of the Rocky Mountains and the Appalachians.

It is, in other words, a classic example of the modern Middle American community. A place Sinclair Lewis would have favored. A place of unexceptional ordinariness, known locally for the making of light machinery, car parts, and rubberware, and surrounded by large and generally family-owned farms where soybeans and corn are grown, and where hogs are raised. Reading between the lines, one can perhaps detect the faintest tone of fretfulness: a concern for the town's future, born of such newfangled developments as the spread of manufacturing to Mexico, the outsourcing to Asia of much of the service economy, and the drumbeat growth of China. No doubt wishing to encourage new businesses, the chamber of commerce makes a claim for Wapakoneta that is shared with many other towns similarly unburdened by excessive splendor: that by virtue of its strategically important location, with all the roads and railway lines that run nearby, it is something called "a transportation hub."

It is a town with a past built on the bedrock of America's previous success, a present that clings by its fingernails to its own notion of stability, and yet a future in which the old Ohio bedrock seems not quite as firm as had initially been supposed, one that most people in consequence do not care to ponder too closely.

However, those who expect Wapakoneta to be only blandly Middle American, and perhaps a little unadventurous and dull, might be surprised to find another side to its history. The astronaut Neil Armstrong, born in the town in 1930, went to the local high school and, quite rightly, no one will let you forget it. (Only two other luminaries of the town are thought worthy of mention, and both are by contrast memorably forgettable: one a heavily mustachioed hero of the Civil War who fought at Vicksburg; the other the screenwriter of *The Bells of St. Mary's,* who also happens to have invented a device allowing naval vessels to lift mines harmlessly from the seabed.)

The town's Web site is where all this is so serendipitously revealed. It opens with a scratchy sound recording of an unidentified baritone reading a launchpad countdown. He follows this with the announcement of the *liftoff,* in July 1969, of the Apollo 11 spacecraft, a ship that is destined, he says gravely, *for the moon.* And while his voice is intoning on what turns out to be an original NASA recording, an image of the moon swirls and grows steadily bigger on the screen—until it is eventually replaced, with a booster's flourish, by an image of a bustling community and, in bold type, the name of the town: Wapakoneta.

It is fitting that this small town should celebrate so eagerly the exploration of space: The worldwide excitement over the samples of lunar rock brought back to earth is just one small indication of the value, in real scientific terms, of America's having sent up a man to get them. But there was an unanticipated and less obvious consequence of the expedition, the effect of which has been, in many ways, rather more enduring.

For it appears now that one field of scientific discovery was changed forever by the journeyings of Neil Armstrong and all those others who have gone to the moon in the years since. This sea change has come about specifically in the science of geology, and it is a change that has its origins in a very simple fact. When Wapakoneta's first citizen was teetering gingerly about up there on the moon, he was able to do something that had never been done before, and that provided science with a profound, paradigm-shifting moment of unforgettable symbolism: He was able to stand on the lunar surface and *look back at the earth.*

To be sure, astronauts who had gone into orbit in the years beforehand were also able to see the totality of the planet; but there was something wholly remarkable in being able to stand upright on one world and gaze back at another, more than 200,000 miles away.

The great American biologist and philosopher Lewis Thomas wrote in 1974 of the symbolic importance of humankind having this new perspective:

> Viewed from the distance of the moon, the astonishing thing about the earth, catching the breath, is that it is alive. The photographs show the dry, pounded surface of the moon in the foreground, dry as an old bone. Aloft, floating free beneath the moist, gleaming, membrane of bright blue sky, is the rising earth, the only exuberant

thing in this part of the cosmos. If you could look long enough, you would see the swirling of the great drifts of white cloud, covering and uncovering the half-hidden masses of land. If you had been looking for a very long, geologic time, you could have seen the continents themselves in motion, drifting apart on their crustal plates, held afloat by the fire beneath. It has the organized, self-contained look of a live creature, full of information, marvelously skilled in handling the sun.

Five years later a British chemist and environmentalist named James Lovelock, thinking along much these same lines, used the moon view of the earth to advance a long-considered idea he called the Gaia hypothesis. The idea—which he christened with the ancient Greeks' name for the earth goddess, Gaia or Ge, and which has been rechristened as the even more plausible-sounding Gaia theory, now that his supporters believe so much of it has been proved—holds that the earth in its totality is very much a living entity. It is alive, it is fragile, and everything that is in it preserves a complex balance with everything else in a state of mutually beneficial equilibrium. It so happens, to the dismay of many present-day scientific philosophers, that humankind's current disharmonious behavior is affecting this careful balance; there is a growing feeling that it must be changed, radically and soon, if life on earth is to continue and to flourish.

This is not an environmental book by any means. It is, more simply, the story of one remarkable and tragic event that befell California a century ago, when a 300-mile-long swath of the earth briefly shifted, wrecking the cities that lay atop it. But, though it is not intended to be a Gaia book, it seems right to tell the story of the events that so ruined the city of San Francisco in 1906 within the *context* of the Gaia idea. There is, for a start, an interesting synchronicity at work: At the moment when Thomas and Lovelock were putting forward their ideas (in the late 1960s, at the same time as the beginning of space travel and, in part, of course, because of it), the geological sciences were also changing very profoundly, as we shall see.

Neil Armstrong was able to gaze across the quarter million miles that separate these two small planetary bodies and look directly toward that area of America from which he had come, at the hills and valleys of the selfsame rocks where he had grown up—rocks that established geology and the fossil record tell us belong to the Silurian Age. And I have no doubt that it was in large measure because of this most extraordinary vision—extraordinary both for Neil Armstrong and, in time, for the rest of us, too—that the birth was signaled of what is now coming to be regarded as an entirely new science. It was a science that was born and then helped to its feet quite simply by virtue of this new perspective that Neil Armstrong's view, even though it had been long anticipated by those who sent him and his colleagues into space, would shed on our planet.

What he saw—and what we saw through his eyes, which we now perhaps take somewhat for granted—was a thing of incredible and fragile beauty. It was a floating near-spherical body, tricked out in deep blue and pale green, with the white of polar ice and mountain summits, with great gray swirls and sheets of clouds and storms, and with the terminator line, that divides darkness and light seeming to sweep slowly across the planet's face as it turned into and out of the sun. It was a lovely aspect to contemplate. And it was a view that in time compelled humankind *to take stock.*

"To see oursels as others see us," as Robert Burns had written. Here and now, all of sudden, we realized that we could do just that—and, with this unanticipated ability to do so, something about us suddenly changed. Almost overnight, and essentially because of this new worldview to which we had access, we discovered a whole raft of new reasons to ponder the oldest of age-old questions: just where we stood in the celestial scheme of things, what the universe and its creation might mean, and how the very earth itself may have first come into being. And such ruminations led, in short order, to the makings of the scientific revolution—and, most specifically, to the geological revolution—that is central to this story.

A Born-Again Science

Like alchemy and the medicine of the leech and the bleeding rod, the Old Geology is a science born long ago (most formally in the eighteenth century): one that, unlike so many of its sister sciences—chemistry, physics, medicine, and astronomy—never truly left the era of its making. Since its beginnings geology has been a field mired in some alluvial quagmire, defined by dusty cases of fossils, barely comprehensible diagrams of crystals, and the different kinds of breaks that were made in the earth's surface (as well as by unlovely Continental words like *graben, gabbro,* and *graywacke*), and explained with cracked-varnish wall roller charts showing how the world may have looked at the time of the Permian Period. To me it remains the most lyrical and romantic of the sciences; but in terms of glamor, and when compared with astrophysics or molecular biology, the Old Geology is somewhat wanting.

The New Geology is, on the other hand, a creature fashioned wholly from the science of the space age, from the attitude that was born when Neil Armstrong first looked back and gazed at the earth. It is a science that now presents us with an entire canon of new ways in which we might look at this planet and at our stellar and solar neighbors.

It seems to me quite fair and proper that the principles of this new science should underpin everything that follows: the terrifying and extraordinary event that enfolded the small but fast-growing western American city of San Francisco one twilit California morning in the middle of April 1906.

Many other scientific disciplines that are revolutionary and dauntingly modern—cosmology, genetic engineering, quantum mechanics—have been formed or founded in recent years, and had no past to hold them back. But geology is different. It is a very old science indeed and hugely proud of its origins: Portraits of the bearded ancients of its founding priesthood invariably hang in esteemed positions in departments from Anchorage to Adelaide. Its antiquity, however, has long been a problem for it, one that has tended to inhibit too many of its practitioners from escaping the glutinous hold of its earliest ideas. Students who remember measuring the umbos of brachiopods or trying to fathom the mysteries of recumbent folding can reach through the centuries and join hands with students who were taught the same topics at the time of George IV and President John Adams. It was only when the professors happened to mention in more modern classes such wonders as the K-T boundary event, with the massive dinosaur extinctions that were mysteriously triggered at the end of the Cretaceous Period (perhaps by a monstrous collision with an immense asteroid), that geology as taught seemed, briefly, to come alive.

Now, however, thanks to a number of recent developments—space travel being one of them, the most spectacular but in terms of science not the most important—geology has suddenly and seriously changed, and at a pace so rapid as to bewilder and astonish all who come up against it anew, or return to it after a while away. It is probably fair to say that never before has any long-existing science been remodeled and reworked so profoundly, so suddenly, and in so short a time. Wholly unimagined visions and possibilities allow us to contemplate our planet in brand-new ways. These means have evolved right before our eyes, and, to the less prescient among us, they have done so well-nigh invisibly and, moreover, in rather less than half a century.

Thanks to the attitudes and instruments and scientific philosophies of the new science, all the events of great geological moment—with chief among them the earthquakes and volcanoes that so plague humankind—can now be seen and interpreted in an entirely fresh context, and in a manner that had rarely before occurred to those who practiced the confusing and cobweb-bound older science with which (from memories of school and university) we are still so vaguely familiar.

IT WAS NOT NEIL ARMSTRONG'S venture alone that brought about this transformation. It is fair to say that geology flowered as rapidly as it did because at almost the exact same moment as the rockets started to soar up through the stratosphere from their bases in Florida (and from the cosmodromes in Baikonur—for this new perspective was one offered to Russian scientists too, of course) something else occurred. A previously little-known professor in Toronto (a man whose very ordinary surname—Wilson—might have kept him marooned in the shadows forever, had not one of his given names—Tuzo—been so strange) drew up the foundations of an entirely new geological subdiscipline, the now all-too-familiar theory known as plate tectonics.

Plate tectonics and space travel each burst onto the world stage at the same time—plate tectonics becoming fully developed by 1967, manned lunar exploring getting under way in 1969—and it is this that led to the unprecedented evolution of the science that was common to both. I shall try to explain the more relevant details of plate tectonics later in the story; but in essence it was a theory that also happened to encourage its believers to stand back, as it were, just as Neil Armstrong was doing at that moment. Plate tectonics allowed us—compelled us, even—to view the world as a complete entity, for the first time to look and to see *the earth entire.*

For it should be remembered that every single one of those Old Geologists—the tweedy figures who, with hammer and lens and acid bottle, had explored and observed and thought and written since the days when it was first realized that the earth is actually very old and that rocks are laid down with some natural purpose and that no deity had anything much to do with the actual manufacture of the planet—found their evidence for the theories and principles of the Old Geology in the rocks, fossils, faults and minerals that were scattered around simply and solely *on the surface of the earth.* They made crucially important discoveries, true; they laid the foundations for this most elemental of disciplines, true; but they did so by examining only the topmost layers—or at most the topmost few miles of thickness, if you will—of the planet.

And that, it is now realized, was a very limiting way indeed of conducting the science—a science that, after all, should more properly be concerned with the nature and history of the earth *in its entirety,* and not with its surface alone. Before the 1970s we had knowledge about the earth's outer cover and not much more. What we wanted to know involved, if we thought about it, much, much more. We wanted to know—and geology was, in its theoretical essence, established purely so as to enable us the better *to* know—about the earth as a whole. And when the intellectual revolution of the sixties came about, we started swiftly to understand that up until that point we had, quite literally, only been scratching the surface; we had never considered the earth as it truly deserved to be considered.

It promptly started to dawn on those sixties geologists who had listened to Tuzo Wilson or his acolytes, or who had seen the spacecraft pictures, that it was somewhat misleading for a science to draw conclusions about the earth entire by examining only those minor features that occurred upon, or just beneath, the planet's outer covering. A fault in Scotland or the relic of a volcano in Montana or the succession of types of trilobite that had been found buried in a shale high on a hillside in British Columbia—such things might be interesting in and of themselves, but only when they were viewed in the context of the big picture, of the planet as a whole, were they able to offer up evidence that allowed the whole-earth portrait to be inked in and made to look something like complete.

So this, then, lies at the heart of the New Geology. The world is these days viewed by most as one entire and immense system, the most refined of its details all interwoven with the biggest of big concepts. It is a living system four and a half billion years old. In a purely physical sense it is an entity warmed up from inside by radioactive decay, with fragments of its fairly recently cooled crust moving about on top of its more mobile inner self, and with solid rocks that have formed (or are still forming) on or beside these fragments creating continents or the floors of oceans. These rafts of solid rock have since been (or are still being) folded or lifted or broken apart as the plates on which they ride move about until they collide and bounce and dive beneath one another. In places, the rocks rise up to great heights; these are eventually eroded, causing the formation of sediment. A geological cycle of creation and decay continues, endlessly. And meanwhile there is life, almost in global terms a brief irrelevance; animals and plants evolve and disappear by turns on the various wet or dry surfaces of the planet according to a series of complex sets of rules that have been laid down by the practical realities of tectonics, of temperature, of pressure, and of almost limitless quantities of time.

The finer details of these things have been studied for decades—such arcane niceties as the suture lines of ammonites (by which one can determine the species and subspecies of this particular beast, which floated gently about in the Mesozoic seas), or the varying degrees of sphericity of the ooliths in a Jurassic limestone, or the patterns of those parts of bivalved creatures that are inelegantly known as muscle scars. But now, in the light of the whole-earth, big-picture view of the science of which they are so infinitesimal a part, they seem tangential to the broad realities of the New Geology, as the pores in an elephant's skin do to a biologist or the volume of sap that courses through the leaves of a live oak from San Antonio does to a forest botanist.

Which is not to say that such things are unworthy of our fascination. Small pieces of puzzles can often lead to grand ideas: The beaks of the Galápagos finches, after all, led Charles Darwin to his big notions about natural selection, the origin of species, and evolution. But it is important to remember that Darwin had at the time all of what was known of earth's biology at his intellectual disposal—every beak and claw, every feather and fin was there, and his journeys took him to far and remote parts of our planet, so that he saw and thought about evidence from all manner of perspectives. When he sat down to write and think at his desk in Down House, he had an immense and almost unimaginable accumulation of information available to him, the finches' beaks being just a scattering of tiles from the great mosaic of biological knowledge.

But, by contrast, geology, at least before the 1960s, was able to lay out before its practitioners only the tiniest portion of available information—very little more than the superficial, the minute, the peripherally relevant. And then, in the nick of time (for without it, where would geology have gone?), everything altered: Along came the astronauts and the unmanned satellites and the space-born magnetometers and gravimeters and mass spectrometers and ion probes, and along came J. Tuzo Wilson and a whole army of like-minded tectonicists. They, combined with the new way of looking at the earth, taught the Old Geological community that there was much, much more to know—and what was once merely a hunch, an inner feeling, became a settled idea. It became abundantly clear that very few grand theories could actually ever be derived from minutiae such as ammonite suture lines and oolith sphericities and relative umbo sizes alone, except forensically; and that nowadays the grand geological ideas are the ones that truly matter.

The View from On High

And seen in that great and glorious context is the earth of the Californian morning of what Western Christian mankind had chosen to call April 18, 1906. Had any geologist at the time been able to look down at the planet in its entirety and witness what took place then, he would at the very least have been utterly amazed by the physical context of the event, even if the event itself, when viewed from on high, appeared less than overwhelming.

For, as context, the planet would have been memorably beautiful. Had he been standing on the moon, say—had he been a 1906 version of Neil Armstrong, scanning with a hugely powerful telescope the surface of the blue and green and white ball that was hanging in his ink black sky—he would have seen illuminated in front of him (assuming that the cloud cover was not too dense) a tract of the world that extended from what some of mankind called India to what others called the Rocky Mountains, all of which would have been bathed in the brilliant white light of sunshine.

He readily could have made out all of Europe and Africa, Asia Minor, and Arabia; he could have seen the deep blue of the Atlantic Ocean, the pure white mass of Greenland to its north, the blinding white immensity of the Antarctic deep below. The corpulent mass of what we now know as Brazil would have been sparkling in the sunlight, with the city-smudged eastern coasts of North America and Patagonia only slightly less so, places peopled with a humanity that was just waking on what many of earth's inhabitants would call a Wednesday, a day that thousands of miles away, in the darkness of China and all points east, was in any case already coming to its end.

At the moment that we find interesting—five o'clock in the morning, give or take—he could have seen the terminator line of western darkness pushing its way rapidly toward the Pacific. The earth would have been moving relentlessly at a speed of some hundreds of miles an hour eastward toward it, opening ever more populated parts of the landmasses to the light of the dawning day.

The line at that very moment would seem to begin in the north near Melville Island in the Canadian Arctic, pass on down through Banks Island and the unpopulated and icebound wilderness of the Northwest Territories and the Yukon, through Saskatchewan and Alberta, raggedly on down through the newly created state of Montana, through the bison-and-Comanche country of Wyoming and Colorado and New Mexico, across the Rio Grande toward Acapulco, and arrive at a point on the coast where it would finally slide off the North American landmass and eventually brighten the still-inky emptiness of the Pacific Ocean.

To the east of the line, all would have been bright and daylight. To the west, an impenetrable dark. And on the line itself, an uncertain penumbra of a few hundred miles of a swath of half dark and half light. On earth this penumbral vagueness would have translated itself into the morning twilights that early risers were experiencing just then in cities and on farms and in small villages all the way from Vancouver Island in the north down to Baja California in the south, where the day designated as April 18 was about to begin.

It is fanciful to suppose that anyone watching so far away, in distance or in time, would have had access to a telescope that was large enough to do the job. But, assuming that such a device did exist, and that the person at this lunar viewing point had its brass and glassware trained precisely on the northern coast of California at that very particular moment, with the terminator line brightening his view inch by inch—what, precisely, would he have seen?

The answer is inevitably dismaying to all of those who like to think that the earth and its inhabitants and the events that occur upon it have any importance at all, in a cosmic sense. For from that distance he would have seen, essentially, nothing.

Yet at a few minutes past five in the morning of that day something did, indeed, happen.

The planet very briefly *shrugged.*

It flexed itself for a few seconds, perhaps a little short of a minute. If our observer had been acutely aware of his geography, and if he had been fortunate enough to have been staring at a very precisely defined spot in the north of California at exactly the right moment, then he would have seen what appeared to be a tiny ripple spurt in toward the coast from the sea. He would, moreover, have seen that spreading ripple as it moved slowly and steadily inshore, and then watched as it moved, fanlike and subtle, up and down the coastline as a tiny *shudder.* It would have seemed to him a momentary loss of focus, something that would have made his vision suddenly blur very slightly, and then just as quickly clear again.

If he had blinked, he would have missed it. Having noticed it, however, he would probably have assumed it was more of a problem with his lens and his telescope than with the surface of the planet below. And even if he had realized that the ripple and the shudder had in fact occurred on the green and blue and white planet that floated serene in the lunar sky, he would have been quick to conclude that whatever it was had been momentary, trivial, and utterly forgettable. No more, for the earth entire, than a gentle and momentary heave of the shoulders.

The Street Before Morning

It was all so very different down on the surface of the planet itself. On earth, in the western part of that great entity called by most English-speakers North America and particularly in and around the fragile and rather delicately constructed young northern city of San Francisco, a number of people grasped all too quickly that something of immense significance was happening. And if they were sensible and observant they took care to note and remember exactly when it all began, and we have their memories set down for us still.

In the city it was a little after five by the local clocks and still not yet light (though some speak of a rosy glow just discernible behind the hill named Mount Diablo, to the east). The air was cold and moist in the way that spring mornings often are in Northern California. But there is a robust heartiness about those who choose to live in this corner of America, and, in spite of the chill and the gloom of this particular morning, a man of middle years, described in the directories simply as a laborer, was already in the sea taking his morning constitutional: This involved swimming through the rough-breaking waters of the Pacific, a few yards off the shore at Ocean Beach.

At the same moment, five city miles away, a young reporter was walking home with two friends, having completed the routine tasks of what newspapermen were in those days starting to call "the graveyard shift." He had stopped on Larkin Street near City Hall to smoke a cigarette and exchange pleasantries with a pair of patrolling policemen, and so further secure these necessary professional connections.

A professor of geology—an immensely eminent man of sixty-three who had been honored around the world for painstakingly exploring and mapping the Rocky Mountains, the Grand Canyon, Death Valley, and a score of other remote and dramatic wildernesses besides—was lying asleep in a room at the Faculty Club at the University of California.

The head of the City Weather Bureau, a future professor at Harvard College, an expert on frost, and at the time an enthusiastic advocate for naming the study of weather "aerography," was also asleep, in his house at 3016½ Clay Street. But, as was his custom, he slept lightly, and kept a flashlight, a watch, an already-date-stamped notebook, and a pencil on the table beside his pillow so that he might be ready for whatever mayhem—aerographic or meteorological or otherwise naturally made—the night might throw at him.

And an elderly English astronomer, the founder of the first real observatory in California, a man of great energy and yet one whose career had precipitated no little controversy and disappointment and who had just retired at the age of eighty-one from the post of professor of geography at the University of California, was lying half awake in his house at 2221 Washington Street, on that fashionable square in Pacific Heights known as Lafayette Park.

AT THE PRECISE MOMENT when the members of this quintet—three of them very distinguished men of science and two others of relatively modest social standing—were undertaking their very mundane activities of swimming or walking or chatting or sleeping or drowsing, with most of them unknown to one another and each certainly unaware of the others' exact circumstances at that second, it was twelve minutes after five o'clock in the morning.

However, this was a matter of provable fact only for the Englishman, so far as the record relates. His name was George Davidson, and he, like his fellow scientists, wrote about the event that was to follow with a certain icy detachment. He took care to mark the time that he first noticed something happening: Suddenly and without warning his room, his house, and the very land all was standing upon began to shake, with a great, ever-increasing, and uncontrollable violence.

It was, he knew full well, an earthquake.

It came, he later reported,

> from north to south, and the only description I am able to give of its effect is that it seemed like a terrier shaking a rat. I was in bed, but was awakened by the first shock. I began to count the seconds as I went towards the table where my watch was, being able through much practice closely to approximate the time in that manner. The shock came at 5.12 o'clock. The first sixty seconds were the most severe. From that time on it decreased gradually for about thirty seconds. There was then the slightest perceptible lull. Then the shock continued for sixty seconds longer, being slighter in degree in this minute than in any part of the preceding minute and a half. There were two slight shocks afterwards which I did not time.

Professor Davidson must have been as terrified as anyone, but he was a man trained to observe, and he knew in an instant what was taking place. So he took painstaking care to note that his watch, as he later reported, stood at 5h 12m 00s. Only he then added the caveat, for safety's sake—and with the sense of caution that was hardwired into his astronomer's mind—that this observation was *subject to an error of plus or minus two seconds.* This reflected, one imagines, any error that he might have made when calculating how long he had spent staggering, his nightshirt awry and his mind still marginally befuddled by sleep, from his bed to the bureau where his watch was ticking and readying itself to slide, along with the pitcher and the shaving mug, onto the redwood floor.

The first full series of hard shocks, say his notes, lasted until 5h 13m 00s. The shocks were slightly less from that point until 5h 13m 30s, then there was a slight lull, and by 5h 14m 30s all was quiet again. The entire event—which was to destroy an American city and leave an indelible imprint on the mind of the entire nation—had lasted for just over two and a half minutes. That, at least, was the considered view of a man so esteemed that three mountains, a glacier, a seamount, an inlet, a bank, and a San Francisco avenue were later named after him. The official report on the earthquake said, in a tone that brooked no dispute, "We shall accept Professor Davidson's time as the most accurate obtainable for San Francisco."

JUST ONE BLOCK SOUTH and eight-tenths of a mile to the west slept the weatherman whose name, still celebrated in meteorological circles (though he had only one mountain named after him), was Alexander George McAdie. A New Yorker, he became a soldier in the Signal Corps after college and made a name for himself by promoting the use of kites rather than balloons for the study of the upper atmosphere (in which signalers were officially interested, since radio waves were affected by what went on there). He became an academic and then joined the government. In 1895, together with his young wife, Mary, he moved out to San Francisco to head the Weather Bureau and to direct the state's Climate and Crop Service, the latter post carrying with it the title of professor. One of his tasks at the bureau was to record, as accurately as possible, any and all seismic events that happened in and around San Francisco.

Professor McAdie was an ambitious and a punctilious man, and at the very moment that he was awakened on that dark and chilly April morning, both his ambition and his scrupulous regard for factual observation—as critical in the world of weather as in the study of the stars—came promptly to the fore. As had been his custom ever since he went through the Great Charleston Earthquake of 1886 ("for twenty years I have timed every earthquake I have felt," he was later to write), the instant he awoke and felt movement he clicked on his flashlight, noted the time on his fob watch, and recorded in his notebook everything that transpired.

> I have lookt up the record in my note-book made on April 18, 1906, while the earthquake was still perceptible. I find the entry "5h 12m" and after that "Severe lasted nearly 40 seconds." As I now remember it the portion "severe, etc" was entered immediately after the shaking.

The only snag was that poor Professor McAdie somehow managed to misread his watch during all the confusion, and he wreathed himself in a magnificent maze of complications as he tried to explain the mistake. He wrote that the day before the earthquake,

> my error was "1 minute slow" at noon by time-ball, or time signals received in Weather Bureau and which my watch has been compared for a number of years. The rate of my watch is 5 seconds loss per day; therefore the corrected time of my entry is 5h 13m 05s AM. This is not of course the beginning of the quake. I would say perhaps 6 or more seconds may have elapsed between the act of waking, realizing, and looking at the watch and making my entry. I remember distinctly getting the minute-hand's position, previous to the most violent portion of the shock. The end of the shock I did not get exactly, as I was watching the second-hand, and the end

> came several seconds before I fully took in that the motion had ceased. The second-hand was somewhere between 40 and 50 when I realized this. I lost the position of the second-hand because of difficulty in keeping my feet, somewhere around the 20-second mark.
>
> However, there is one uncertainty. I may have read my watch wrong. I have no reason to think I did; but I know from experiment such things are possible. I have the original entries untouched since the time they were made.

The official report accepts that the unfortunate man did effect an error in making what was probably the most critical observation of his career—but, out of courtesy, adds that such a mistake would have been very easy to make. The one-minute error is, then, officially compensated for, and Alexander McAdie enters the lists as having, essentially, timed the Great San Francisco Earthquake as beginning at 5h 12m 05s, recorded that it became *extremely severe* at 5h 12m 25s, and noted that it tailed off into bearable oblivion at 5h 12m 50s. The whole event, in McAdie's eyes, extended over little more than forty seconds—about half the time that Davidson had computed, from his observations that were made a little bit closer to town.

NINE MILES ACROSS the Bay in Berkeley slept Grove Karl Gilbert, one of the lions of early American geology and a figure still revered today as one of the greatest scientists of the nineteenth century. He was in the closing years of his career when he arrived at the University of California—appointed ostensibly to investigate whether miners should be given permission to resume the environmentally harsh (and, for half a century, totally banned) hydraulic method of mining, in which incredibly powerful jets of water were played on an exposed rock face to unseat any minerals (gold especially) that might be lodged there. At twelve minutes past five on that Wednesday morning Gilbert was rudely awakened by a sudden fierce vibration. The floors creaked and swayed below him. The light fixture swung in an arc above him—its swing aligned, he noticed, along an imaginary north-south line on the ceiling—and the water in a pitcher on the washstand splashed out on the container's southern side. He, like everyone else, was briefly alarmed, but then that feeling was rapidly overtaken in his particular case by, of all things, pleasure. His account, no doubt written with an eye cocked to posterity, begins as if he had only recently devoured the best-known work by Jane Austen:

> It is the natural and legitimate ambition of a properly constituted geologist to see a glacier, witness an eruption and feel an earthquake. The glacier is always ready, awaiting his visit; the eruption has a course to run, and alacrity is always needed to catch its more important phases; but the earthquake, unheralded and brief, may elude him through his entire lifetime. It has been my fortune to experience only a single weak tremor, and I had, moreover, been tantalized by narrowly missing the great Inyo earthquake of 1872 and the Alaska earthquake of 1899. When, therefore, I was awakened in Berkeley on the eighteenth of April last by a tumult of motions and noises, it was with unalloyed pleasure that I became aware that a vigorous earthquake was in progress.

The net result of Gilbert's enthusiasm was probably much to his liking. Within hours he was co-opted into all of the inevitable investigations and seismic postmortems; his name appears on every masthead of every official publication about the event for the next decade. Such fame as he had already enjoyed—from his writings on topics as diverse as the erosion of Niagara Falls, the recognizable features of the moon's face, the Bonneville Salt Flats, and the philosophy of scientific hypothesis—was magnified a hundredfold by his authoritative account of the San Francisco events; and his reputation was in no way besmirched when he began to speculate (as legions of others have since, often unwisely) on how earthquakes might successfully be predicted.

FRED HEWITT WAS walking home, together with two colleagues, after his night shift as a reporter on the *Examiner.* He lived at 500 Fillmore Street, and so to get home each morning he would turn north up

Larkin Street, cross Golden Gate Avenue, Turk Street, Eddy Street, and Ellis Street, before turning west onto O'Farrell and walking up a steep hill and then down again to the valley, at the bottom of which huddled the small brick houses and shops (now all most fashionable) of Fillmore Street. It was some minutes after five o'clock when he and his two friends crossed Golden Gate Avenue, spent five minutes talking to a pair of policemen—"blue coated guardians" as he later wrote for his paper—and said their farewells. Hewitt had turned north, the policemen back south down Larkin, when suddenly:

> The hand of an avenging God fell upon San Francisco. The ground rose and fell like an ocean at ebb tide. Then came the crash . . . I saw those policemen enveloped by a shower of falling stone.
>
> It is impossible to judge the length of that shock. To me it seemed like an eternity. I was thrown prone on my back and the pavement pulsated like a living thing. Around me the huge buildings, looking more terrible because of the queer dance they were performing, wobbled and veered. Crash followed crash and resounded on all sides. Screeches rent the air as terrified humanity streamed out into the open in agony of despair.
>
> Affrighted horses dashed headlong into ruins as they raced away in their abject fear. Then there was a lull. The most terrible was yet to come.
>
> The first portion of the shock was just a mild forerunning of what was to follow. The pause in the action of the earth's surface couldn't have been more than a fraction of a second. It was sufficient, however, to allow me to collect myself. In the center of two streets I rose to my feet. Then came the second and more terrific crash. . . .

AT THE VERY SAME MOMENT Clarence Judson was taking his early and very cold awakening bath in the Pacific, off Ocean Beach. He lived on Forty-seventh Avenue, less than a hundred yards from the spume and spray of the wild Pacific, and in a part of the city that was still half sand dunes and home to a few hardy souls only, he could happily put on a bathrobe and shoes and stroll to the ocean. Looking back on his short walk, he remained convinced that something was already happening somewhere under the sea. The breakers, he noted, did not advance toward the beach that morning as they usually did, in a series of parallels, each one hitting the strand with a ferocious roar of breaking water (which is what he liked, offering his work-weary bones a cold massage of highly oxygenated spray). Instead the waves were roaring in to shore "cross-wise and in broken lines, in a vicious, snappy sort of rip-and-tear fashion."

Nevertheless, Judson discarded his robe, hat, and shoes, and went into the water, walking out to sea until he was up to his shoulders. Then, surprising but not unduly alarming him, he was hit by one enormous breaker, a wave that he could then see pounding its way high up the beach. It was a freak wave, he supposed, and, while a little odd, was nothing to bother about. There was so much water, and the wave crest was so high, that he was lifted quite off his feet, and a few seconds later the undertow began to pull him out to sea. He was a strong man and a good swimmer. He was still not especially concerned. But then out of nowhere came an immense shock, a silent detonation that cannonaded through the water like an explosion. Even while he was floating in the sea Judson was knocked down, as he described it, to his knees.

> I got up and was down again. I was dazed and stunned, and being tossed about by the breakers, my ears full of salt water and about a gallon in my stomach. I was thrown down three times, and only by desperate fighting did I get out at all. It was a close call.
>
> I tried to run to where my shoes, hat and bathrobe lay, but I guess I must have described all kinds of figures in the sand. I thought I was paralyzed. Then I thought of lightning, as the beach was full of phosphorescence. Every step I took left a brilliant incandescent streak. I jumped on my bathrobe to save me.

Neither Judson the swimmer nor Hewitt the reporter nor the sleeping great explorer Grove Karl Gilbert was in a convenient position that morning to enjoy either the luxury of immediate note taking or the congenial precision of knowing the exact time of day when the crisis struck. But, upon later interrogation, having collected their thoughts and subjected themselves to forensic self-examination, they agreed that twelve minutes past five in the cool dark morning was *exactly* the time of day when each saw, heard, felt, otherwise experienced, or was rudely shaken awake by the singular event that these next pages will try to recount.

ONE

Chronicle: A Year of Living Dangerously

April, April,
Laugh thy golden laughter;
Then, the moment after,
Weep thy golden tears!

SIR WILLIAM WATSON, "April," 1903

SO FAR AS THE ANCIENTS OF CHINA ARE CONCERNED, 1906 was a year of the Fire Horse—a time of grave unpredictability that comes along every six decades, and a time when all manner of strange events are inclined to occur. So to the seers and the hermits in their faraway mountain aeries such events as unrolled during the year would have come as no surprise. The rest of humankind was less well prepared, however, and were caught unawares. And what instruments we have agree that, so far as matters of the earth were concerned, 1906 was, yes, a very bad year indeed.

At least it was bad seismically speaking, being a very violent and a very lethal year. And the flurry of activity that marked what the numbers show to have been among the most ill behaved of times of the entire century began in the morning of the last day of January, when there was an enormous earthquake under the seabed of the Pacific Ocean.

It is said today to have been the greatest and most powerful earthquake that had until that moment ever been registered by the machines of humankind, and it struck a score of communities along the South American coast, devastating towns, inundating fields, and causing huge waves to tear out into the open ocean. Its shaking lasted for more than four minutes, and as many as 2,000 people are thought to have died in the disaster. Scores of thousands were injured and made homeless, and countless villages and at least one major port city were totally destroyed. The effects of the huge traveling sea waves from the event were felt as far away as San Diego, and in Honolulu Harbor in Hawaii all the steamboats waiting at anchor were spun around and carried upward on an enormous tsunami, which ebbed and flowed like a tide every few minutes, bringing confusion and alarm in its wake.

The epicenter of this earthquake, whose details are still pored over, is now calculated to have been some eighty miles due west of a prominent headland known as El Cabo de San Francisco, in Ecuador.

The town that was all but destroyed—but which has since been rebuilt, only to be damaged many times subsequently—was the island port of Tumaco, now a prominent oil terminal. But in 1906 it was a place where fishermen brought in sizable catches of tuna and sardines, and where traders hawked bales of rubber and pallets of cinchona bark, ready to be pressed for quinine. Tumaco is some thirty miles north of the Ecuadoran frontier, in Colombia.

Both Ecuador and Colombia suffered grievously from the earthquake, and even today people in the villages by the mangrove swamps of the estuaries speak fearfully of the morning when several hundred miles of their coastline, from the port of Guayaquil in the south to Buenaventura in the north, were devastated by the power of the water and the four minutes of ground shaking. Seismologists working in the 1930s, when Charles Richter* created his scale of magnitude, esti-

---

*Richter, who spent most of his career in Pasadena, at the California Institute of Technology, was a somewhat unusual man; an avid nudist and vegetarian who, to judge from his correspondence and his diaries, enjoyed a prodigious sexual appetite. More detail on his famous scale, together with an explanation of earthquake magnitudes and intensities, can be found in the appendix.

mated that the Ecuadoran-Colombian Earthquake of 1906 had a magnitude of 8.4, as high as anything then known; new calculations today suggest an even greater magnitude, of anything approaching 8.8—as bad a disaster as could possibly be imagined, whether it rated 8.4 or 8.8 or somewhere in between, for the two young republics struggling to their feet.

But the earth wasn't done yet. Sixteen days later there was another very large earthquake, this time on the island of St. Lucia, one of the four specks of Caribbean limestone, sand, and coral that make up what was then the British crown colony of the Windward Islands. According to interpretations of the damage data made in the 1970s, it rated somewhere between VII and VIII on the magnificently named Medvedev-Sponheuer-Karnik Earthquake Intensity Scale. Just as with the Ecuadoran event of the month before, this February earthquake had its epicenter in the sea too, somewhere off the northeastern tip of St. Lucia, and about twenty miles south of the French possession of Martinique.

This event, which collapsed buildings on both St. Lucia and Martinique, and which was felt by the populations of other islands in the eastern Caribbean, including Dominica and Grenada and St. Vincent, did not kill anyone. But it triggered a burst of smaller earthquakes—probably a swarm of so-called volcanic earthquakes, which tend to occur when spurts of magma force their way up into the earth's upper crust, after the crust has been weakened by a deeper earthquake that has been caused by the movement of tectonic plates. This wave of lesser earth movements went on for two or three weeks, and for a while the placid life of an island whose people produced, according to the Colonial Office report of the time, a heavenly confection of "sugar, rum, cocoa, coconuts, bananas, bay oil, bay rum, spices and sea island cotton" was dangerously interrupted. The colonial governor, who had his headquarters in Grenada, was alerted, and a Royal Navy warship was dispatched from the squadron in Bermuda. Assistance was offered, assessments were made, and St. Lucia was from that moment on formally designated an earthquake-prone territory, risky enough to be of note but not sufficiently dangerous to be abandoned.

Still it was not over. Five days later a tremendous outbreak of ground shaking occurred in Shemakha, an ancient town of mosques and temples in the Caucasus Mountains, and a place, given its location, that was long accustomed to seismic happenings. This was not an especially large earthquake—not by normal Caucasus standards, at least—and there seem to have been no reports of deaths. The same could not be said, however, of a truly enormous quake that ripped through the island of Formosa four weeks later, on March 17. This historic event, known variously as the Chiayi or the Meishan Earthquake, tore along a nine-mile fault in the west of the island, displacing the land on each side of the fault by six or seven feet horizontally and three or four feet vertically. At least 9,000 houses were destroyed, 2,000 islanders injured, and no fewer than 1,228 people killed. The Japanese, who ten years before had taken control of the island from China, organized a formidable rescue operation; but the fact remains that this was one of the largest and most terrifying quakes to have struck Taiwan for many years, and for some weeks following the disaster the situation overwhelmed all efforts to contain it.

And then Vesuvius erupted. For ten terrifying days, beginning with a cannonade of rocks that was hurled 40,000 feet into the air above Naples on April 6, the only volcano on the European mainland underwent its most severe eruption for 300 years; some vulcanologists at the time said it may even have been greater in drama and strength than the legendary eruption of A.D. 79, when Pompeii and Herculaneum were destroyed by rivers of gas and glowing ashes.

In 1906 just 150 people are thought to have died. The villages of Bosco Trecase, San Giuseppe, Ottajano (the destruction of which, said a newspaper of the time, "appalled the civilized world"), Poggiomarino, and Somma were all covered in several feet of ash, and some of them had to be hastily abandoned. When the market in the town of Oliveto collapsed under the weight of hundreds of tons of eruptive material, scores of shoppers were trapped inside, with dozens killed. Moreover, the very shape of Vesuvius was drastically changed by the explosions. The summit crater's edges were shaved to an almost perfect horizontality—the shape the mountain has to this day, in fact. Beforehand it had been ragged and untidy—with cliffs more than 1,000 feet high around the steaming, smoking, and seemingly bottomless crater itself.

It was not until April 16 that the explosions subsided and the eruptions stilled. The seismometers that had been measuring this extraordinary and, in terms of its power, nearly unprecedented display fell silent later that night. After ten days of malevolent unpredictability, the needles on the instruments that were monitoring matters in southern Italy all suddenly ceased their vibrating, at last. This was the beginning of the week, a Monday, and, with the damping down of Vesuvius, there seemed some small reason to suppose that the worst might be past. To more than a few on that evening there was, no doubt, a sense of relief—a sense that perhaps the world had now done its worst, and that it would lapse into a steady quiet once again, reverting to that blessed state in which the rocks stay where they are, the earth calms itself, and peace returns.

That night was quiet. All of the succeeding Tuesday passed without incident. The world and its wife slept peacefully in their beds through much of Tuesday night.

But on the Wednesday morning everything would suddenly change. The surface of the earth was to be ripped apart yet again—and in an instant it became abundantly clear that there would be no time for relief, relaxation, or reflection. Another earthquake was to strike, and, in terms of lives lost and structures ruined, this would be the worst, by far.

Half a world away from Italy, on the cool morning of 18 April of that same very bad year, seventy-seven days after the earthquake in the Pacific off Ecuador, which might have signaled to some that all was not well with this quarter of the world, the earth in San Francisco and in a score of places nearby suddenly went dramatically and terrifyingly berserk.

And though it is specifically San Francisco's tragedy that forms the core of this account, so far as 1906 was concerned there was yet more to come. Before the year was done a further 20,000 Chileans would die when an enormous quake, of magnitude 8.3 at least, struck the port city of Valparaiso in mid-August, essentially wiping it out. The restlessness of the earth seemed to know no bounds. Ruin was being visited on humankind on a titanic scale. The destructive appetite of an uneasy planet seems from this vantage point to have been almost inexhaustible.

The figures collected since show with perfect clarity that 1906 was indeed the most dreadfully active of years. Only twice in the twentieth century were there more big earthquakes in any twelve-month period—but those of 1906 happened, by chance, to strike at great cities and so killed tens of thousands of people, rendering the year the most seismically dangerous of the century.

Much the same now appears to have happened in 2004, by far the most dangerous year of the young century that has followed. It began with an immense earthquake in Iran; a series of shocks and volcanoes then shuddered all around the world for much of the twelve months following; and then the catastrophic Sumatran tsunami struck precisely one year after it all began. Geologists looking at the statistics have lately started to wonder if some cruel butterfly effect might be at work—a pattern that might permit a ferocious event on one side of the planet to trigger a similar disaster far away on the other. Those who believe in the ideas of Gaia think this might be so: As the plates shifting against one another are all interconnected, jostlings on one part of the planet's surface might well create sympathetic movements elsewhere. Thus far there is no firm evidence—only the numbers, and the anecdotes, that show incontrovertibly that some years are seismically very much more dangerous than others.

And 1906, it appears, was one of the very worst years of all time.

TWO

The Temporary City

Fourmillante cité, cité pleine de rêves
Où le spectre en plein jour raccroche le passant!

BAUDELAIRE, *"Les Sept Vieillards,"* 1857

I FIRST SAW SAN FRANCISCO IN THE EARLY SEVENTIES, at the end of a long westward drive that had taken me clear across the North American continent. The moment I glimpsed it, its cluster of skyscrapers glinting on the far horizon, was on a cool blue late summer's dawn—and it was a sight every bit as breathtaking as I had imagined. And yet somehow, when I look back on that journey, it was the night before that remains more firmly etched in my memory. It haunts me still, I think, because the place where I decided to put up for the last night on the road—a place I had chosen by the purest chance, with just a pinprick on the map—turned out to be so very remarkable, so steeped in Northern California history, lore, and myth, that it managed to put my destination city into a context that might otherwise have taken years for me to understand.

It was the end of a long and tiring day, and dusk was gathering over the peach orchards, the walnut groves, and the asparagus fields of the Central Valley of California. I had been driving westward from a city on the plains for the better part of a week, over prairies, ranges, salt flats, and deserts. All that now lay between me and the vast emptiness of the Pacific Ocean was the fast-darkening silhouette of a very large mountain. It rose up ahead of the car, slightly to my right, its bulk filling the windshield, the brown of its dry August grasses purpling in the coming twilight, its shape broad hipped and brooding, like the arching back of an enormous whale.

The map said it was Mount Diablo, which I first thought came from the Spanish for "Devil's Mountain," and the reference books said it had once been the most important peak in the entire American West. "The great central landmark of the state," someone wrote. It was the summit from which all Californian distances and directions were once formally triangulated and measured—and are still computed, by some, to this day. It was a mountain that everyone in the north of the state could see (at least they could before the invention of the haze-making automobile ruined so many Californian views), and from its summit, it was said, just about all of the north of the state could be seen in return, along with tantalizing glimpses of parts of Oregon and Nevada, too.

I decided I would spend the night there, making my final miles to San Francisco in the morning. I was in no particular hurry, and besides, I had never seen my destination city from afar. From up there, or so it appeared from the map, in the morning I ought to have a spectacular view.

I turned off the freeway and made my way through the corona of newly built bungalows sprouting around the mountain's flanks—settlements like Clayton, Concord, and Walnut Creek, large and quite respectable Starbucks-and-Saks suburbs these days, but rather more sparse outposts back then. I threaded my way through a maze of back lanes, and soon the houses thinned and the road became red earth. I saw a gate and beyond it a small hut, from which stepped a ranger, all freshly pressed khaki, with a peaked hat and a gun. He issued me a ticket and a yellow chit that allowed me to pitch my tent. He asked if I might like to buy some firewood, which the state park authorized him to sell. He told me I was the last one in for the day. But then he

The great landmark peak of Mount Diablo, with a rare winter dusting making it visible for scores of miles. At the summit is a beacon, illuminated only on the anniversary of the Japanese attack on Pearl Harbor.

stepped back into his hut and riffled through his log. In fact not just the last, he said; I was the *only* person staying on the mountain that night. It was a Wednesday, and in the middle of the week few people bothered to stay. So now he'd be closing the gate and going home. I'd be all alone on Devil's Mountain, with no risk of anyone disturbing my sleep until the park opened again at dawn. Plenty of animals, though, he added. Most of them friendly enough. But it would be peaceful, that'd be for sure.

The road wound steeply up the north side of the mountain, and, with the sun sinking fast behind the low ranges of the distant coast, the sycamores and the pines, the junipers and the great old oaks, were cast into sharp relief, their shadows ever lengthening on the meadows and, as I climbed higher, against the canyon walls. By the time I had found somewhere to set up camp, in a glade floored with a carpet of soft pine needles, well above the 3,000-foot line—the mountain's summit is 3,850 feet above sea level—the night had entirely closed in, and I had to use the car's headlights to show me where I could hammer in the tent and lay the fire for dinner and tea. While I was cooking, a small battalion of raccoons marched in out of the night, their eyes blazing; I lobbed a stone at one of them to stop him from stealing what little food I had left. It hit him square on the nose, and he left howling. Neither he nor his friends returned.

I turned in early and found it difficult to sleep. I may have been fretting over the vengeful potential of members of the raccoon family. Might they come in angry droves and try to hound me from their turf? But whatever it was that kept me awake that night, I do not remember minding much. I had a canvas bag of books with me, and a gas lantern. It seems to me today, so many years later, that much of what would eventually come to fascinate me about California—its extraordinary history, its phenomenal wealth, its lyrically complex topography and geology, and its transcendent loveliness—was learned that night or seen the next morning, things that were concentrated, as tinctures of experience and reality, into those few square miles of rugged upland around this very remarkable mountain.

THE NAMING OF MOUNT DIABLO, for example, has a much more curious complexity about it than might be supposed. Naturally it begins with the Spanish, who in the sixteenth century had extended their empire of New Spain northward from Mexico into what was eventually to be named California.*

One of the Spanish expeditions to the northern interior in the late

*Like "The Seven Cities of Cibola," "Quivira," and "El Dorado," the word "California" is fanciful, an invented name for an imagined utopia. The word was coined by a sixteenth-century writer named García Ordóñez de Montalvo, who wrote a short story about an island filled with gold and diamonds, populated by black Amazons who rode griffins and were ruled by a queen named Califia. From the first Spanish contact in 1542 until an expedition in 1701, the real California was thought to be an island, too. The first physical features to be given the name were

eighteenth century discovered the magnificence of the mountain: "The view from south to north is beautiful, for its end cannot be seen," wrote one young army lieutenant. Shortly thereafter priests opened a mission nearby (that of San José, fifty miles to the southwest, the fourteenth of the eventual twenty-one by which the Spaniards intended to conquer, convert, and "missionize" the native peoples). Farmers arrived from Castile to settle on the mountain's fertile flanks (the town of Concord was founded around this time), and the local governor let it be known that the fields on the lower slopes might be used for winter grazing; and then, by the beginning of the nineteenth century, Spain's policy of subduing the native peoples of the region and bringing them to God got properly under way—with predictably sour initial results.

During the spring of 1806 a platoon of Spanish soldiers were sent out from their small adobe fortress, the Presidio, near the Golden Gate, to the northern side of the mountain, their allotted task to hunt down some local Miwok Indians. These natives were members of a group the Spaniards liked to call the Carquinez—said to be a local expression meaning "trader"—and the soldiers were supposed to welcome them into the civilizing embrace of Madrid as well as Rome (the Spanish, of course, were not in the Americas simply to proselytize on behalf of the papacy). The Miwok were having none of this, however, and after running for some days eventually stood their ground in what was then a thicket of willow and bay laurel trees, at a spot that today is near an airport for the suburb of Pacheco. To frighten or intimidate the Spaniards, the Miwok sent out a medicine man so weirdly dressed and made up as to persuade the Europeans that the Indians were under the spell of the very devil himself. "An unknown personage, decorated with the most extraordinary plumage and making divers movements, suddenly appeared," wrote one trooper, clearly terrified. He and his colleagues promptly took flight, the Indians escaped north across the Sacramento River (the narrows where they did so remain the Car-

---

a gulf (still so named, and into which the Colorado River would flow if it flowed into the sea at all) and a cape, both identified as such on a Spanish map of 1562.

quinez Strait to this day, with two important toll bridges the expensive scourge of commuters), and the returning soldiers reported to their commander back at the Presidio that they had lost the natives in what they in consequence christened "the thicket of the devil"—*Monte del Diablo.*

Later cartographers assumed that the word *monte*—which in this context meant "thicket"—actually identified a mountain, although had the soldiers wanted to describe a mountain they would probably have used the less ambiguous *montaña.* But, in any case, the later mapmakers mistakenly co-opted the name to describe the peak itself—and the error has endured for the better part of two centuries, memorialized in a pile of rock two-thirds of a mile high and now perhaps the best-known wrongly described mountain in the Bay Area. In 1865 the new California legislature tried to change the name to the memorably unromantic Coal Hill; unsurprisingly, in a state for which romance would eventually become a byword, the attempt failed.

Bret Harte, the Albany Yankee who went out west to write about the miners and the Chinese and other exotica of the Gold Rush days (and who would hire Mark Twain and later write in collaboration with him), added spice to the legend with a famous short story. He had come to know the mountain well. His first job after leaving Brooklyn in 1854 was as tutor to the sons of a local farmer, and his fascination with and support for the Indians and Mexicans—which would later render him highly unpopular with the less racially sensitive of the white Californian settlers—was essentially born out of his encounters with members of the Bay Miwok tribes on and around the mountain. The story, which he wrote in 1863, concerns an eighteenth-century Spanish padre who, while walking on the mountainside, is confronted by the devil. The satanic figure warns him of an intolerable and seemingly inevitable future, the loss of Spanish California to the Americans. But, the devil insists, this Yankee invasion can still be reversed—all he has to do is abandon God and adopt the agreeable diversions of Lucifer. The priest demurs; there is a fight. The priest awakens from a dream—and the high hill where all this low drama played out wins a name, the Devil's Mountain, for all time.

Spain, religion, soldiering, rambunctious literature, and the tribes of the Miwok—so much of early California distilled into one small Mount Diablo story. Small wonder that I kept my Coleman lamp guttering, and my pages turning, well into that breezy summer night.

THE DULLARDS WHO WANTED to christen the mountain Coal Hill had a point, however. The rocks of Diablo were, by the middle of the nineteenth century, throwing up all manner of mineral wealth, and not a few people were beginning to make modest fortunes.

Throughout Northern California there had been a lively interest in mining—one might almost say a quietly fanatical obsession—ever since that January morning in 1848 when a misanthropic carpenter from Lambertville, New Jersey, named James Wilson Marshall, discovered a scattering, a *blossom,* of gold flakes in the millrace of John Sutter's lumber works near Sacramento, inadvertently triggering the Gold Rush, which in essence created* the state of California. In these fairly metropolitan parts—Mount Diablo is only thirty miles from San Francisco, and rather less wild than where other claims were being staked farther afield—prospectors in the hundreds were a-roving in search of minerals of one kind or another. And though maybe in the pleasant purlieus of the hills close to San Francisco Bay there was less of a need for the burro and bedroll, for the claim stake and the slouch hat—the accoutrements of the grizzled caricature of the prospecting type—there was nonetheless a great deal of very energetic exploration of Diablo's riverbeds and the canyon walls, mainly conducted by new immigrants who had come by ship to see what the Gold Rush fuss was all about and stopped at Diablo on their way. Many lingered, and found lodes, outcrops, deposits, and ore bodies; and by the 1860s, as a result of their energies, a series of substantial discoveries had been made that

*And named: California has officially been the "Golden State" since 1968; gold has been the official state mineral since 1965, and "Eureka" the state motto since 1963. There was an attempt to make "In God We Trust" the motto, but that ploy went the same way as Coal Hill and slid into respectable oblivion.

would turn Mount Diablo and its slopes into a place as spectacular for its buried wealth as it was already magical for its aspect and topographic power.

Coal—dull, brown, dirty, unromantic coal—was the first mineral to be discovered there, peeking seductively out of a thick group of fairly recent sediments known as the Domengine Formation, a band of which lies just to the north of the mountain, in the flattening fields that spread down to the Sacramento and San Joaquin Rivers. Gold and silver, mercury ore and copper, were nearby, too—but their quantities were small, and the difficulty of extracting them was prodigious and at first far too expensive. But coal, on the other hand, was abundant, and soon miners and mine owners realized how they could make their fortunes from it.

For it happened that the discovery of coal on the slopes of Mount Diablo coincided almost precisely with the ending of placer gold mining in Sacramento, forty miles to the east. River sand was washed and the lighter minerals tipped back into the stream, leaving the sparkles of gold behind; even Bret Harte found a nugget worth twelve dollars on his first day of experimenting. But the relative ease with which gold had initially been extracted from the ground was as evanescent as it was seductive. By 1852, a mere four years after Marshall's discovery (which was announced formally by the president, James Polk, thereby suddenly redirecting westward migration toward California and away from the more immediately enticing farmlands of Oregon), the glistening nuggets had all, it seemed, been found. Such gold and silver as might be lurking in the rocks had from then on to be won by more traditional mining methods, with the gold hosed out by the use of enormous jets of water or the hard igneous rocks crushed and hammered and pulverized and blown to smithereens in order to release their bounty.

To do this, the mine owners required machines, big machines, nearly all of them made of iron. They needed rock crushers and pumps, stamp mills and conveyor trestles. They had to lay railway lines and acquire railway cars, as well as immense tracked engines that could claw the rock from the cliffs and tear down whole mountains. All of these great iron machines and manufactures had to be created. But the East was far away, and the costs and perils of bringing such enormous goods overland to California prohibitive, so a good deal had to be made locally. There had to be foundries and forges and rolling mills and factories in Northern California, with their primary function to keep the gold and silver mining industries alive—to help make the growing dream of California into a reality, and not let it die prematurely, a chimera.

All of these new factories depended upon a source of heat—a source that helped to smelt and puddle and make the steam and turn the wheels and fans and conveyor belts. And that source, of course, was coal. Not the eastern and foreign coal that was then being brought in by ship for $30 and $40 a ton—but the newly won and homegrown Californian coal, carted in from the Mount Diablo mines that were, come the 1860s, being dug, drilled, and blasted into the foothills between the peak and the two rivers to the north. Using locally mined coal would make the factories of San Francisco and Sacramento hum, and hum for a sound return on the backers' investments.

It was a short-lived experiment. Only 4 million tons were ever extracted from the Diablo mines; competition from cheaper and more accessible coal farther east put paid to the California coalfields. Before long such pithead communities as had sprung up were abandoned, doors left open and swinging on their rusting hinges, stores emptied and their windows broken; wild grasses and chaparral engulfed the outhouses. Some say all that remains today are the trees that miners brought with them to cheer up their dismal work camps, among them black locust trees, almond bushes, and a fast-growing weed known as ailanthus, or tree of heaven, which was brought in by the Chinese shopkeepers and grew in such wild profusion that it is now regarded as a blight, fit only to be slashed and burned and carted away by the ton.

There is a tendency common to most of us to take the more modest of our landscapes for granted. We see a wide and fertile plain, and we drive across it, as fast as its flatness allows, rarely pondering what might have brought it into being. We come across a valley, and, though we might take pleasure in its appearance, we give it all too little thought, other than perhaps to assume there is probably a river somewhere within its folds. And, while we are generally awestruck by the more spectacular mountain ranges, it seems true to say that those hills that are simply hills, or those mountains that are simply mountains, rarely prompt us to ask: Just why are they there? What forces first made them and set them down here, in this particular place?

Mount Diablo, though its isolation makes it somewhat more dramatic than most, is just such a place. It is a mountain generally outside the orbit of popular consideration, big enough to be of note, too seemingly ordinary to be puzzled over. To its neighbors in towns like Concord and Antioch and Walnut Creek, who can so readily see its bulk from their picture windows, it is simply a piece of scenery, an eternal and unyielding part of the view. Very few people ever stop to wonder why this particular mountain is where it is; what forces caused the land to slope upward as it does. And yet those forces, their complicated workings all encapsulated in the geologic history of this one mountain, are part of the same set of forces that caused the destruction of San Francisco. To understand this, to understand Mount Diablo, is to begin to understand why California, perched precariously at the edge of the North American world, has been destined by its geology to be both so beautiful and so dangerous.

Like most, I had given precious little thought to such things on that summer's evening. As I drove up through the gloaming along the winding mountain roads, my interest in the hills, such as it was, related simply to my need to muster sufficient horsepower to overcome them. To me that night—and to countless others who were seeing the hillside every day, and who might have paused to wonder why it might be and what it might be—*volcano* was perhaps the explanation that most readily came to mind, especially in the American West, where such things are much more commonly seen. The shape of the mountain when it is viewed from afar—a low double cone, which from the Central Valley side seems to rise spectacularly alone out of the plains like an Etna, a Vesuvius, or a Mount St. Helens—rather reinforces the impression.

But Mount Diablo is most certainly not a volcano. Few places in the world are as geologically complicated as California, and few parts of California's underside are as raggedly confused as Mount Diablo. And though the story that will unfold in later chapters has to do with geological structures and happenstances that involve San Francisco, Santa Rosa, San Jose, the San Andreas Fault, and a host of other faults besides that are some distance away from this rather peaceful-looking mountaintop, the saga of why Diablo is where it is and what it is has in fact great relevance to the geology and the geological processes that once destroyed San Francisco and that may yet destroy it again.

THE GEOLOGY OF the northern half of California—whether we are talking about San Francisco Bay or the Central Valley, the Coast Range or the Sierra, the Monterey headlands or the coast of Humboldt County, or Mount Diablo itself—is all interlinked, subtly, confusingly, and, for the geological mapmakers, often maddeningly. These links go far beyond the borders of the state—political lines that pay no heed, in this case, to the absolutes of geology.* They spread far, far beyond—as we shall discover, they reach up to Alaska, they percolate across to Wyoming and Montana, they reach back west across two oceans as far, in fact, as India and Australia. One might say, indeed, that the story of what makes California so complex and so interesting and so dangerous—and what makes Diablo so similarly geologically alluring—has implications for, and connections to, the planet in its entirety.

About 170 million years ago—in the early to middle part of the Jurassic, when dinosaurs were the dominant large land creatures in other parts of the world—the floor of what we now call the Pacific Ocean began to spread outward, eastward and westward, from a cen-

---

*Few are the world's political frontiers that mark tectonic boundaries as well. Within America one might argue that where North Carolina marches with Tennessee is the demarcation line of an ancient orogeny, evidence of a period of mountain building and folding. But beyond America, it seems clear that only the well-guarded border between Nepal and Tibet happens also to mark the divide between two plates, the Indian and the Asian, each pushing firmly against the other, and forming the Himalayas as a result.

tral suture line. The section of the floor that moved east did so with such force and speed that about 50 million years later—120 million years ago—it collided, gently but powerfully, with the cliffs and mountains at the western edge of what we now call North America. When that almighty crash occurred, two things appear to have taken place.

First, a sliver of the ocean floor, which happened to be made of rocks that were somewhat heavier than those of the cliffs and hills of North America (thus causing it to lie so low that the ocean was able to accumulate above it, while the American hills seemed to float, almost ethereally, in the air above), was forced downward, sliding under the cliffs and hills like an envelope being pushed surreptitiously underneath a carpet. And second, while it was being slipped underneath, it dragged down with it all the sand and soft rock that had accumulated on top of it while it had been moving eastward from the center of the ocean. It had traveled a long way to the point of collision—perhaps 5,000 miles—and it had taken a long time to arrive—perhaps 50 million years. The result was an ever-moving floor with an unimaginably large thickness of accumulated material on top—and some of it dipped under, carrying with it the floor material, while some stayed offshore, like surplus froth scraped off a cappuccino. It remained like a great barrier of islands, well to the west of the place where one plate was sliding under the other.

Thus a basin was created between the hills at the point of collision and the hills of the offshore islands—with the former (to reiterate) being a complicated and multilayered arrangement of ocean floor together with mixed-up and younger sediments, the latter being rather more entirely youngish mixed-up sediments with little or no ocean floor anywhere around, other than deep, deep down below. And over the next many millions of years, more sediment accumulated in the basin, the makeup of which was determined by how it was accumulating (whether in silt-rich rivers, on some sandy shoreline, by settling on the bottoms of deep oceans, or born from onetime marshes or long-buried sand dunes) and what the weather was like while it was doing so (warm and humid or freezing and dry).

There must have been one period, about 40 million years ago, when huge ferns and soft trees crashed, dying, into warm and fetid swamps. The layers of material in the swamps were then compressed and buried and heated, and after time and maturation they produced the coal of the Domengine Formation for which the miners hunted so assiduously in the 1860s. But there were other periods when there were deep blue seas instead of botanically rich swamplands, and they were alive with shellfish, sharks, and other noble creatures; and there were other more recent times when mastodons and saber-toothed cats trekked over windy grasslands and left their bodies to rot and become skeletons in forests of trees that seem not dissimilar to those growing in California today.

And then, 20 million years or so ago, the oceanic plate suddenly changed direction, for reasons that will be made clear (or as clear as the science will allow, since much remains unexplained) in a later chapter. Instead of pushing eastward, to smash head-on into the cliffs and hills of North America, the oceanic material began to move *northwestward,* proceeding smartly up alongside the coast, scraping past it instead of plunging underneath it.

The connection with San Francisco now becomes clear, for this new movement is exactly the same movement between plates that would go on to produce the cracks in the earth's surface that nowadays trigger the myriad earthquakes that occur farther to the west. At Mount Diablo, though, the movement did something rather different: It caused all the material that had plunged below the North American Plate to be ripped northward, bringing it hard up against the newer sediments that had been accumulating behind where it had been diving downward. The newly arrived material began to wrap itself around the older crustal and downthrust rocks, as if it were an immense shell of pastry. In time it buried the older material entirely and made it more or less invisible, coated in a thick covering.

And there this material would have remained, except that about 4 million years ago, for reasons that will also be explained later, the whole mixture was dramatically folded upward. The old, hard rocks were in the center of the fold; the new, young, soft sediments were on the outside. Under the influence of weather the young, soft rocks were swiftly worn away, thus exposing, as a range of steep and dramatic hills, the core of ancient ocean floor and dragged-down sediments from the dinosaur-era age of the Middle Jurassic. Mount Diablo was born. The soft rocks, which in one particular case were thick with abundant layers of soft coal, were to be found on the flanks of the hills; the unyieldingly hard and relatively ancient rocks, which were good for road stone but not at all rich in the kind of minerals that make men wealthy, were left in the middle.

The crucial element of this long and complicated story is the relatively uncomplicated but still somewhat mysterious event that took place almost exactly 20 million years ago: the moment when the onward press of the ocean crust suddenly, and for a reason that long remained a mystery, translated itself into a northward, sliding motion—as when an army suddenly wearies of charging head-on at the enemy and begins to execute a mysterious, somewhat cunning motion to one side that appears to be an attempt to outflank the foe. At this point—which took place in the middle of the period of world history known as the Miocene—everything that now in essence defines seismic California was brought into being.

This was the moment of making—when the earthshaking, city-killing, history-creating, epoch-changing linear system, 750 miles in length, and known since the beginning of the nineteenth century, broadly and generically, as the San Andreas Fault, was created. That the oncoming plate's change of direction also helped to create this hill, the massif that rises so formidably in the picture windows of the residents of the towns of Clayton, Pittsburg, and Pleasant Hill and their like, reinforces the view that I started to hold in the tent that breezy night: that Mount Diablo is more connected and interlinked with the events of the San Francisco tragedy than almost all those who live beside it and below it have ever properly supposed.

BY NOW I WAS READY to sleep. I stepped briefly out of the tent and looked up at a sky ablaze with stars: Cassiopeia and Gemini unusually bright, Castor and Pollux winking down from the roof of the universe. There seemed to be a gathering of clouds rolling onshore from over the faraway Pacific. The loom of lights from the cities spread a blush of orange-pink on the underside, making them glow bright against the velvet of the coastal night.

DAWN CAME UP all too early that next morning, and from where I was camped, close to the top on the western side of the mountain, the day and its morning sky were pale blue and clear as crystal. There had evidently been a shower in the small hours: The fire was out, the ashes were damp and cold, the pine needles glistened with more than the usual dew. I walked up along the empty road (the gates below would not open for another hour, and so I could revel in the knowledge that I still had the mountaintop entirely to myself) to the summit. At the top is a immense octagonal stone building—a onetime aircraft beacon,* a uniquely visible landmark and, as it happens, a memorial to one of the defining periods in recent American history, the Great Depression. The building, fashioned from a highly fossiliferous sandstone quarried locally, had been erected in the 1930s, using the brawn and muscle of scores of out-of-work men who had been organized into a local chapter of the federally funded Civilian Conservation Corps, and who had

*The beacon was erected in 1928 by Standard Oil, to help guide in eastbound planes coming from over the Pacific, and westbound planes arriving at the Bay from the difficult and stormy skies over the Sierras. Pilots from New York said they could see its beam while they were still in the relatively safe skies over Nevada, more than a hundred miles away. The invention of radar and radio beacons—and the fact that airliners flew at ever-greater altitudes—soon rendered the searchlight all but worthless, and the Diablo Beacon is these days lit only once a year, each December 7, as a memorial to the attack on Pearl Harbor. During the last days of 1941 it had become quite literally a beacon of safety for all those aircraft fleeing the Hawaiian bombardment, and a valedictory light for those others heading out to wage war.

lived for two years in a camp close to where I had slept the night before. I climbed the steps up to the parapet and gazed, silently and dumbstruck, at one of the most stupendous views in all America.

William Brewer, when he first surveyed the mountain in 1862, estimated that 80,000 square miles could be seen from the summit. "I made an estimate," he wrote in his report for the Geological Survey, for which he was principal assistant, "that in tolerably plain view the extent of land and sea embraced between the extreme limits of vision range over 300 miles from north to south, and 260 to 280 miles from east to west. Probably but few views in North America are more extensive, and certainly nothing in Europe."

There was no haze, no smoke from forest fires, no fog, no pollution—at least, not where I first looked, toward the southeast, almost directly into the rising sun. The dry flatlands of the Central Valley spread out brown and cornrowed, with their arrow-straight irrigation canals gleaming like tinsel. Beyond them in the distance, forming the horizon, rose the jagged and snow-topped wall of the Sierra Nevadas, a range that stretched 250 miles from the Cascades and Lassen Peak in the north to China Lake and the fringes of the Mojave Desert in the south.

This immense wall of rock was where California's two most precious assets were created: its water and its electricity (though more properly only one of these was a real asset, since the latter was created by the former, provided that the water was above sea level and thus possessed of potential energy). The moist westerly winds that blew in from the ocean were stopped in their tracks by this great granite massif, which, happily for the state (though less so for that part of California known as Death Valley as well as the entire western part of the state of Nevada, which lies in the mountains' rain shadow), proved too high for the winds to climb over, too elongated for them to weave their way around. The snows that then fell so constantly in due course melted and became rivers, with the waters then employed either to slake thirsts or to create power.

Most of the range I could see, and, had I had a detailed chart, I imagine I could have recognized the noblest peaks: Mammoth, Dana, Gibbs, Parker, Darwin, Lamarck, and the great summits around Yosemite. The very highest, Mount Whitney—named for the first head of the Geological Survey—was too far away to sight, but, had I known exactly what I was looking at that morning, I daresay I could have made out far to the southeast Mount Brewer, the 13,500-foot Sierra peak that had been named for the remarkable and heroic William Brewer, an explorer of (according to his biographer, Francis Farquhar) "the strongest fiber, of unflagging energy, the soundest judgement, the utmost tact, and of unequivocal honesty and loyalty."* He was also accorded the honor of having the California spruce, *Picea breweriana,* named after him; but he is otherwise too little remembered, like so many of those who opened up this staggeringly beautiful country, and who beckoned to outsiders to come west for reasons aesthetic and not purely mercantile. John Muir and Ansel Adams were first among those who heard this call, but many of the rest of early Californians, one is often tempted to suppose, belonged to the more strictly practical school of migrants.

I turned away from the sun, to my left, and gazed up to the north—a view dominated initially by the most spectacularly severe of California's peaks, the ice-covered, glacier-strewn, and symmetrically perfect cone of Mount Shasta. Shasta's standing, like that of Mount Kailash in Tibet, stems from its magnificent isolation: It is a double volcano, and it rises, without warning, from where the Central Valley peters out, close to the logging town of Redding, which was also just visible from that morning's vantage point.

Much closer, and just below Diablo's slopes, were the two great rivers that most properly delineate the geography of this part of California; the nearer of the two was the San Joaquin, which I could see

---

*Brewer climbed his mountain, twice. He left the records of his second ascent, in 1895, in a bottle on the summit—a bottle that was later found and taken to the small museum in the offices of the Sierra Club in San Francisco. But, irony of ironies, it was later destroyed in the fire that followed the earthquake in 1906.

faintly, squirreling its way through the great sprawling inland port city of Stockton;* the more distant was the Sacramento River (the Río Sacramento, or the "River of the Sacrament"), which coursed down from the north, and from the mountainsides behind the state's present-day capital. The two joined in plain sight below me in a wide bay and glinting marshlands; and then, after squeezing their combined way through the Carquinez Strait—named after the Indians who had given the missionizing Spanish soldiers such a hard time with their diabolical medicine man—and beneath the iron highway bridges on which I could already see the crawl of morning motor traffic, their waters spread out into a wide and flat declivity: the immense, 1,600-square-mile tidal harbor that unfolds from the hills, valleys, wineries, sheep farms, and flower meadows of Napa and Sonoma in the north, down to the plants, factories, airfields, think tanks, and next-century-paradigm-shifting industries around Palo Alto, Stanford, and San Jose in the south.

There the rivers widened at last and evolved lazily and muddily into the feature that, once its value as a harbor was realized, was enthusiastically named to honor Saint Francis of Assisi, Bahía San Francisco, San Francisco Bay.

And center stage, seeming almost to be floating out on the wide waters, stood the city, the obvious and very evident capital of it all. There, at last, was my goal, the first sight of my destination, of the great city that had always been the Bay's principal port. It had at first been called after the pretty shrub that grew there in such abundance, yerba buena. But in 1847, in commemoration of the universally venerated twelfth-century priest from Assisi in whose kindly principles its founders so ardently believed, it had been given the name it bears today: San Francisco.

---

*It was named for Commodore Robert Stockton, who, in 1846, captured the Mexican military outpost of Los Angeles and four days later formally annexed the Mexican province of Alta California in the name of the United States. Stockton is, essentially, the founder of American California.

What I remember most about the city, which was spread out beyond the low hills and clustered like a jewel box of gleaming spires and glittering windows on its tiny thumbnail of a peninsula, was just how astonishingly *delicate* it all looked. It was quite unlike New York or Chicago or Boston. Those places were gray, massive, battleship-like cities, cities that were indelibly written into, and indestructibly welded onto, their landscapes, each fully a part of the topography that once shaped it. London, nestled among its own enclosing hills, had looked and felt much the same for centuries. As had Berlin, Paris, Moscow. And Rome, of course, "the eternal city." Tokyo, so ancient and so modern and so *always there,* regardless of the fires and wars that had scourged it. Even Hong Kong, Sydney—even such defiantly modern cities as these had acquired a look of settled permanence.

But not, it seemed to me, this preternaturally beautiful city of San Francisco. I squinted through a big brass telescope that had been obligingly placed on the parapet. My feeling that this was a confection of untoward and only half-urban-looking delicacy was confirmed by the magnifying lenses. How tightly San Francisco appeared to cling on to its hillsides: One could imagine knuckles whitened, sinews straining, teeth gritted.

I walked back downhill, enchanted and fascinated by all I had seen. And then there came to me the one word that, more than any other, has stayed with me and has haunted me since that morning—a word whose relevance was only compounded by what I knew, by what all of us knew, of San Francisco's history.

As I folded my tent, cleaned up my campsite, and packed the car for the last thirty-odd miles back down the hillside and onto the freeway that would eventually take me over that enormous series of iron bridges and into the city itself, one word kept running around and around in my mind. This fragile, enchanting-looking place had also appeared, more than anything else, most terribly and fatally *vulnerable.*

Perhaps all cities, like all the creations of humankind, lack the real permanence they often seem to seek. But this was something more. Because of where San Francisco was built, and because of the febrile and uncertain nature of the world that underpins its foundations, it has a unique vulnerability and suffers under a greater sense of edgy impermanence than any other great city anywhere. San Francisco that morning seemed, in the context of the landscape spread around it, a city more temporary than any other great urban creation that humans have ever made.

THREE

Chronicle: Such Almost Modern Times

Time has no divisions to mark its passage, there is never a thunderstorm or blare of trumpets to announce the beginning of a new month or year. Even when a new century begins it is only we mortals who ring bells and fire off pistols.

Thomas Mann, The Magic Mountain, *1924*

Nature, and Nature's laws lay hid in night
God said Let Newton Be, *and all was light.*

Alexander Pope, 1730

IT WAS A HYBRID YEAR, A YEAR BETWEEN ERAS, ONE THAT still balanced on the cusp. It was a year that was pinioned by two centuries, held in equipoise between the comfortable and apparently innocent ways that echoed down from the nineteenth century just gone, and the infinitely more complex and challenging ways starting to well up from the all-too-modern twentieth century just begun.

Science—much of it born in direct response to the events of that April in San Francisco—was behind a great deal of the coming change. One harbinger of these transitions was a modest-looking paper that had been published in Leipzig six months before the earthquake. It was written in German and appeared in the monthly journal *Annalen der Physik.* It was titled "Does the Inertia of a Body Depend upon Its Energy Content?" The author of the paper—it was the fourth he had written for the *Annalen* during 1905, a year that would later come to be seen as the annus mirabilis of his entire career—was a young clerk named Albert Einstein, then working in the Swiss Patent Office in Bern. He conceived of the paper as a footnote, an afterthought to papers that more fully described what would later come to be regarded as his special theory of relativity. But the paper has an enduring fame among scientists, one that derives from a single sentence written just seven lines from the end. It presented Einstein's very simple conclusion, at which he arrived after working his patient way through a series of less-than-simple calculations, by stating that "It directly follows that if a body gives off energy L in the form of radiation, its mass diminishes by L/c^2." This, after making allowances for the German nomenclature of the time, would become the best-known equation of all time: $E = mc^2$

And, though it would take a while before the paper was circulated and its significance fully realized, the year that followed its publication—1906, the year of the seismic mayhem that culminated in the destruction of San Francisco—unarguably marks the true beginning of an era that is still broadly recognizable today. It was the start of the atomic age, and the San Francisco Earthquake was the first internationally recognized disaster to be logged into its history.

In a score of other ways, by way of a welter of major technological achievements and a blizzard of less obvious scientific advances, it was as if the world was sloughing off its old and wrinkled skin and a glistening new replacement was slithering smoothly into being. Psychoanalysis was recognized as a science at about the same time, with the publication of Sigmund Freud's seminal *Die Traumdeutung* (*The Interpretation of Dreams*) at the turn of the century; as W. H. Auden said of Freud, in creating the new science he ceased in an instant to be merely a man and became instead "a whole climate of opinion."

Mass production of automobiles began in 1900 (at the Olds Company in Detroit—Ford was not founded until 1904). And the motion-picture industry was on the verge of being born by 1901, thanks to inventions being made at an almost feverish rate by the Edison Company and to the opening of scores of what were known as Film Exchanges, which led in turn to the opening of public cinemas (and which also allowed the 1906 San Francisco Earthquake to become the first natural disaster to be captured in moving pictures; hundreds of minutes of disaster film can still be found in archives).

Marconi sent his first Morse code signal by radio across the Atlantic that same year. The Atlantic Ocean had long before been crossed by telegraph wires; it was to be the turn of the Pacific in 1903, which acquired its first submarine cable in the same year that Orville and Wilbur Wright launched their tiny biplane beside the sea at Kitty Hawk. A message sent from the White House took only twelve minutes to come back to the transmitting telegraphers, having circumnavigated the planet via a now fully connected skein of cables, and at what was then a barely imaginable speed. The diesel engine was introduced to the United States in 1904. The rambunctious and boisterously enthusiastic radio engineer Lee De Forest created the triode, known as the Audion, around the same time and with it sent the first voice broadcasts across the oceans; later he transmitted music from the top of the Eiffel Tower during his honeymoon (the second of the four he would enjoy in his lifetime), the signal being picked up 500 miles away. And finally, though Elisha Otis had invented the principle of the elevator as early as 1852, the first such device having been installed in a New York department store in 1857, and though the Bessemer process allowed steel girders to be used (in place of iron columns) in construction, thus permitting taller and taller buildings to be made, it was not until these early-twentieth-century years that a combination of engineering, ambition, and architectural enthusiasm coalesced sufficiently to allow for the building in New York of the world's first true skyscrapers. They were to be far taller than the big buildings of Chicago that are usually taken to be the ur-structures of the breed, and yet in both cities they were ornate and rhapsodical and flamboyant—and for the next two decades many were modeled on classical structures from ancient Greece and the Roman Empire, or from Venice, Spain, and the work of the Moors.

The ornate appearance of Manhattan's first pair of tall buildings, the Metropolitan Life and the Woolworth, which were being built around the time of the events in San Francisco, suggests the lingering hesitancy of the time. Their look implies that, despite the rush of scientific discovery and progress, people were not ready—as they rarely are—to abandon completely the attitudes of those more comfortable years that were just beginning to slip into history. The British poet John Betjeman, who was born in London in 1906 (six months after the earthquake), wrote much later, in the thirties, a brief farewell to King George V, a slight poem that reflects the conflicted view of these earlier times as well. It offers a sentimental grace note to an era that the poet knew only as a child, but that he saw as having been fully and finally extinguished by the ways of the modern:

Old men in country houses hear clocks ticking
Over thick carpets with a deadened force;
Old men who never cheated, never doubted,
Communicated monthly, sit and stare . . .

In the San Francisco that existed at the time of its greatest tragedy the modern was being half embraced, half disdained. Many of the more conservative Americans of the day clung to the attitudes and customs of the time that Betjeman, across the Atlantic in England, would soon so sorely miss—and to these people it scarcely mattered that Einstein had just pried a nugget of understanding from the universe, or that aircraft had been invented, or that mass production was beginning, or that the ether was starting to chatter with radio transmissions, or that immense commercial buildings were being erected to be filled by men who barked into telephones and betrayed one another in newly cutthroat ways that would in due course be exposed by muckraking newspapers (this last term was introduced in 1906, a month after the earthquake).

NO, THE MODERN was not going to be embraced without a struggle. There was too much disregard for certain aspects of the new. The 78,000 registered cars on U.S. roads in 1905 (up from just 300 in 1895) were still considered, for example, to be little more than barely useful toys; the 51-day journey taken in 1903 by the drivers of a 20-horsepower Winton automobile from San Francisco to New York, the first-ever cross-country road trip, was initially denounced as a fraud. Then again, social change was slow to be accepted, too: A woman was arrested for smoking a cigarette in an open car in New York. "You can't do that on Fifth Avenue," the arresting officer reportedly said. And an Illinois congressman loudly attacked the brand-new practice of adding strange confections to hitherto natural foods, to make them taste better, or to eke out the ingredients and make their supply more profitable.*

Despite all of science's froth and exuberance, and despite a growing perception of the brave new world it promised, 1906 was in many ways a year still hesitant, partly pinioned in the Edwardian era by its mannered ways. And though this hesitancy was perhaps more apparent back in England—which still had an empire to act as bulwark to its more reactionary values—it permeated America, California, and San Francisco, too, though inevitably in America it was colored by a bolder and brasher style.

Teddy Roosevelt, a man who epitomized these conflicts in attitude and style—"Speak softly and carry a big stick" was the adage for which he remains best known—had been president since 1901, assuming the post after William McKinley was assassinated at the Pan-American Exposition in Buffalo, New York. The tone he then set reflected the duality of the times. He had been a frail and sickly child, but as a young man he was determined to conquer his limitations by strenuous exercise, becoming a rancher and a volunteer cavalryman. When he eventually came to lead his country he did so in a rambunctious, almost defiantly physical way, bringing explorers and soldiers and boxers to the White House, making stirring speeches advocating valor and national sturdiness, and promoting American prominence throughout

*Representative James R. Mann had long entertained a particular dislike for the adulteration of coffee with, he wrote, "Scheele's green, iron oxide, yellow ocher, chrome yellow, burnt umber, Venetian red, turmeric, Prussian blue and indigo . . . roasted peas, beans, wheat, rye, oats, chicory, brown bread, charcoal, red slate, bark and date stones." The legislation that he sponsored, the Pure Food and Drug Act, passed into law on June 30, 1906.

the world. His memorials are legion, the most notable (aside from his having won the Nobel Peace Prize in 1906 for brokering the peace treaty ending the war between Russia and Japan) being the Panama Canal, the building of which he secured by acquiring the Canal Zone from Colombia in 1903.

His dream was for Americans to dominate in particular the Pacific Ocean, to stand firm against the encroaching ambitions of any "Orientals" who might entertain similar hopes, and to create in San Francisco a base for a naval force that would secure America's supremacy forever. He may have been an Easterner of the old school—a graduate of Harvard and Columbia, married to a Bostonian—but he was a man who valued America's blue-water West, in part because of the buccaneering spirit of those who settled there, but also because of the coast's utility as a base for his own imperial ambitions. When he went to San Francisco in 1903 to dedicate the monument to Admiral George Dewey and the fleet that had so roundly defeated the Spanish in Manila Bay, he thrilled the immense crowd by declaring that the proper place for all Americans was "with the great expanding peoples, with the people that dare to be great." San Francisco was, in Teddy Roosevelt's eyes, very much the Imperial City, a gateway to great fortunes won on the far side of its vast ocean. The city could not have had a more enthusiastic champion in the nation's capital when it suffered its greatest calamity.

A YEAR, AND A COUNTRY, and a president, all of them in balance, all expectant and optimistic and apprehensive by turns as a whole world of changes—changes political, psychological, social, and, most of all, scientific—began to sweep in from the future. The year, in a state of such fine equilibrium, was unusually vulnerable to the unexpected, causing events like the eruption of Vesuvius and the destruction of San Francisco to cast a disproportionately long shadow on science, society, philosophy, religion, and art.

Yet there is a difference in the way that Americans of those half-modern times reacted to the events—a difference in the way that the

Churches, in particular, are shown being ruined by the 1755 Lisbon Earthquake, an event seen by most locals as the act of a cruel and capricious Divinity.

vulnerability of the people manifested itself, compared with that of earlier years.

Back in 1755, for example, on November 1, All Saints' Day, the city of Lisbon experienced a truly enormous earthquake—some say it reached 9 on the Richter-Scale-to-be—and as many as 60,000 people died. It triggered total mayhem. A few wise men in eighteenth-century Portugal, most notably the prime minister of the day, Sebastião de Melo, reacted to the event with a cool rationality: They ordered countrywide surveys to be made and replacement buildings erected only after engineers had marched soldiers around models of the proposed structures to ensure they would not collapse as a result of vibration, and so on. But, generally, rational reaction to that earthquake was minimal: Most of Lisbon displayed the kind of wild primitivism that characterizes a people who are shocked and unprepared and intellectually ill equipped to be able to offer answers as to why a catastrophe like this might have happened. *God was responsible,* it was widely assumed. Catholic priests roved around the ruins, selecting at random those they believed guilty of heresy and thus to blame for annoying the Divine, who in turn had ordered up the disaster. The priests had them hanged on the spot.

In San Francisco, a century and a half later, it was all very different. When the San Andreas Fault ruptured just before dawn on that April Wednesday, the new-forming appreciation of science meant that a good number of the city's inhabitants understood, at least basically, what had just taken place. Many of them speculated sensibly and rationally as to why. The official reaction to the disaster was generally swift and measured, ordered and rational—and the consequences of that reaction were far-reaching and remain with us to this day. True, some few lingering attachments to less sophisticated times did also leave their mark—which is why the hybrid nature of the year 1906 has rendered the earthquake so resonant and peculiarly interesting an event in American history, and in the history of the world.

Lisbon's disaster, widely regarded as an unstoppable act of a cruel and capricious God, is now largely forgotten. The San Francisco catastrophe, recognized, on the other hand, as having been the act of a perhaps not wholly unpredictable nature, never will be. San Francisco will not be forgotten because, thanks to the growing understanding of science, it became the first seismic event to awaken mankind to the realization that nature's whims could perhaps be measured, perhaps one day anticipated, then met and even overcome. The tragedy led scientists to begin studying the earth with far greater vigor than ever before. It offered the first opportunity for humans to imagine what it might be like if they, and not God or nature, were ever to be *in control.* To that extent, the fact that the earthquake occurred in this specific changeling year of 1906 was more than a little fortuitous.

FOUR

From Plate to Shining Plate

I am lost in wonder and amazement. It is not a country, but a world.

Oscar Wilde, quoted in the St. Louis Daily Globe-Democrat, *1882*

A Simple Plan

SINCE THE MID-1960S, WHEN THE PLATE TECTONIC THEORIES were first adduced and observations proved them to be correct (to all but a few hard-core skeptics), it has been realized that the brittle exterior surface of the earth, broken into a number of enormous slab-like fragments generally known as plates, is in constant motion because of the upwelling and downthrusting of convection currents in the material immediately below that crust—rather like the motions of the creamy scum that forms on the surface of a soup that is boiling merrily away underneath it. The upwelling and downward plunging of these currents takes place in the earth's very plastic or highly molten mantle, which itself exists as a thick band—1,800 miles thick, in fact—just below the solid crust and above the molten-ingot sphere of heavy metals that spins at the heart of the earth and is called, somewhat unpoetically, the planet's core.

The convection currents exist because the inner earth is frighteningly hot—and it is hot for two primary reasons: first, because of the kinetic energy that was released when all the space debris combined, in bodies that are called planetesimals, to make the protoplanet that we now call earth; and second, because the inner earth contains huge quantities of radioactive materials, isotopes of potassium, thorium, and radium in the main, that emit heat because they have been decaying over the millennia. But there is a secondary reason for the earth's being so very hot and having a crust in a constant state of motion (in a way that the crustal surfaces of Mercury, Venus, Mars, and earth's moon do not seem to move at all), and this relates to the earth's size.

The earth is much bigger than the other close-in planetary bodies like Venus and Mercury and, because of its size, it stores a lot of heat in its interior. But this cannot be radiated away into space at the same rate as heat from the smaller bodies, because of the simple realities of solid geometry: The larger a sphere, the smaller its surface area when compared with its volume. Earth has a very large volume and contains a great deal of heat within that volume—and yet its surface area is too small to radiate away all that energy heat and decay heat with the kind of efficiency and dispatch that other, smaller bodies enjoy. So the core burns; the mantle bubbles and boils and moves up and down in convection curves and plumes, just as the aforesaid soup might do on a stovetop; and the scum of cream that floats on the top, the many-miles-thick earth's crust, which bears all of the seabeds and all of the continents in the solid and broken-up plates of which it is now known to be composed, moves. It slips and slides about under the influence of those convection currents; the plates confront one another, interact with one another, jostle against one another; and a chain of consequences—some of them dramatic, some majestic, some terrifying, and all of them of singular importance for the humankind that clings to existence on top of some of the plates—is visible on all sides. The consequences of the plates' interactions with one another are what we see and feel and know: the topography of the world that rises and falls all around us.

DEPENDING ON HOW they are classified and counted, there are between six and thirty-six major plates recognized as wrapping themselves around the entirety of our planet. The edges of these plates are where the geological business of the world is most dramatically conducted. Where plates of the same kind collide with each other head-on, their edges hitting each other, mountain ranges rise up. Where they pull away from each other in midocean, volcanic islands ooze slow-moving lava streams onto the seafloor. Where plates of different kinds smash into each other, and one plate rides up over the other, as if in some kind of bizarre traffic accident, violently explosive volcanoes and other frightening manifestations of the earth's power are thrown up. And where plates slide alongside each other, like ships passing too closely in the night, there are other kinds of mayhem and havoc on display—not the least being violent earthquakes, of the kind most frequently experienced in California.

The biggest of the plates have suitably big names—the African Plate, the Eurasian (by far the largest of all), the Antarctic. There is the Pacific Plate, for example, the only large plate that does not sport a continent. Lesser plates have more romantic, less familiar appellations: the Caroline Plate, for example, the Cocos, the Nazca, the Gorda, and the Juan de Fuca. And new plates are still being identified and classified, as more observations lead to an ever-greater understanding of the complexity of the earth's crust. Recently, for example, there has been the christening of the Resurrection Plate, a few thousand square miles of real estate that lies mostly below the sea, tucked up against the southern coast of Alaska and itself bounded by a pair of hitherto unfamiliar lesser plates known, exotically, as the Kula and the Farallon.

These last three small plates nudge up against the western edge of one truly enormous neighbor: the North American Plate. And it was across this immense tectonic entity that I decided in 2004 to make a journey. I decided that to better understand what had happened in San Francisco a century ago, and to place it all in its appropriate geological and historical context, I would get myself there by driving. I would drive my car from where I live on the Atlantic Coast all the way to this

Major Tectonic Plates and Fault Lines

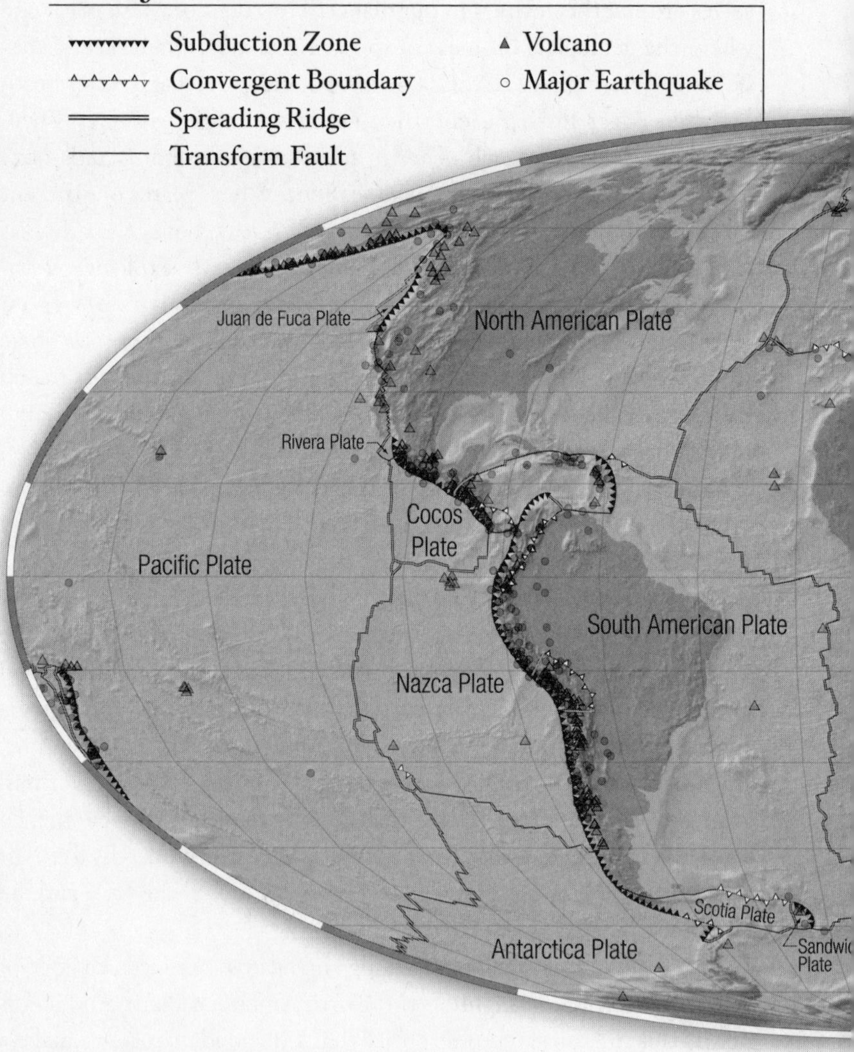

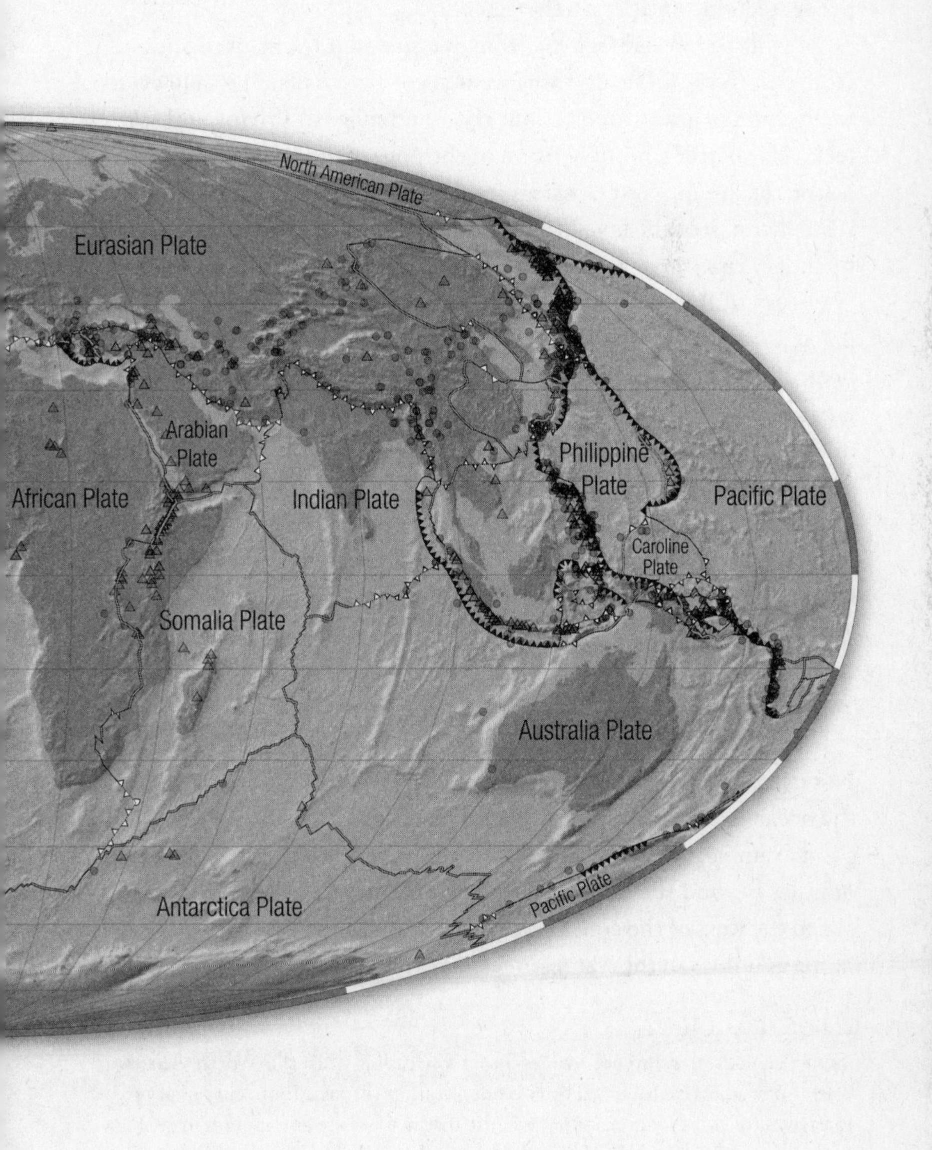
North American Plate
Eurasian Plate
Arabian Plate
African Plate
Indian Plate
Philippine Plate
Pacific Plate
Caroline Plate
Somalia Plate
Australia Plate
Antarctica Plate
Pacific Plate

once benighted city on the Pacific and, in doing so, traverse the breadth of the plate, one of the largest and most complex and bewildering tectonic entities on the planet.*

My journey would take me from one side of it to the other, driving from east to west. If everything went according to plan, I would eventually end up, after three or four thousand miles of driving and, with luck, a far better comprehension of the undersurface of North America, on the distant West Coast and at the very place—that ragged but essentially die-straight line—where the North American Plate's western boundary rubs up against and slides along, quite literally, the eastern boundary of the adjoining Pacific Plate. I would eventually reach the place, in other words, where all this rubbing and sliding, which has been going on for the better part of the last 15 million years, produces great and often terrible earthquakes, of which the events of April 1906 in San Francisco remain the best known.

It seemed a simple plan. But then I looked more closely at such maps as exist displaying the plate boundaries and came face-to-face with a small geographical snag—a snag that caused some not inconsiderable complications, just as I was planning the trip.

The Plate Entire

Not one of the major tectonic plates exactly overlies the continent or the ocean from which it derives its name. The African Plate, to take a good example, spills out all over the place, extending some hundreds of miles beyond the boundaries of continental Africa itself. It surges out to the west of the continent and into the Atlantic Ocean, incorporating as it does so the Azores and the Canary Islands, and such British

*Some geological reference works insist on calling this the North America Plate—one adjective followed by two nouns rather than the more customary two adjectives followed by one noun. Pedantic distinctions are as common in geology as they are in lexicography: I remain terminologically agnostic in this case, but will keep the usage "American."

colonies as St. Helena, Tristan da Cunha, and Ascension Island. At its eastern margin it embraces Diego Garcia, Mauritius, and Socotra; and to the north it stretches into the Greek Islands and the southern coast of Turkey—ecumenical in its politics, but hardly confined to the Africa that is merely carried along on the plate's ample middle.

Such indistinct boundaries are a feature of all the big plates, with the North American Plate being probably the most geographically undisciplined of all. It runs in the east from Iceland clear across the American continent to the Russian peninsula of Kamchatka. Sixteen million square miles, depending on how the less certain edges of the plate are drawn, are incorporated within this immense storehouse of geology. A journey across it, from precise edge to precise edge, would in theory take me all the way from somewhere near Reykjavík in the freezing North Atlantic to the muskeg and larch forests near Magadan in the Russian Far East.

And something else became readily apparent as I trawled my way through the staggering amassment of papers and journals and tomes that chronicle the newest developments in the astonishingly fast-moving world of modern geology. Much of the new science—most of the new discoveries and new realizations that have turned geology so comprehensively on its head in the last two decades or so—have been made by examining the rocks and their structures that lie to the *east* of the shores of modern North America. It would be idle to say that other rocks in other places have not brought forth devastating revelations, too. But, as it happens, the central story of what has taken place in the world—and, by extension, the central story of what has taken place in California, what is taking place there now, and what may well take place there in the near future—is one that has largely been derived from discoveries made by geologists and geophysicists and theorists not in the American West or anywhere else, but off the American East Coast.

The discoveries that have been made, the theories that have been born, the fieldwork that has confirmed them—all originated, for the most part, in the rocks and the rock structures that unroll between the Appalachian Mountains and the Highlands of Scotland. The New Geology provides a picture of the way the world works, and answers, in one splendid moment, all the questions that can be distilled into the perfectly simple: *Why San Francisco?* And the New Geology—it is worth the capitals, so different is it from the geology of Lyell and Murchison and Smith and all its gray-bearded founders—that begins it all itself, began 2,000 miles east of the eastern American coastline, and 6,000 miles from the western American coastline and San Francisco, in the middle of the 40,000 square miles of fire and snow and black, black North Atlantic rock known as Iceland.

THE EASTERN FRONT

Iceland is still a center of creation, the place where the North American Plate is being born. It is known, and has been known since the late 1960s, that Iceland sits astride—indeed, is created by—a ridge, where material from the mantle wells up and spills out between two plates that are moving apart from each other, under the influence of the almighty convection currents that drive the earth's outer engine. It is in Iceland that one can see—as I did, back in the mid-1960s—the thick lava, the raw material of the inner planet, belching up from deep within the earth. It piles itself up and helps to shift the already-west-moving North American Plate ever westward; and, as it continues to well up, it squeezes and compresses parts of the plate to such a degree that earthquakes, among a whole host of other phenomena, are triggered.

This is why Iceland, which lies so far from California as to appear quite irrelevant to the geological goings-on in the state (and most certainly it must appear so to most Californians, for when did a surfer at Malibu or a vintner in Sonoma or a camper on Mount Diablo last contemplate his or her connection to the Vatnajökull icecap?), is not, in fact, irrelevant at all. It is a crucial first part of the jigsaw puzzle, important to visit, essential to understand.

I first went there in 1965 as a student, as a climber (of modest ability only), and notionally, since I was a member of the Oxford University Explorers' Club, on an expedition. We went up into the lava fields and onto the icecaps, and saw spread out before us a landscape quite unlike any other in the world—a place where (though no one knew it then) the planet is being steadily and spectacularly reborn.

Essential to my plan for crossing the North American Plate was that in due course I would reach the place where it met its western neighbor, the Pacific Plate. Once I got there, I would in all probability want to place one foot on each of the two plates, as a tourist trophy—much as one likes to have snapshots taken with a foot on each side of the equator (one of them thus experiencing summer, the other technically winter), or to capture the moment of being briefly astride the Greenwich meridian, or (while at sea) the international date line. And then, as I looked through the field notes I had kept from that long-ago trip to Iceland, something similar occurred to me. Even though we had no idea that an entity called the North American Plate even existed (a measure of how new plate tectonic theory is), was there any chance, I wondered, that during that expedition I had managed to stand on the place where that plate met its eastern neighbor, the Eurasian Plate? Had I done so, it would have given my intended American trip a pleasing sense of symmetry, if nothing else.

So, briefly excited, I examined my old notes even more closely. I reread our expedition report; I pored over our yellowed maps. And yes, I thought after a while, there was indeed one place in Iceland, one place to which at the time I had attached little importance, where I might in fact have done that very thing.

My search for that place on the eastern edge began with the reddest of red herrings. Shortly before we went off to Iceland, a brand-new island had been formed, amid much fanfare from the world's press, off the country's southwestern coast. The locals named the island Surtsey, after the Nordic fire god Surtur; and since we were there, and curious, and ostensibly interested in a new topography, we found ourselves a small fishing smack and an intrepid young skipper, and went off to have a look. The sea was the color of iron and heaving; the weather was foul, and Surtsey lay twenty miles off the coast. Eventually we reached it—a gray, smoking hulk of fresh lava, by then a good 300 feet high and perhaps half a mile across. There was a dusting of green against the gray, as the first plants had already been established (notably a cluster of something called sea rocket); the first animal had arrived (a housefly, not easily visible from the deck of a rolling smack); and the first of the innumerable seabirds that would stop there, at what for them was a newly convenient resting place, were wheeling down from the clouds—herring gulls, kittiwakes, fulmars, and (from the sound of their chattering growl) murres were clustered happily on the craggy shore.

From the look of the contour lines on the adjoining islands and on the main island itself, and from the *son et lumière* of all this fresh volcanic activity, it seemed to us very much as though it was somewhere around this point that the Mid-Atlantic Ridge was welling up from the mantle below, with the coast that lay to the west of where Surtsey rose technically American tectonic real estate, and the cliffs and hills that stretched away to the east—toward the Faeroes, Shetlands, Scotland, and France—European territory.

But today's scientific literature says otherwise. True, if an imaginary line is drawn from Surtsey northeastward to Heimaey and the other Westmannaeyjar Islands (which have to be evacuated with dismaying frequency because of their very active volcanoes, the last time being in 1973), and extended farther northward through the peak of Iceland's most infamously dangerous volcano, Hekla (whose eruption in 1783 caused terribly cold winters all over northern Europe), it does look beguilingly like a ridgeline.

But geophysicists who are currently monitoring the line—and few places in the world are as closely monitored as Iceland (which is hardly surprising, given that it marks a place where the planet is tearing itself open)—are doubtful. Surtsey is very interesting, of course; but, though the region often behaves in a quite spectacular way and so is noticed by the outside world (as with the birth of the island, for instance, which was greeted with the enthusiasm usually reserved for pandas or heirs to royal thrones), it appears from close examination that nothing much is going on there other than that a plume of superhot material is welling up from within the earth and spilling out along this fissure track.

Most important, the sides of this fissure are not moving away from each other—they are not spreading as they would be if this were the point of the plate junction. At least, the movement is nothing like the one that is occurring sixty miles east of this line, sixty miles closer to the capital city of Reykjavík. It is there that geophysicists now believe a spreading ridge exists—and most specifically at a place to which I had been that sixties summer. I had visited it not because of any great interest in its topography but because it was where Iceland's ancient parliament had first met more than a thousand years before.

The parliament was called the Althing, and it sat for many years, from the tenth century until the thirteenth; then Iceland entered into a treaty with Norway and for a while lost its independence. The Althing met in a natural rocky amphitheater northeast of where the present capital lies, at an old town called Thingvellir that is a shrine to all Icelandic people. The structure can be reached by traveling—as I had done—a road that runs along the western edge of a lake called Thingvallavatn. It now turns out that this road, which passes through a canyon cut through cliffs of layered basaltic rock, follows exactly the spreading center of the ridge. For the canyon is cut not by a river but by a series of faults, caused by the two sides of the canyon pulling away from each other, with the valley floor between dropping down because its supports have been stolen from it.

The cliffs on the east of the canyon are in Europe; those on the west are American. They are pulling apart at a rate of about one-tenth of an inch every year; and the floor between the cliffs—where the roadway runs today and where I drove back in 1965—is dropping at about the same rate.

This, then, is the true eastern edge of the North American Plate, and I had indeed stood there forty years before, perhaps with a foot on both it and on its Eurasian Plate neighbor, even though I didn't know it at the time. I was pleasantly intrigued when I realized that one of the world's first structures made to house the fledgling idea that later evolved into a form of rudimentary democracy had been sited, centuries ago, at the very edge of one of the world's most crucially important geological pivot points. Synchronicities between geology and expressions of humans' physical achievements are legion, of course—roads and railways run along valleys, cities tend to be built at river crossings or by estuaries, national boundaries follow mountain chains. But few are the coincidences, so far as I know, between the underpinnings of the earth and the foundation of ideas; and I found it elegant and satisfying to imagine Iceland as a place where this seems to be true. Not least, perhaps, because of its reciprocal: that a wealth of ideas of quite another kind is being produced at the other end of the same plate, in California. Tectonic plates may have more of an effect on those who inhabit their livelier parts than anyone cares to notice.

The eastern edge of the North American Plate—and, coincidentally, the site of one of the world's oldest parliamentary democracies: Thingvellir, Iceland. The plate's western edge, which is infinitely less stable, is pictured on page 170.

FROM SURTSEY we had to head west, first across the Denmark Strait to Greenland and then on down to Newfoundland. As we did so, the rocks that composed the landscape became steadily older, and the real character of the plate started, in spectacular fashion, to assert its identity and personality.

Not that this aging of the rocks was dramatically apparent at first. Our icebreaker rammed its way slowly, steadily, and very noisily across the strait, punching leads in the floes for three full days before finally emerging below the curtain wall of tall and embrasured black cliffs with which the East Greenland coast is fortified. We knew a fair amount about these cliffs, and the nunataks, the black mountains that speared through the ice cap behind. We knew they were basalt, the same fine-grained frozen lava that made up the canyon walls that rose beside Thingvellir, and we knew that they were older, though in the geological scheme of things only marginally so. Those back at Surtsey were brand-new—rocks of the entirely modern Holocene Epoch, which we had seen being fashioned before our eyes. Those in the cliffs above the Althing, on the other hand, were a little older—the simple fact of their solid existence being proof of that—by a few million years (comparing the amount of the decay products of rubidium and strontium and other once-radioactive-marker elements would easily give an accurate figure). And the rocks here in Greenland were older still—perhaps 30 million years old, maybe a little more. They were nowhere near as old as rocks in the island's center—but these were buried beneath miles of ice and for now were barely visible. The East Greenland basalts merely hinted at the age of things to come.

But the East Greenland basalts interested us for another reason, when we first went there in the mid-sixties. Back then no one could be certain that the continents were spreading apart—and only a scattering of well-connected scientists had any idea of the existence of such entities as tectonic plates—but there was, nonetheless, a widespread feeling that the continents might not always have been where they were today. At the beginning of the century Alfred Wegener had said that a phenomenon he called continental drift had occurred, and that what we think of as a solid earth was not solid at all. These ideas, though they were for decades derided by many in the scientific establishment, were in later postwar years tempting some believers to look for evidence that might prove that Wegener was right—the continents had moved. The apparent "fit" across the Atlantic—the bulge of Brazil looking as though it might fit handily into the bight of West Africa, for example—was proving impossible to ignore. And so student expeditions galore were being sent out from sixties Britain around the world, often organized for reasons of biological or anthropological inquiry that had nothing, ostensibly, to do with the possibilities of continental movement; the expedition leaders, however, were taken aside before setting sail and politely asked if they would mind keeping a weather eye open for any compelling evidence, for any jigsaw puzzle pieces that looked as if they might fit.

The expedition that we took to the Blosseville Coast, as it was called, of East Greenland was just one such search for evidence: We were there to examine the basalts in a very specific way, to try to establish if there was indeed anything about their makeup that might suggest, conclusively, that their position had shifted in the 30 million years since they had poured and oozed and erupted out of the ground. So we spent several weeks taking core samples of the rocks, which we looked at back in the laboratories in Oxford. When we compared the direction of the magnetism in the tiny crystals of hematite, which had frozen themselves into the basalt like microscopic compasses at the very moment they were laid down, we proved, in what came to be our own private (if small) *Eureka!* moment in the affirmation of tectonic plate theory building, that the Atlantic Ocean had indeed become fifteen degrees of longitude wider at this point* during the 30 million years since the basalts were laid down. Greenland had moved westward during those years—whether it had shifted or drifted, been shoved or pulled, or accelerated away we could not tell. But those little

*Latitude 70° N—the implication being that the ocean could have widened itself very much more, by several thousand miles, in the more southerly latitudes.

magnets told us incontrovertibly that it had moved, and the way was now left open to others to explain why.*

There were other Greenland rocks nearby that were fascinating, too, and that remain a subject of much study still (in a way the basalts do not, as their story is now familiar to everyone in the geological community). These were located about a hundred miles farther south, close to the community of Angmagssalik: They are to be found in what is generally known as the Skaergaard Layered Igneous Intrusion, and constitute one of the more remarkable geological phenomena known.

When they were first spotted from the deck of a ship, all on board thought, seeing the horizontal layers that were such a prominent feature, that they were sediments, like sandstones, with bands showing traces of the history of the period when they were deposited. But when geologists clambered ashore they quickly discovered that the rocks were not sedimentary at all but in fact volcanic—which was shocking, to say the very least. And these volcanic rocks had the outward appearance of having been laid down, instead of having been spurted out of a volcano, which was something that had never been seen before. Fieldwork began in earnest, and eventually it was discovered that the layering had been the result of a pod, or a lens, of trapped volcanic magma having been allowed, thanks to a coincidence of unusual circumstances, to cool very slowly indeed—to cool so slowly as to allow all the heavier crystals and minerals that formed during the cooling to sink downward, and all the lighter minerals and crystals to float upward, until the moment when the rock froze, or solidified, and layers were locked into place for all eternity. There are, for example, places where crystals of chromium salts have floated down to one particular place in the pluton—the tear-shaped body of rock, a mile or so in circumference—to form a layer of chromium so solid and metallic and thick that when hit by a hammer it rings.

---

*A full account of the expedition and its findings appears in chapter 3 of my *Krakatoa: The Day the World Exploded: August 27, 1883*.

Geologists from the world over visit the Skaergaard—and its two sister intrusions, one known as the Stillwater, in western Montana, and the other in the bushveld in South Africa. To see it is a compulsion common to many, to be there, to marinate oneself in its millions of unsolved chemical and physical mysteries. For an igneous geologist—the kind of scientist whom I have long thought of as an inquirer into a science of an elemental purity, a figure largely unbothered by the baser concerns of commerce—there is perhaps no more hallowed a place than the Skaergaard. Reputations have been won and lost there, theories have advanced and retreated like the glaciers that spill down to the sea. To the outside world the place is little known (and it was very little known indeed back in the sixties), but today if one whispers names like Kangerlugssuaq and Basistoppen and Aputiteq and Mount Wager (named after the Oxford professor, mine for a while, who first worked on it) in certain corners of university common rooms around the world, men in beards and wearing oiled-wool sweaters and corduroys will nod knowingly and smile.*

Curious and interesting though the Skaergaard is, its relevance to the North American Plate and, by extension, to the mechanics of the San Francisco Earthquake might seem just a little tenuous. It is not tenuous at all, however, and for two reasons.

The Skaergaard, it turns out, sits at the base of a four-mile-thick layer of basalts; during the millions of years that the Atlantic has been opening, untold trillions of cubic feet of lava have spewed from vents in what was then the center of Iceland, depositing on each spreading side layers of basalt that are miles and miles thick. But it is not quite so

---

*However pure and unmoved by commerce most igneous geologists may be, the growing familiarity of the Skaergaard and intrusions like it have not escaped the notice of those commercial companies that like to mine the planet for treasure. The chromium layers are just one of a number of metal-rich zones, and there is gold and platinum and palladium in showy abundance, seemingly there for the taking. One supposes that the Inuit Greenlanders may wish to preserve their corner of the world from plunder—but how long, with riches like these on hand, will they be able to hold out?

important for this story to know what lies on top of the Skaergaard. More important is what the intrusion itself lies on—and it so happens that it rests on top of two layers of rock, each of which, crucially, is of much, much greater age.

It sits first on a massively thick layer of sediments, filled with fossils, that was laid down in Cretaceous times, as much as 145 million years ago. These Cretaceous sediments in turn sit on a basement of rocks—highly contorted schists and gneisses, formerly made up of an entire spectrum of more recognizable sedimentary and igneous rocks but now hopelessly contorted under the influence of intense pressure, extreme heat, and eons of time—that are so ancient that some scholars of radiochemical-dating techniques believe they are the oldest in the world. It was the rocks of the Isua Formation in West Greenland that had, until 2002, been positively dated at 3,500 million years old, thus claiming the record; the fact that more recent studies have said a group of rocks at Porpoise Cove on Hudson Bay in Canada is 350 million years older still—before that the earth seems to have been an inchoate blob of hot gases and slithery supermelt—does not make the central question posed by the rocks' existence any less relevant.

And that question is: How is it that such very old rocks lie cheek by jowl with an array of rocks of relatively recent age? What mechanism allowed the world's youngest country of any extent—which is what Iceland is—to stand directly beside an island that has within its geological suite some of the oldest rocks on the planet? How did it happen—and why, for that matter?

The World Beyond Ur

Ever since 1915, when the quietly charming and tragically misunderstood German Alfred Wegener—a meteorologist, an explorer, and a pipe-smoking theorist of the first water—proposed the idea that the continents had not always been where they are today, there has been a grudging familiarity abroad with five hitherto entirely unfamiliar words: Panthalassa, Gondwanaland, the Tethys, Pangaea, and Laurasia. All of these words entered the language at Wegener's urging, though the first had been invented in the late nineteenth century by an Austrian polymath named Eduard Seuss, who thought continents floated and sank, popping up out of the abyssal dark on the command of some heavenly genie. Seuss, regarded by his supporters as an early seer of the science of geology, wrote an impenetrable four-volume book, *The Face of the Earth,* setting out his theories.

Wegener's basic notion, vaguely familiar in a back-of-the-mind kind of way today, held that there had at some stage on our planet been the one huge supercontinent, Pan-*gaea,* which was surrounded by the vast proto-sea, Pan-*thalassa.* Part of this continent then foundered or split up or in some other exotic way broke apart to leave behind an immense scattering of bodies of land, of which Gondwanaland*—an assemblage that contained today's India, Madagascar, Australia, southern Africa, and Patagonia—is the best known, together with a sea that surrounded the remaining continental islands and that Seuss was moved to call, after an ancient Greek sea giant, the Tethys.

This basic model has survived ever since. The birth of the plate tectonic theory has done little to dampen the enthusiasm for the story of Pangaea, and it is now reckoned with some certainty that such a supercontinent did indeed exist between about 200 and 300 million years ago. Its plates did indeed break up more or less as Wegener surmised and, in doing so, formed mountain ranges and oceans, many of which still exist today. Movement is still going on: The Atlantic widens, Australia shifts northward, plates hurtle slowly toward Alaska, the Pacific folds and buckles itself beneath California. This much is certain: Its processes are complex and its effects dramatic and often catastrophic.

But what of the time *before* Pangaea—what of the time before the cutoff period of 300 million years that is commonly ascribed to the

*The Gonds are a southern Indian people known to the imperial British for their protruding navels and, in more recent times, for their remarkable mural paintings.

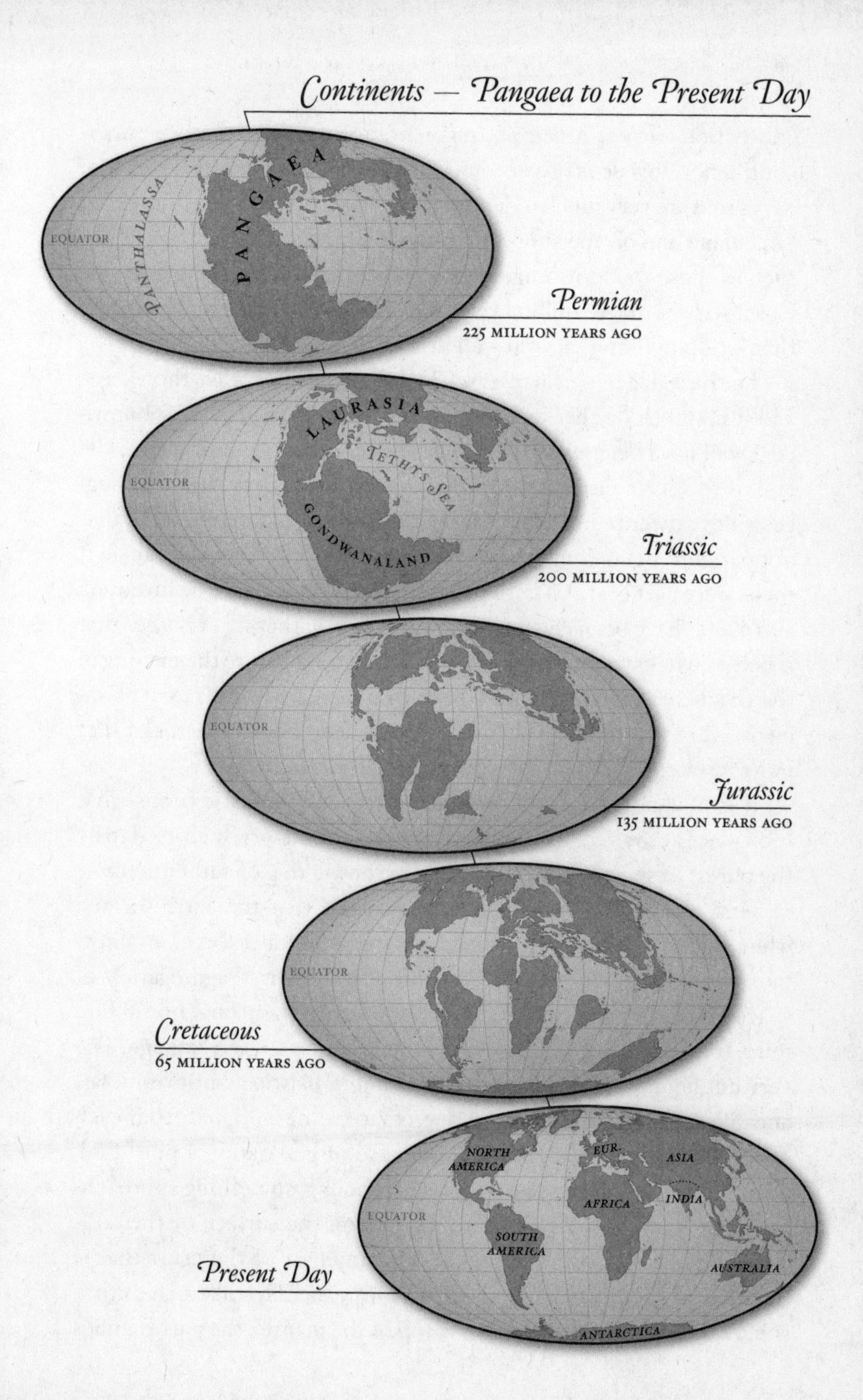
Continents — Pangaea to the Present Day
PANGAEA
PANTHALASSA
EQUATOR
Permian
225 MILLION YEARS AGO
LAURASIA
TETHYS SEA
GONDWANALAND
EQUATOR
Triassic
200 MILLION YEARS AGO
EQUATOR
Jurassic
135 MILLION YEARS AGO
EQUATOR
Cretaceous
65 MILLION YEARS AGO
NORTH AMERICA
EUR.
ASIA
AFRICA
INDIA
EQUATOR
SOUTH AMERICA
AUSTRALIA
ANTARCTICA
Present Day

convection-current-ordained fragmentation of the Pangaean supercontinent? How does one account for the existence and the placing of rocks that are very much older than this—such as the ancient rocks of Greenland and on the shores of Hudson Bay? How do juxtapositions such as those found in Angmagssalik and inland from the Blosseville Coast—rocks not 50 million years old sitting atop rocks that are more than 3,000 million years old—all fit into the picture?

The pace of geological research is alarmingly impressive these days, and it may well be that by the time this appears in print new knowledge will have been uncovered and new models drawn. But since the end of the last century information has come to light that now suggests the existence of a great extended family of supercontinents that popped into being long before Pangaea, and that these very ancient rocks were part of it. An entire new taxonomy has had to be invented, and names have been given to an entirely new gathering of bodies that is believed to have existed in the world long, long before the making of the fragments of today's plates. Parts of those early worlds exist as ancient echoes in today's Greenland, lying deep below the rocks that more modern processes have created to lie above them.

It is now possible to imagine in increasingly realistic terms what took place at the very beginning of the planet's history. It all used to be the purest of speculation; now it has a growing ring of authenticity.

First things first. It is generally accepted that the earth formed when, under the bonding influence of gravitational force, an enormous liquid or solid mass coalesced out of a myriad aggregation of space-borne components that were drawn together some 90 million miles from the star we call the sun, some 4,550,000,000 years ago. The very hot liquid-metal core and the hot liquid-plastic mantle of metals and silicate magma became, in due course, differentiated from each other, also under the pervasive influence of gravitational pull. And, after about 100 million years of gradual cooling, something approaching a stable and solidified scum formed on the surface of the still-boiling or simmering planet. (The scummy part of the crust that is involved in tectonics is these days more generally called the lithosphere, and the plastic layer at the top of the mantle, the part that lies above an important line that is recognized only by the high priests of physics, is known as the asthenosphere.)

It was at this point in the planet's history that the earth's eggshell-like crust, which was slowly forming on the surface from this cooling scum, began to stop doing what up to that point it was prone to do, and that is to keep on remelting itself. For eons it kept sinking back into the mantle just a few millennia after it had formed, utterly wrecking itself in the process—and then it would pop up out of the molten ocean of lava and be reborn in a totally different guise. Instead, all of a sudden, large chunks of crust were staying afloat, more or less permanently. In cooling, the crust was forming itself into rocks that would themselves be permanent—if only the external forces permitted them to remain at the surface and did not try to drag them or push them down toward the heat again.

As they slowly cooled , some of these rocks-to-be separated themselves out, according to perfectly understandable laws of physics: The lighter materials of the scum rose to the surface, the heavier ones passed downward in one enormous fractionating column—a little like the Skaergaard, though over infinitely longer periods of time and under very different physical conditions. The lighter materials generally formed themselves into those rocks we now call granites—the coarse-*grained* rocks that tend to be prettily light in color as well as in constitution. The heavier fractions created layers of rocks like basalt and diorite and gabbro, which were darker and tended to sag downward under the force of gravity, forming sloughs, whereas the granites tended to form uplands. The darker and heavier slabs lay sluglike and low on the earth's surface, and in time they began both to accumulate and to accommodate water that fell from the skies; over many millions of years, this resulted in the creation of oceans. Dark rocks underlay the seas; granites made up the new continents. And this law of basic igneous geology has remained a verifiable truth ever since.

The new crust, as it spread and wafted itself around the surface of the sphere, also became cracked, as cooling crusts of clinker and furnace slag are wont to do, and the plates, or rafts, or slabs of floating or sagging clinker that were then formed between the cracks began to swirl about, thanks to the currents of terrifyingly hot material that were (as they still are today) upwelling and sinking back underneath. No doubt the slaggy scum came under the influence of other forces: There was gravity, there were great gyrations in the planet's magnetism, there was its spinning motion, the occasionally too-close-for-comfort proximity of the moon and other planets, and the tilting and wobbling of the earth's own axis of rotation. The third planet from the sun, it must be remembered, is in geological terms a comparatively small ball of material, subject to all manner of kinetic and thermal influences; and the first continent-in-the-making was turned this way and that for millions of years, as it struggled gamely to get a grip on itself and remain more or less in place on the ever-changing molten mantle that underlay it.

And finally, about 3,000 Ma,* it emerged as a fully fledged entity, sizable and solid and stable enough to be given a new name, to be classified as something else entirely. Enough of the crustal material floating about had now gathered itself together. Scores of islands of granitic material, floating on a basement of darker rock, had agglomerated, like raindrops on a windshield, to produce ever-larger accretions, which themselves met and married and did so again and again—until, midway through that period of geological times now called the Archaean, they combined to form one very large body, one covering sufficient of the earth's otherwise still-molten surface to be classified as, and called, a continent.

Those with fanciful imaginations might say it was shaped rather like a bird, an albatross with outstretched and enormous wings. It was small, compared to the immensity of the earth—it seems to have been

---

*This is the current internationally accepted abbreviation for Million Years Before Present. The lowercase *a* is a creation of the Système International d'Unités, which sets down standards for all manner of measurements, including the meter, the kilogram, the second, the ampere, the candela, the Kelvin, and the mole. It is the accepted abbreviation for the Latin *anno.* It serves also as replacement, in the geological context, for another abbreviation, that of the archaic word "agone": *ago.*

about 5,000 miles from birdlike wingtip to wingtip and maybe no more than a thousand miles from north to south across the thickness of its bird body. It seems to have lain close to the notional equator of the early earth, a little to the western side of where the meridian would eventually be.* And then it broke up, and its granitelike rocks were, in the fullness of geologic time, scattered to the four winds; they have since spread themselves liberally all over the planet. Gigantic amassments of rock from this strange little protocontinent are visible, and perfectly recognizable, in places like Zimbabwe, southern Australia, central India, and Madagascar.

This frail-looking, tiny, and delicate thing is in truth the *fons et origo* of everything that is solid and habitable about our earth today. Which is why, when it was named, shortly before the twentieth century was ending, it did not take much of an effort of mind or spirit to decide to christen it, most appropriately, the continent of Ur.†

However, the supposed creation of Ur prompts a question: If Ur is 3,000 million years ago, how can it be that the rocks of Greenland and around Hudson Bay are 3,500 and 3,850 million years old respectively? Why were they not a part of Ur?

The answer is that for some reason—and at the time of this writing it is still an unfathomed reason—the first aggregation of small precontinental bodies occurred in the planet's Southern Hemisphere. Since the current configuration of the world has by far the greater amount of its landmass to the north of the equator, this presents something of a poser. But the evidence is unassailable: It is very clear that a number of modest-size rock masses had formed and become more or less stable in a variety of places around the world; in particu-

---

*After John Flamsteed designated it as running through Greenwich, near London, in 1675 and then twenty-five countries formally adopted it, or another very close by, at a conference in Washington, D.C., in 1884.

†Ur was named by a geophysicist at the University of North Carolina, John Rogers. His paper announcing the idea, and suggesting it take the name of the ancient capital of the Chaldees, appeared in the *Journal of the Geological Society of India* in 1993.

Pre-Pangaea Continents

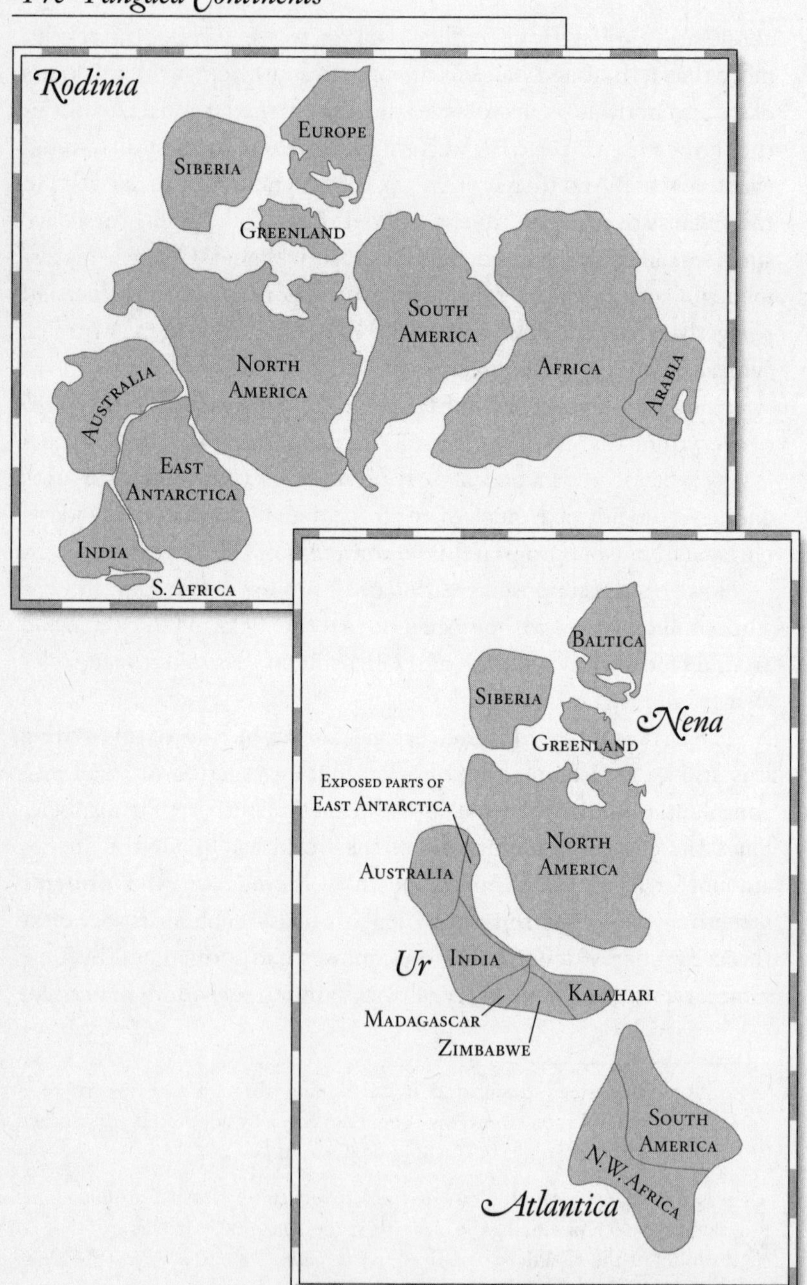

lar, some 850 million years before the formation of Ur, there was a significant quantity of small bodies of granite floating around in the Northern Hemisphere, thousands of miles away from where Ur would eventually form. But none of these northern bodies had met up with any of their neighbors and massed to form a continent-size body. That did not happen north of the equator until about 2,500 Ma—500 million years after Ur was created in the south and fully 1,350 million years after the rocks themselves had been created out of that crystallizing and fractionating column of cooling magma.

The first continent that is believed to have existed in the north has in recent years been christened Arctica. It is a gathering of granite islands that includes the very rocks of Greenland and Hudson Bay that we have been considering; it includes also a vast amount of material of what would later become Siberia; it has a smaller body of very old rock that in billions of years would become Wyoming; and it enfolds hundreds of thousands of square miles of what would later be northern and northwestern Canada.

That said, it needs to be noted that a small plume of national chauvinism intrudes at this point into the story. Canadian geologists have long claimed that this conglomeration of granites and other very old and stable "shield" rocks that exists in the Canadian North and Northwest was already large enough by 2,500 Ma to be called a continent in its own right. The most typical granites of this region occur in and around Kenora, a town on the Lake of the Woods close to the border between Ontario and Manitoba. Much earlier research showed that the Kenora series of rocks displayed evidence of a major episode of ancient mountain building, a so-called orogeny, which had taken place all over Canada, as well as in Wyoming, the Dakotas, and the Outer Hebrides of Scotland (geology knowing no national boundaries, of course, and the distances between these "places" of yesterday having no relation whatsoever to the distances that we know of today: Wyoming and Scotland lapped up so close to each other then as to be one place, making the very concept of "place" more than a little surreal). In recognition of the importance of the Kenoran rocks and the Kenoran Orogeny, Canadians have proudly christened the huge body that they suppose to have existed Kenorland,* and they think of it as having a presence quite as valid and provable as that of Ur and of Arctica. Non-Canadians are not so sure, however, and wonder whether it is much more than a piece of an enormous and very ancient jigsaw puzzle.

The world became steadily more complicated as time wore on. There were two further coalescences about 500 million years later on, when the continents now known as Baltica and Atlantica emerged, also then in the Northern Hemisphere. Baltica held much of what is now northern Europe as far south as today's Ukraine. Atlantica, on the other hand, encompassed what is today's West Africa, Congo, Guyana, Brazil, and the region around the river Plate, all landmasses that would eventually shift south of the equator.

The world was now possessed of four continents—or five, if Kenorland is counted—Ur, Arctica, Baltica, and Atlantica, which sound as if they come from the title of a short story by Borges. It seems that these bodies, all of them massive, cool, and quite stable platforms, probably represent the totality of continental material that would be on the fledgling earth for a very long while.

And, having been fully made, these bodies then began a complex dance—a dance that, in the phrase of some fascinated geophysicists who are brave enough to mix metaphors, seems an accordion-like process, in which continents clang into one another, sometimes joining up, then separating, most colliding and separating once again for the remainder of their existence, right up to the present day. Knowledge of what happened and the sequence in which it did so is still ragged, and since the creations that resulted from the marriage of the various bodies all have new names, too, the whole *megillah* adds up to a delicious tectonic confusion. In essence, though, it seems to have unrolled itself approximately thus:

---

*Whatever dignity might be conferred by naming a continent "Kenorland" is marginally lessened by knowing that Kenora itself is a somewhat less-than-stellar acronym. The word comes from the first two letters of each of the names of the neighboring settlements of Keewatin, Norman, and Rat Portage, the last so called because of the annual migration of muskrats along the local river.

First, about 1,800 million years ago, or maybe a little before, Arctica (which, it will be remembered, contained Siberia, Canada, Greenland, a bit of Wyoming, and the majestic monster called by some Kenorland) collided with Baltica, which held a great deal of what is now far northern Europe down to the Ukraine, and the two subsumed for good measure a small and noncontinental part of what is now the Antarctic. The resulting supercontinent has been called Nena (or Nuna to some, who claim it as an Inuit word that means "the land around us," as in the recently formed Canadian province of Nunavut). Nena then collided with and promptly gobbled up Atlantica and became a truly vast body, which its namers (Americans) have seen fit to call Columbia (or, according to some others, Hudsonland). This enormous continent enfolded all of the world's continental blocks for about 400 million years, then broke asunder once more around 1,300 Ma, whereupon its parts—behaving for the next 300 million years or so with the accordion-like back-and-forth squeezing and pulling apart that seems to have marked the world's early progress—in due course collided with a re-formed Atlantica and a reinvigorated Ur and created yet another extrasupercontinent that has been called Rodinia, after the Russian word for "homeland."

Rodinia then stayed stable, despite being so immense, for the next 300 million years—from about 1,000 Ma to 700 Ma, or the period known as the late Precambrian, when there was a fairly healthy amount of multicellular life to be found on the planet (fossils of which are scattered about the world's outcrops today, in patterns that indicate where Rodinia had been stitched together). But then, and one hesitates to say *once again,* it fractured into a number of slightly different constituent parts—Laurentia, East and West Gondwana, bits of Ur, bits of Atlantica, all of which careened across the millions of square miles of crust before finally reassembling themselves into yet another body, which has been named Pannotia (which essentially means "all southern continents"). Another breakup, another reconciliation—and then, with some parts eventually drifting back together about 550 million years ago, and after a quarter of a billion more years of miniseparations and microdivorces, all were rejoined in an unholy tectonic matrimony almost exactly 280 million years ago, forming what we now know, with all the familiarity of its relative chronological closeness to us (a mere 250 million years, after all), as Pangaea.

From this point on, the world's historical story becomes relatively plain sailing. It is generally very well known, but it belongs in its details to a later part of this story. The weird, almost biblical passages of complexity that speak of these places of a much greater antiquity—of Ur, and then of Atlantica, which in turn begat Rodinia; of Baltica, which begat Nena; and of Rodinia, which ultimately spawned Pannotia, which then in turn begat Pangaea—are daunting, to say the very least. But after the making of Pangaea 250 million years ago, all suddenly becomes much clearer. Pangaea ultimately split into two of its original building blocks, a northern massif called Laurentia and the more southerly and somewhat-better-known Gondwana. The two bodies then themselves began to fracture into a myriad pieces, the present-day oceans opened up, the present-day continents formed from the aforesaid myriad pieces, and these in turn—Africa, Europe, the Americas, Australia, Asia—started to pull apart from one another, or to travel toward one another, or to slide by one another; until eventually, with the exhausted arrival of all the continents, broken up, rewelded, reconfigured, and reconstituted in the approximate configuration of the present, geological history ended. The Triassic became the Miocene, then the Pliocene and the Pleistocene and the Holocene, and the present day arrived with its culminating finality.

For some, other explanations will become clear also. Once the awful rote of the supercontinents' new names has been assimilated, and if all of this theorizing is borne out by fieldwork and the facts and proves to be more or less correct, then the perceived development of the earth that emerges assumes a simple elegance, a certain cosmic symmetry.

Punarapi jananam
Punarapi maranam
Punarapi janaijatare
Sayanam.

Adi Sankara's famous sixth-century Sanskrit song "Bhaja Govindam" seems to have more than a little relevance here. *Birth again; death again; repose in mother's womb again.*

Moreover, the simple elegance of the picture that emerges manages in an instant to answer, as plate tectonics answered in its early days, a lot of the very largest questions.

Why are there rocks both terribly ancient and extraordinarily modern smashed together in modern Greenland? Because, as the history above easily explains, what is now Greenland was once a group of freshly made granites that formed into something big enough to be called the continent of Arctica, and eventually Arctica became Columbia, Columbia evolved into Pangaea—and once Pangaea broke up in fairly modern times, the Atlantic Ocean opened up, the Mid-Atlantic Ridge fissured, and thick black basalts welled up—as they are still doing today on Surtsey. These spilled over and laid themselves down on the old Pangaea-Columbia-Arctica-Ur granitelike continental rocks that exist underneath, making the confection of geology that—in juxtaposition with all the ice and snow of climate and the storms and winds of weather, the polar bears and lichens of biology, and the Eskimo and Inuit and Danes and American soldiers of anthropology—constitutes the great and mysterious island known today as Greenland.

And the added beauty of the history is this: that this very same simple and elegant explanation can be wheeled out to offer valid answers to any of the major geological questions that scientists are currently pondering. What, for example, caused the terrible undersea earthquake in northern Sumatra in 2004, which sent up waves that killed hundreds of thousands in the Bay of Bengal? Why do large and very dangerous volcanoes threaten parts of the Congo? Why are there more than twenty huge volcanoes in the main islands of Indonesia? Why is the geology beneath Moscow, or around Hudson Bay, or southern Queensland, so disarmingly stable? What lies behind the creation of the newly found diamond deposits in the Yukon that have so excited Canada? Why are the Cyclades where they are? And Lipari, Tristan da Cunha, and Kerguelen? To answer any one of these questions one trundles out the same litany of proper names, the same descriptions of motion and rising and falling, and the same visions of long-vanished places: Ur, Pannotia, Gondwana, Arctica.

And why was there an earthquake on Wednesday, April 18, 1906, in San Francisco? The ultimate reason is just the same. The rocks shift, one set against the other, in California as everywhere else—for reasons that, at bottom, have everything to do with the behavior of plates that are the legatees of Gondwana, Arctica, and Ur.

A Certain Nervous Shaking

It is a very long way, both in space and in time, from the continent of Ur to the city of Charleston, South Carolina. But Charleston, the pretty and some might say ostentatiously gracious colonial town that has lazed in the soft, wet heat of the Carolina coast since English settlers founded it in the late seventeenth century, is a place of great seismic interest, and so it is a place that has connections, and to geologists very interesting connections, with the making of the world's ancient continents.

Not that anyone would have suspected such a thing, back when the town was founded. Charleston had just been sitting quietly and prettily on its coastline, its citizens unperturbed and their manners genteel—until the very last day of August 1886. On that day, however, and according to a resident named Paul Pinckney, something happened. There came a sudden slight vibration, he wrote, "as when a cat trots across the floor."

This was the harbinger of what a few moments later would be the greatest eastern earthquake in American history—a powerful, deadly, and highly destructive quake that brought ruin and chaos in its wake.

But it was far from being the only eastern earthquake. My journey across the North American Plate was well under way; Iceland was far behind me, Greenland had faded into the background, and so had Newfoundland and Nova Scotia and the states of upper New England.

Now I was driving in the area around my home in the Berkshire Hills of western Massachusetts,* and all the talk seemed to be of earthquakes. Stories about past earthquakes abounded. There seemed, all of a sudden, to be a surge of seismic activity in this particular corner of the plate. Many of my neighbors in the Berkshires, for instance, claimed to have felt earthquakes—one friend had been woken a few weeks before by a tremor that turned out to have had its epicenter in Amherst, Massachusetts, thirty miles from my home. The local seismic observatories say they record an average of forty earthquakes each year in New England, of which five are strong enough to be felt by any reasonably aware person.

In the American Northeast the geological factors that trigger earthquakes seem to be all around. Just to the west of New York City, for example, there is a fault system that runs in the same direction as the Appalachian Mountains that rise up to its west. These faults, running along what is known as the Ramapo Fault System, occasionally break along this northeasterly line, and, when they do, throw the torn sides upward, or downward, toward the southeast. Earthquakes, which are generally fairly minor and are felt only by a very few people, have thus been recorded sporadically all along the system for centuries.

Commuters will find it droll to realize that some of their rather commonplace hometowns are actually very well known in seismologists' circles, because of historically important earthquakes that have taken place there: The Wappingers Falls Sequence of June 1974* and the Annsville Event of January 1980, for example, are two fairly recent events with considerable standing among those who catalog such things. Some of the states in the region are beginning to take the possibility of earthquakes more seriously than before—it used to be thought that quakes were more or less exclusive to California and the

---

*My journey took me by way of Manhattan, Richmond, Fayetteville, and a small South Carolina town called Florence.

very active West—and Massachusetts, for one, has now added seismic requirements, albeit rather limited in their extent, in local building regulations.

The northeastern earthquakes may be more numerous than most outsiders suppose, but they are quite minor even at their worst. The same cannot be said, however, of what happened in Charleston: That was an almighty disturbance, well worthy of inclusion in the lists of the great and the dreadful.

But though what took place in Charleston was certainly terrifying, it did not come entirely without warning. Four days earlier, on Friday, August 27, up north in Summerville, a hamlet about twenty-five miles from Charleston, the entire population was stunned at about 8:00 A.M. by the sound of what they all thought was a tremendous explosion: A train boiler had blown up, some said; or the phosphorus plant on the far side of the river had caught fire; or a hidden stash of gunpowder had been detonated. But no, there was no ready explanation for something everyone had heard and felt. Though very few knew anything about earthquakes, a number of Summerville's old-timers warned that perhaps this had been one. But the rest of the day passed without incident. Ah, well, they said collectively. Back to the porch and the evening julep. The morning papers would explain.

But next morning, Saturday, at 5:00 A.M., it all happened again. This time people were awakened, and many were tumbled from their beds by a sudden violent crash. For the rest of the day the earth below continued to tremble: Sometimes a series of inexplicable vibrations was accompanied by a slight rumbling sound, and at other times not. Anxious and afraid, the local burghers telegraphed their observations and their concerns to the state authorities. The newspapers across the state line in Georgia picked up on the fretful condition of their neighbors, did some perfunctory research, and began to worry themselves:

---

*It is believed that the Wappingers Falls Earthquake, which occurred in a section of the Ramapo Fault already long overdue for a rupture, was in fact triggered by drilling from a nearby quarry.

An editorial in the *Atlanta Constitution* the next day warned that perhaps the area, which was usually seismically peaceful, was standing into danger.

On Tuesday evening the tension suddenly broke. Thomas Turner, the president of the Charleston Gas Light Company, had arrived home in Summerville in his carriage, weary from a day's work at the office:

> I had been out in the garden admiring the beauty of the evening and was entering the door of the hall of my house when, without any rumble or warning, the floor seemed to sink under me. I seized the door jambs to steady myself, when the floor seemed to go down in front of me at an angle of twenty-five or thirty degrees. It was so sudden and unexpected, that I was thrown forward into the hall about ten feet, and as quickly thrown backwards, and before I could fall upon the piazza I was again thrown forward into the house. At this moment I observed my sister-in-law crawling on all fours, she having been thrown from the door of her chamber, which she was just entering, into the middle of the sitting room. Amidst the rolling and rocking of the building she managed to reach the hall, but was unable to regain her feet. At this moment we observed the upper part of a lamp, which had been jerked off its stand, to fall upon the floor and burst. The oil took fire, and amidst the roaring and violent motions of the house, we succeeded in extinguishing it with pieces of carpet and rugs. Immediately after we received another shock, which threw us from side to side of the hall. Having gotten the members of my family together, and supporting my niece, who was in a fainting condition, we endeavored to leave the house amidst the crash of falling chimneys and plaster; but at every attempt to reach the door we were hurled backwards and forwards and from side to side, as if we had been in the gangway of a steamer in a heavy cross sea. After some delay we reached the garden. I then returned to get wraps and chairs for the ladies, and again experienced severe shocks and rumblings. These were repeated at intervals of several minutes during the night, but were not of so violent a nature, although the earth waves were very perceptible and several times upset a small lantern placed upon the ground.

Summerville seems to have been very close to the epicenter of what seismologists today believe was a magnitude 7 earthquake,* with a very concentrated episode of ground shaking in a corridor extending between the hamlet and Charleston itself.

Paul Pinckney then picks up the story. He was, as he wrote, "just a lad at the time," but his memory of what occurred in the city remained with him for the rest of his life.

The Tuesday had been hot and sultry, and people were remarking on the unusual stillness of the day. Willis Carrier, the so-called Father of Cool, would not invent air-conditioning for another twenty years, and so the streets were thronged with people doing their best to escape the oppressive indoor heat. Many were returning from church: Even on a Tuesday, and even in a city better known for its soirées, black tobacco, and strong drink, the town maintained a veneer of southern piety. Thomas Turner's gas lamps had just been lit, and groups of people gathered under their dim illumination to gossip, gawk, and ready themselves for home and bed. It was 9:51 P.M. Suddenly, something happened deep within the earth. And all the town clocks stopped dead.

It must have begun with a barely perceptible force. Then the deep-seated murmurings quickly became stronger, and within seconds the vibrations started to come, wrote Paul Pinckney,

> in sharp jolts and shocks which grew momentarily more violent until buildings were shaken as toys. Frantic with terror, the people rushed from their houses, and in so doing many lost their lives from falling chimneys or walls. With one mighty wrench, which did most of the damage, the shock passed. It had been accompanied by a low, rumbling noise, unlike anything ever heard before, and its duration was about one minute.

---

*Charleston's earthquake was large compared to that at Bam in 2003, which had a magnitude of 6.6 but resulted in the deaths of more than 30,000 people; both of these events were modest when ranged against the Sumatran event of December 26, 2004, which had a magnitude of 9.0, killed about 275,000, and is one of the biggest quakes of all time.

Carl McKinley, who was a reporter on the *Charleston News and Courier,* was working in his second-floor office that evening. He was writing at his desk when his attention was "vaguely attracted," as he later wrote, by a noise that seemed to come from the office below. It sounded like something rolling past down on Broad Street—an iron safe, he imagined, or maybe a heavily laden dray:

> For perhaps two or three seconds the occurrence excited no surprise or comment. Then by swift degrees, or all at once—it is difficult to say which—the sound deepened in volume, the tremor became more decided, the ear caught the rattle of window-sashes, gas-fixtures, and other movable objects; the men in the office, with perhaps a simultaneous flash of recollection of the disturbance of the Friday before at Summerville, glanced hurriedly at each other and sprang to their feet with the startled question and answer, "What is that?" "An earthquake!" And then all was bewilderment and confusion.
>
> The long roll deepened and spread into an awful roar, that seemed to pervade at once the troubled earth and the still air above and around. The tremor was now a rude, rapid quiver, that agitated the whole lofty, strong-walled building as though it were being shaken—shaken by the hand of immeasurable power, with intent to tear its joints asunder and scatter its stones and bricks abroad, as a tree casts its over-ripened fruit before the breath of a gale.
>
> There was no intermission in the vibration of the mighty subterranean engine. From the first to the last it was a continuous jar, adding force with every moment, and, as it approached and reached the climax of its manifestation, it seemed for a few terrible seconds that no work of human hands could possibly survive the shocks. The floors were heaving underfoot, the surrounding walls and partitions visibly swayed to and fro, the crash of falling masses of stone and brick and mortar was heard overhead and without, the terrible roar filled the ears and seemed to fill the mind and heart, dazing perception, arresting thought and, for a few panting breaths, or while you held your breath in dreadful anticipation of immediate and cruel death, you felt that life was already past and waited for the end, as a victim with his head on the block awaits the fall of the uplifted ax.

For a second or two it seemed that the worst had passed, and that the violent motion was subsiding. It increased again and became as severe as before. None expected to escape. A sudden rush was simultaneously made to endeavor to attain the open air and fly to a place of safety but before the door was reached all stopped short, as by a common impulse, feeling that hope was vain—that it was only a question of death within the building or without, of being buried beneath the sinking roof or crushed by the falling walls. The uproar slowly died away in seeming distance. The earth was still and oh! the blessed relief of that stillness!

But how rudely the silence was broken! As we dashed down the stairway and out into the street, from every quarter rose the shrieks, the cries of pain and fear.

Almost a hundred people died in the Great Charleston Earthquake of 1886, thousands, out of a population of 49,000, were injured, and it was said that seven in every eight houses—most of them substantial and very beautiful houses—were damaged, some a total ruin. Paul Pinckney, who as a young man had witnessed the destruction of Charleston, had the ill fortune to be in San Francisco twenty years later when it was destroyed by its own great earthquake; he wrote, "I do not hesitate to assert that the temblor which wrecked Charleston was more severe than that of April 18 last, and in relative destruction considerably worse."

In technical terms he was quite wrong—though there seems to have been an extraordinary degree of ground shaking in Charleston, caused by locally peculiar factors: wet ground, compaction of the soil, and so forth. But what did happen was very big indeed, and very dangerous.

A comprehensive study of the Charleston event, published in 1977 by the U.S. Geological Survey, and which included rigorous analysis of all the anecdotal evidence from around the country, showed that the city had suffered an extraordinary and unprecedented seismic assault. The reverberations were felt as far away as New York, Chicago, Milwaukee (where "a lady living on Huron Street was so frightened she

Damage from the Charleston Earthquake of 1886 offered late-nineteenth-century America a hard lesson about the need to build good and strong.

---

fainted," reported the *Sentinel*), Nashville, and Baton Rouge. In the immediate neighborhood of Charleston there had been at least a dozen foreshocks—like the two experienced in Summerville that had prompted the warning in the Atlanta newspaper—and several hundred aftershocks, which can be dispiriting and demoralizing to a population already terrified by what seismologists refer to, with the clinical brutality of their trade, as the mainshock. The aftershocks were as powerful as magnitude 5.3, though they tailed off after a week, allowing the city to get back on its feet and to rebuild.

It did so most nobly. Within four years the population had increased by a tenth, and masons and bricklayers and carpenters were busily removing all the scars of an event that some supposed might ruin the city forever. Some of the original colonial buildings were demolished, but most were shored up and repaired, and modern Charleston is one of the architectural gems of the South, and indeed of all America, because of this wonderful stock of eighteenth-century mansions.* As most of the citizens were suffering under the blight of absolute poverty that had followed the Confederacy's defeat in the Civil War—Charleston, it should be remembered, is where the War between the States began, with the Confederates' capture of Fort Sumter in 1861—the majority of earthquake victims chose to patch up and preserve their damaged houses rather than tear them down and rebuild. "Too poor to paint and too proud to whitewash," was a saying of the time—and it led to a recovery that was based on preservation, not wholesale renewal. The legacy is a city of beautiful houses—gracious homes, as they would say locally—and so a blessing for which all who visit today must be thankful.

But why an earthquake? A look at any map will prompt the question—for Charleston, a town built at the confluence of three sluggish rivers on a muddy coastline, is hundreds of miles from anything that looks even remotely geologically active. The Appalachian hills are too distant to be even blue smudges on the city's western horizon. There is not a volcano in sight, not a hot spring, nothing but mud and gravel and low cliffs and beaches of white sand. The eastern edge of the North American Plate is pulling itself away from the western edge of its European neighbor 1,500 miles away at least, somewhere beneath the waters of the Atlantic. The western edge of the North American Plate is 3,000 miles away, close to the San Francisco toward which we are slowly driving. There are no colliding tectonic plates anywhere near here, not crashing into each other, not easing themselves beneath

---

*Standard Oil, it is often said, did far more lasting damage in Charleston than any earthquake when, in the fifties, it demolished the magnificent mansion that had belonged to a grandee and architect named Gabriel Manigault, and built two gas stations on the property in its place.

each other, not slyly rubbing against each other, edge against edge.*

It can be said without too much of a challenge that all of these places—Charleston, Peekskill, Wappingers Falls, Annsville, and Summerville—stand more or less in the geographical middle of the North American Plate (though not in the middle of *North America*) and so in a place where earthquakes are simply not supposed to happen. The more agitated business of the world is conducted at the edges of the plates: 90 percent of the world's earthquakes occur there, while barely anything goes on in the plates' centers.

So what mechanism caused such a great disturbance in so improbable a center-plate location as Charleston, South Carolina? Why the Wappingers Falls Sequence of 1974, in an equally seismically unpropitious place? Why Peekskill, New York? And why does the Ramapo Fault System exist where it does, with its swarm of minor but irritatingly mobile faults—causing puzzling vibrations throughout the suburban lots of Scarsdale and ripples on the surfaces of all the Hockney-blue swimming pools between Montclair and Bedford Hills?

To begin to answer that—and geologists have been grappling for years with the vexing problem of those earthquakes that occur where they ought not to—I first had to drive some 600 miles west, to the site of one of the most remarkable earthquakes that America has ever known—by some accounts the biggest ever experienced in the Lower

---

*The earth remains very unstable in these parts, whatever the reason. On my computer is a map, brought up-to-date every five minutes, that shows all the earthquakes that occur in the United States and its possessions, every hour of every day. On any given map—which preserves a week's worth of recorded events—there may be as many as 550 separate items, each one a record of some seismic happening, so long as it is greater than 1.0 on the Richter scale. As it happens, at the very moment I was completing work on the 1886 Charleston Earthquake, a small red square appeared on the map, in southeastern South Carolina. I zoomed in and found that at 09:13:14 Universal Time on this particular summer's day there had been a magnitude 3.0 event six miles below the earth's surface, just outside the hamlet of Summerville—where it had all started a century and a quarter before.

Forty-eight states. I then had to travel another 400 miles westward, to a village set deep in the midwestern plains, where a seismograph is mounted inside the general store. My second destination was a somewhat obscure and all-but-forgotten place, though one of some importance in explaining why America suffers earthquakes so far from the edges of the plate on which it stands. My first intended stop, however, was at a town that suffered an event that took place over a series of weeks during the winter between 1811 to 1812—a hitherto unremarkable Mississippi riverside town that has since entered the lore and the lexicon of seismologists around the world: New Madrid, Missouri.

The Unstill Center

Hundreds of anecdotes and newspaper reports suggest that there was something very unusual about the violent shaking that spread from the three great winter earthquakes that were centered on the village of New Madrid, Missouri, around the end of 1811. To widespread contemporary astonishment, the quake was felt in places as far away as New York, Toronto, Montreal, Boston, Chicago, Milwaukee, New Orleans, and Charleston (where it rang the bells of Saint Philip's Episcopal Church, later to be destroyed by the 1886 quake). And it was more than merely *felt:* The quake caused damage far away, together with panic and despondency in large measure.

Indeed, the kind of shaking (or *rocking,* which was the word used in the American Southeast at the time) that can bring down poorly made buildings appears to present-day researchers to have extended in 1811 over almost a quarter of a million square miles of territory—and the earthquake was felt, was *detectable,* over a million square miles. By way of comparison the San Francisco Earthquake, though it was much more intense and more lethally destructive, was recorded over an area of just 60,000 square miles, and its damaging heart extended to no more than a fifth of that. Clearly there is something strange about the subterranean makeup of Middle America that allows such earthquakes as begin there to radiate far more energetically beyond their epicenters than do those in the American West.

NOT THAT MANY Americans today are especially aware of the New Madrid event. Nor, it seems, are many even aware of the existence of New Madrid, in spite of the name being so familiar to the scientific world. I confess I had some slight difficulty finding the place, tucked away as it is in the flat and featureless boot-heel territory, the cotton and soybean country where Missouri marches confusingly alongside three other southern states—Arkansas, Kentucky, Tennessee—as well as a part of Illinois that was once so decidedly Secessionist in outlook that it was hard to imagine it being part of the same state that also embraced Chicago. Illinois is supposed to be a place of steers and hogs and railroads and dry goods—not of cotton bolls and corncob pipes and the lazy drawl of the pecan groves. Myth has it that President Lincoln, when making speeches in Cairo, Illinois—a town a few miles across and upriver from New Madrid, and about as far south as one can go in the state—tended to play down the Abolitionist and emancipation rhetoric he was wont to employ upstate: Here, despite the name of the state, he was already deep in Dixie.

I had been driving westward from Charleston for the previous two days. I had sped over the alluvial plains of the Carolinas for all too many tedious hours before climbing up over the ancient furrows of the Appalachians in eastern Tennessee, a state so very wide it seems as though it might have no end at all. But eventually, after descending steadily for 200 miles or so, after turning northwest near a town called Humboldt, after crossing the immense brown winding sheet of the Mississippi River near Caruthersville, and after driving north past a scattering of hardscrabble hamlets called Wardell and Braggadocio and Peach Orchard and Portage, I saw the signpost for my destination. I turned back toward the east at a junction where I was warned that I would find myself on my way to Cape Girardeau and St. Louis if I were careless enough to carry on up north. It was an icy January morning, and I drove on a narrow road through a deserted and frozen cotton field and past battalions of dead cornstalks, and in due course came to the undistinguished sprawl of double-wide trailers and small frame houses and grain elevators and a cotton-gathering plant and a scruffy little main street that is the community of New Madrid. I had been, I thought, to many more lovely towns than this.

There was a café, a store, a frowzy-looking bar, a museum—and then the community ended abruptly, stopping against a long, low hill of frost-browned grass that was topped by a grove of locust trees. I climbed up a flight of steps, and, from the hill's flat summit, along which ran a gravel track, I could suddenly see the Mississippi River. Spread out before me, it was vast, pale coffee brown, and slow-moving, with gigantic trains of barges being pushed silently and steadily along the navigation channel. The hill turned out to be a levee, a curtain wall built to protect the town and the cotton fields from the snowmelt floods of spring. I was looking east: There was the river, its buoys bending in the stream. Back west lay the town, and beyond, low and waterlogged meadows with wisps of gray mephitic mists.

Almost two centuries ago this place had been Spanish land—and the Kentucky that I could see on the distant horizon, where an aluminum smelter was belching its wind-whipped plume of smoke from behind the sedges, shallow marshes, and the brown grass slopes of the river's left-bank levee, was once the farthest extent of the United States, the western edge of the America that had existed before the Louisiana Purchase. The river had in those days been the formal frontier between America and New Spain—and the Spaniards briefly engineered a cunning and Machiavellian way of securing it, by encouraging Americans from the east side of the river to settle on Spanish land on the west side, thus forming a protective and insulating barrier against more militant Americans rather farther away.

One man who agreed to the Spanish scheme, and thus who, in doing so, briefly turned his back on his own country, was a Revolutionary War hero named George Morgan. He took the Spanish land grant, accepted Spanish citizenship, and, along with seventy like-minded men and women who traveled downstream in four heavily armed boats, established a town on the west bank that he named, in honor of his new masters, New Madrid. It was an arrangement that had all the ingredients of a disaster. Everything soon unraveled.

Colonel Morgan himself had fierce rows with the local governor, soon abandoned his claim, and went off to become a farmer in Pennsylvania. The riverbank where Morgan had established a neatly planned little grid of a town was soon eaten away by the ferocious currents, and the town began to vanish as swiftly as it was settled. Only when it was moved northward and eastward could it flourish, to the extent that it ever did. Then, eleven years after its foundation, the Spaniards gave up their land to the French—early maps of the settlement would call it *Nouvelle Madrid*—and finally, three years later, the French sold the property to the United States, and what had been known officially as New Madrid, Louisiana, New Spain, in due course became the grimy little frontier town of New Madrid, Missouri, USA.

Seven years later—rather more than two decades after this town was established—the wrath of God descended. It was to smite—as some locals imagined*—a place that was already suffering under the burdens of its strange history. It struck first in pitch dark, in the small

*In 1811 there was a large catalog of particularly ominous auguries, with which so many earthquakes seem to be invested. Presiding ominously over all was the Great Comet, discovered that spring by Honoré Flaugergues, seen in Europe by Herschel and Humboldt and then witnessed in South Africa—where there was also an earthquake. When it appeared in America, it spawned much wailing and lamentation. The Shawnee chief Tecumseh, feared as much as Geronimo was in the West, is said to have predicted a dread disaster. There had been floods on the Ohio and Mississippi Rivers, hurricanes off Cape Hatteras, a hugely lethal theater fire in Virginia, plagues of pigeons and Carolina parakeets, and congregations of depressed-looking squirrels that drowned themselves in a number of Midwestern lakes. Such portents as these last—which also had hogs biting each other, outbreaks of strange fevers, lightning storms without thunder, strangely colored sunsets, "electrical columns" appearing in the sky, and wolves going sensationally mad—are doubtless the kind of *post hoc ergo propter hoc* observations that attend just about any great disaster.

hours of the ice-cold morning of Monday, December 16. A letter written by some long-forgotten correspondent in the town to a distant friend is taken as a fairly accurate reflection of what happened:

> About two o'clock this morning we were awakened by a most tremendous noise, while the house danced about and seemed as if it would fall on our heads. I soon conjectured the cause of our troubles, and cried out it was an Earthquake, and for the family to leave the house; which we found very difficult to do, owing to its rolling and jostling about. The shock was soon over and no injury was sustained, except the loss of the chimney, and the exposure of my family to the cold of the night. At the time of this shock the heavens were very clear and serene, not a breath of air stirring; but in five minutes it became very dark, and a vapor which seemed to impregnate the atmosphere, had a very disagreeable smell, and produced a difficult of respiration. I knew not how to account for this at the time, but when I saw, in the morning, the situation of my neighbors' houses, all of them more or less injured. I attributed it to the dust . . . the darkness continued until daybreak; during this time we had eight more shocks, none of them as violent as the first.
>
> At half past six o'clock in the morning it cleared up, and believing the danger over I left home, to see what injury my neighbors had sustained. A few minutes after my departure there was another shock, extremely violent. I hurried home as fast as I could, but the agitation of the earth was so great that it was with much difficulty that I kept my balance—the motion of the earth was about twelve inches to and fro. I cannot give you an accurate description of this moment; the earth seemed convulsed—the houses shook very much—chimnies [*sic*] falling in every direction. The loud, hoarse roaring which attended the earthquake, together with the cries, screams and yells of the people, seems still ringing in my ears.
>
> Fifteen minutes after seven o'clock we had another shock. This one was the most severe one we have yet had—the darkness returned, and the noise was remarkably loud. The first motions of the earth were similar to the preceding shocks, but before they ceased we rebounded up and down, and it was with difficulty we kept our seats. At this instant I expected a dreadful catastrophe. . . .

The disturbance that night was the beginning of a five-month-long nightmare, a seemingly unending succession of sudden movements and severely destructive earthquakes to which seismologists still refer as the New Madrid Sequence. At times the earth just shook and shook intermittently, like "the flesh of a beef just killed," as one local farmer put it; at other times it undulated, like a ship on the sea; or it shuddered so violently that horses were knocked to the ground and trees collapsed into rents torn in the soil. Crevasses opened with no warning, emitting strange hissing sounds as they did so. Curious lights danced in the branches of trees—like the static electricity of St. Elmo's Fire, seen on the masts of ships caught in powerful storms. Strange and generally foul smells oozed up from pores in the ground. What looked like putrefying boils suddenly bubbled out onto the surface of fields, and sand volcanoes or fountains of mud covered the

The turbulence in both the earth and the Mississippi River after the 1811 earthquake centered at New Madrid was formidable, the stuff of legend.

meadows, briefly rendering them (had anyone been in the mood for farming) useless.

And the Mississippi, mighty as it was—in places more than half a mile across—began to behave in ways that defied credulity. After one particularly dramatic sequence of shaking, it developed violent overfalls and began to flow *backward,* its waters surging in borelike shudders up toward the mouth of the Ohio, and spuming as they battled against the furious downstream torrent. A Scotsman named John Bradbury, sent over from Liverpool to catalog the botanical life of the American West, happened to be on the river during one of these episodes. His boat, he later wrote, suddenly and for no obvious reason, lurched upward, so rudely that everyone imagined an imminent capsize. Bradbury was knocked off his feet, but, when he grabbed hold of the railing and tried to stand, he saw that the river was covered with foam, and thick with broken branches and whole trees, the entire surface whipped up into a lather as if during a fearful storm—and yet there was no wind, no lightning, no rain. "The noise was inconceivably loud and terrific," he said, although he could "distinctly hear the crash of falling trees, and the screaming of the wild fowl on the river . . . all nature was in a state of dissolution." The riverbanks were caving in with enormous crashes, jagged chasms were being torn in the low cliffs, and dozens of boats—all empty and forlorn—were drifting past in the frenzy, evidence, so Bradbury supposed, of the deaths of all those caught up in the terror.

The botanist, who was clearly well trained in scientific observation, was able to count twenty-seven distinct additional shocks as the night wore on, while his party's little cutter was hove to, waiting for an end to whatever was happening. In time he noticed a distinct pattern: First there was a sudden sound, usually from "a little northwards of east," and then there was a shock; and once that was over, the sound disappeared, quietening itself to a whisper as it did so, in what seemed to Bradbury was a generally westerly direction. In time it diminished entirely, and Bradbury's party made its way downriver—eventually reaching Natchez in the new year, 1812. There he saw the arrival of the steam-driven stern-wheeler *New Orleans,* the first steam-powered vessel to navigate the Mississippi: It had passed through New Madrid while the water was foaming and the surviving inhabitants were wailing in terror and, he supposed, demanding to be picked up. But, he reported, the captain had decided not to stop—partly because he was afraid of berthing while the earthquake was going on, and partly because he soon realized from the wailing of the town's inhabitants that they were as terrified by his mighty new vessel, belching smoke and making its fearsome engine noise, as they were by what was happening beneath their feet.

The number of events in the sequence was truly prodigious. Jared Brooks, of Louisville, Kentucky, counted no fewer than 1,874 separate earthshaking episodes in and around New Madrid over the next few weeks. Shocks like the first enormous one of December 16 occurred two more times—once on January 23 and again on February 7, this last being the mightiest of all. And then the world fell silent again. New Madrid rebuilt itself to the extent that it ever would, and quietly resumed its place as an otherwise forgettable little town in the middle of an otherwise most forgettable corner of the American Midwest.

Except that it remains a very seismically active place today, a place where the needles on seismographs mounted in the tiny city museum down by the levee wave furiously all the time, like antennae of some excitable insects. A number of faults have recently been identified, running deep and hidden below the alluvial soil that today nourishes the cotton, corn, and soybeans up above. Two of these faults trend from the northeast to the southwest, just as the Appalachian Mountains, now hundreds of miles away back east, do also; and the other two run at ninety degrees to them, appearing on the map as lines of cross-hatching.

A map that shows the estimated epicenters of the thousands of earthquakes that occurred in 1811, and of the tens of thousands that have caused the land here to chatter away noisily ever since, follows the traces exactly. A line of black dots and other symbols of varying size and shape runs from the hamlet of Marked Tree in Arkansas up to the tiny town of Vienna, Illinois, northeast to southwest, illustrating occurrences of one of the sets of quakes; and another line seems to form a crosspiece over the first one, running as it does between Fisk, Missouri, and Trenton, Tennessee, and sporting another array of dots, triangles, and circles, revealing how the quakes followed that fault line, too.

The two lines intersect in New Madrid, at a place shown on the seismic map as a wildly confusing maze of dots, dashes, lines, stars, circles, and other coded markings. These indicate that this is perhaps the most active, and in time possibly the most potentially deadly, location in all of America. The relatively new science of paleoseismology—the looking for evidence of early earthquakes—tells us that this intersection has not only been plagued by small and occasional quakes in the past, but also suffered from sequences of truly enormous earthquakes. It seems to have experienced a major sequence around A.D. 900, and again around A.D. 1500. If these are taken into consideration with the 1811 event, it now looks as though New Madrid is being visited by chaos and lethality every three or four centuries—with the corollary that the end of the twenty-first century, perhaps, will be the time for the intersection of the fault lines to act up again.

If it does so, matters will probably be infinitely more terrible. Not so much because of New Madrid's future as a hub of commerce and population—such a thing will probably never happen. It will be terrible because the epicenter exists, poised to rupture, at the still center of a full one million square miles of American territory that has now within its borders cities, factories, waterways, military bases, suburbs, and slums that would all be seriously, devastatingly affected by a series of quakes of the same size as those that took place in 1811. It is likely that cities will be bigger, with populations more densely settled, roads and railway lines more important, factories more sophisticated, and—unless lessons are learned quickly—all settled and built with only a limited awareness of what the earth below may be readying itself to do.

When the faults ruptured in 1811, the western half of the earthquake's penumbra was to all intents and purposes uninhabited, with the other half peopled by pioneers and farmers and men and women who had made for themselves only modest dwellings in equally modest and still-growing towns. Now, however, from Chicago to Memphis,* from Oklahoma City to Nashville, there are giant conurbations crammed with people and with valuable buildings and priceless businesses. Few of the inhabitants spend much of their time today remembering that they have chosen or chanced to be in a notoriously dangerous earthquake zone. And if they ever do, then they generally reassure themselves by saying, without any statistical justification for doing so, that the winter of 1811 was when the region got its pent-up seismicity out of its system, and all is likely to be peaceful from now on.

I HAD ONE MORE PLACE to visit before I entered the American West proper, before I began to travel into that part of the country that came under the direct influence of the specific geological peculiarities that generated and triggered—or, as geology likes to put it, were the ultimate and proximate causes of—the San Francisco Earthquake. My next immediate destination lay just a few hundred miles farther on from Missouri, at a general store in a tiny Oklahoma village with the distinctly unmemorable name of Meers.

It had been named after a prospector, but nothing of substance had ever been found anywhere near it, by Mr. Meers or by anyone else. Back in the 1890s there had been all kinds of rumors of gold being found in the Wichita Mountains, including one tale of a housewife finding a nugget as big as a piece of buckshot in the craw of a fowl she was preparing for Christmas dinner. Such stories were often part of the western myth, tales circulated to encourage settlers to come out from back east. And some gullible, greedy or bored adventurers did indeed go there and build a clutch of shacks beside a creek. For a while their hopes were kept alive by town hucksters who had brought in some real gold miners from Colorado and salted a local shaft with im-

---

*A city with soul, it was remarked, but, since it is so close to the New Madrid seismic zone, a city with an Achilles' heel as well.

ported nuggets. There was something of an Oklahoma gold rush, but then the mines petered out, or were just abandoned as having been worthless all along.

Before long such townsfolk as remained closed the mines, bought longhorn cattle from dealers in Texas, and began to ranch. Cattle ranching remains the staple of Meers today, though there is a lake, and a wildlife reserve with herds of buffalo, elk, and deer. The place positively hums at weekends when artillerymen from the huge base nearby at Fort Sill stop by to play. The first man I met at the Meers general store was an army colonel just back from serving two years in Bosnia. He was training on a new kind of howitzer and was expecting to be sent off to Iraq any day. Remote though the town may be, it has managed to forge some kinds of links with the world beyond.*

This was amply displayed in 1985, when Meers began to excite, if only in a modest way, a keen interest in the worldwide community of seismologists. Teams of surveyors suddenly became fixated upon a strange escarpment that had apparently popped out of nowhere close by the town, and that ran for fifteen miles or so to the north, then ended as abruptly as it had begun. They couldn't find any good reason for its existence—but once it was examined in detail, it was discovered that it marked the trace of a very distinct fault line. Moreover, it was a fault line that showed evidence of having moved, swiftly and violently, at least twice since it had been created. Two big earthquakes had apparently hit Meers—or rather, where Meers would eventually be, the more recent of the two quakes having taken place just over a thousand years ago. Exactly why this might be, no one at the time could tell.

And so the Geological Survey of Oklahoma promptly put one of the mine shafts, abandoned for three-quarters of a century, to good use again. Technicians placed an array of seismometers deep inside the mine, then ran a coaxial cable up to a seismograph, and placed this in the only convenient, safe, and constantly monitored location they

---

*Though not with France, which most townsfolk regard with contempt since learning of the Gallic disdain for the Iraq War. They serve only "freedom fries" at the general store, and the Fort Sill artillerymen approve.

could find at the time, which was the Meers general store. It sits there still today, close by the main entrance, glanced at only occasionally by the cowboys coming in for their Meers Burgers and shakes and bottles of beer. The instrument's recording drum turns slowly behind a protective plastic cover, its dials flicker, and its celebrity status is underscored by a vanity wall of yellowed press clippings that tell of its importance to the seismic world, and of the sensitivity of this particular machine. Earthquakes that happened on Diego Garcia, more than 10,000 miles away, were recorded on the drum at Meers, it says on the wall; but also, lest any visiting artilleryman try to show off, the guns firing out on the butts twenty miles away at Fort Sill never seem to cause the machine to record the slightest sympathetic tremor.

So far the seismograph has recorded only one earthquake of any size that was definitely caused by movement on the Meers Fault—and since monitoring what everyone agreed was an alarmingly large and unpredictable fault was the purpose of the machine, some might regard this as a problem. This one movement was very small and took place in April 1997: since then, total seismic silence. Which makes seismologists all the more curious: Just when, the geological community wants to know, is the fault going to rupture?

It is a classic of its kind. It is primed, tense, filled with stress, and ready to pop at any moment. But it seems to want to do so only every millennium or two—alarming for those who live around Meers (though in truth no one seems to mind one whit; and the owners of the store appreciate the curious travelers who drop by to inspect the machine), and alarming, too, because there are perhaps many thousands of faults just like those at Meers waiting to break and cause who knows how much mayhem, at unpredictable times of their own choosing.

Knowing a little about Meers makes one far less confident, far more skeptical, about the newfangled science of earthquake prediction. The notion that we might one day be able to forecast quakes is a quest that ultimately motivates all seismic research—or the funding for it, at least—even if few seismologists care to admit it. And the sheer capriciousness of deeply hidden fault systems like those in Oklahoma serves mainly to remind researchers that it will be a long,

The general store in Meers, Oklahoma, features enormous hamburgers in addition to a seismograph, which sits, ticking away imperturbably, beside one of the refrigerators.

---

long time before a predictive technique can be devised that will offer an anxious general public any useful degree of certitude. For although earthquakes do tend to happen in seismically obvious places—like San Francisco and Sumatra—they do also take place in the less predictable places, just out of seismic orneriness.

HERE IN THE MIDDLE of the North American Plate, far from the unstable edges where all the action takes place, are four locations where earthquakes simply should not happen but do: the Ramapo Fault in New York; the Summerville-Charleston axis in South Carolina; the New Madrid and Wabash Valley nexus in Missouri and its neighbor states; and the Meers Fault in Oklahoma. All of these are classified as intraplate events—they occur within the stable-sounding centers of tectonic plates. There are similar examples from around the world: Ungava in western Canada, Christmas Day 1989; Newcastle, Australia, in 1989, three days later; Killari, India, 1993.

And in America, too, there have been yet others. Some, such as the huge quakes of central New Hampshire and Cape Ann, Massachusetts, are only dimly recorded and only in the geological record, since they happened long before there were Americans on hand to write accounts. More recently, there are records from sudden single events that puzzle and perplex to this day: the Grand Banks Earthquake of 1929 (which broke submarine cables and spawned waves that killed dozens in the outports of Newfoundland); and three still-notorious New York State events: Massena in 1944, Goodnow in 1983, Ardsley in 1985. They are sporadic, intermittent, *unpredictable*—and as a consequence they are much more dangerous to those living innocently near them than are the more generally expected earthquakes of California, where the cities make sure that buildings are made superstrong and the public is kept endlessly aware.

The specific causes of these intraplate earthquakes are various. The proximate cause of events at Meers, for example, is explained away as the existence of, and occasional movement along, one very particular fault. The subterranean machine that causes the New Madrid quakes is driven by the reaction that occurs when two very different kinds of faults intersect and collide with each other, right under the town.

The ultimate cause of events like these is more interesting, however—as well as being disarmingly simple.

The best theory that anyone has these days for the underlying cause of all intraplate earthquakes is that they represent the relief of stresses built up eons ago, when the mountains or valleys or areas of basin and range were themselves being created. It is eminently reasonable to suppose that the plates themselves, being buckled and elevated and twisted and compressed at their edges, were subject to stress far away from where the buckling and elevation most dramatically occurred.

A piecrust, as it is heated in the oven, will bow upward in the middle, where the fruit bulges and expands in the heat; it will also buckle up at the edges, where the heat conducted through the pan becomes most intense. In the area between the bow and the buckle, the crust, the second it emerges from the oven, lacks the nearly smooth and unblemished appearance it had when first placed inside: It now has myriad cracks and crevices, bulges and breaks. All of this may add to the pie's charm, but it also suggests something of the stresses that built up in the crust as it was being cooked. Once the pie leaves the oven, and the temperature and pressure begin to ease, the cracks change and the aspect of the surface alters: The cracks start to widen or narrow, the surface perhaps tries to revert to its original unblemished and undistorted self, and the stresses that mounted during the cooking process are generally relieved—until a kind of stasis is achieved and the pie becomes what it will be, until it is eaten.

And that, more or less, is what has happened and is still happening in the stable-sounding central parts of all tectonic plates. Almost all of the events just mentioned—Meers, Cape Ann, the Grand Banks, Ardsley—seem to have resulted from the relief of the kind of stress patterns that are illustrated by the cooling piecrust.

But New Madrid, as it happens, is rather different. Some researchers now believe that the origins of the Mississippi Valley seismicity in this place have a rather more dramatic cause. They believe that here the North American Plate may be trying to split itself into two new plates along the line of major seismicity. Explorations of the deep underneath of the region suggest that material might be welling up out of the mantle well below the earthquakes' epicenters. And, while this does not detract from the analogy of the pie—piecrusts can split into two if suitably stressed—it does have awesome implications for the region. Missouri would then become in a geological sense rather like Iceland, but covered with soybeans, cotton, and corn: It would turn into a place peppered with slow-moving volcanoes and hot springs and cliffs that move apart from one another. And there would in time be two Americas, drifting apart at the rate that fingernails grow, and with an as yet unnamed ocean in between, and somewhere drowned deep below, New Madrid, in what was once called Missouri.

No one can tell how much stress has built up in the millions of years since the Appalachians and the Rockies and all the other plate-edge mountains were created; nor can anyone say how long it will be before the middle of the continent achieves the kind of stasis that rules out all future earthquakes, forever. The only way one can make any attempt at rationally planning for earthquakes in places like this, where, generally speaking, earthquakes do not happen, is to look very closely at those places where they have, albeit very infrequently, taken place. By doing this, one has a faint hope of imagining what could take place at some infuriatingly unspecifiable time in the future: It is only by looking at what has occurred in years gone by that one can imagine what might yet occur.

The most venerable of the guiding mantras of geology is that *the present is the key to the past.* But in the very different world of seismicity it seems more prudent to suggest the converse, that *the past is the key to the present.* And, of course, to the future.

FROM HERE it was on westward, to Amarillo and the wide-open skies of the far western prairies. I drove, for hour after hour, through a perfectly dry landscape of buttes and mesas, of arroyos and tumbleweed, of lonely gas stations and wind pumps that clacked emptily in the hot breeze.

And then, just before sunset, there rose ahead the silhouette of a range of enormous hills, and behind them a jagged line of range upon range of mountains, their summits capped with snow. This, at last, was the range once called the Stone Mountains, now much more familiarly known as the Rockies. This was not yet the end of the North American Plate—but it was the beginning of the end, the part of the North American continent that came most directly under the influence of the forces that converged and coalesced at the end, out at the edge. This was now, in the strictest sense, the American West. It was a place where the geology was all so very different from what had gone before—a place where the mechanisms that caused the mightiest of all America's earthquakes started to become tangible, and visible.

The car began to labor up the first of the fronting slopes. The sky was filled with stars, and the night air suddenly began to feel brisk, and then very cold.

FIVE

Chronicle: The State of the Golden State

Give me men to match my mountains.

Inscription in the state capitol, Sacramento, said variously to be the message sent back east by the first arriving Mormons, or an old Australian settler poem

California, more than any other part of the Union, is a country by itself, and San Francisco a capital.

JAMES BRYCE,
The American Commonwealth, *1901*

AT THE TIME OF THE SAN FRANCISCO EARTHQUAKE, America was a country well on in maturity, having freed itself from the grip of the British crown some 130 years earlier. But California was still young, and many who lived in San Francisco in 1906 would have been alive and alert when the territory was granted statehood and first welcomed into the nation. California had at the time been American territory for just fifty-eight years, formally a state for fifty-six years, and had enjoyed the benefits of a formal constitution for not much more than a quarter of a century.

America's newness as a nation is very much more evident in the West than in those eastern states where the country had its beginnings. And America's ownership of this western half is the consequence of the five separate occasions since independence when the government in Washington conducted negotiations and signed treaties that vastly increased the country's size.

There was the astonishing Louisiana Purchase in 1803, for instance, when Thomas Jefferson persuaded France to sell its 530 million acres between the Mississippi and the Rockies, which overnight doubled the size of the country. Then, in 1846, after a series of bellicose threats, London was finally persuaded to hand over the few hundred thousand acres of the Oregon Territory, which the two countries had hitherto administered together. And in 1867 Secretary of State William Seward bought from Russia, in what many at the time thought an improvident waste of funds, the nearly 600,000 acres of Alaska.*

Most relevant here, though, is the treaty signed in February 1848 in the northern Mexico City neighborhood of Villa de Guadalupe Hidalgo—a town better known today for the basilica built in remembrance of the two apparitions of the Virgin said to have been seen by a sixteenth-century Indian convert. Today the basilica is by far the holiest place in all Mexico and a destination as important for all the properly pious of Latin America as Mecca is for Muslims.

The Treaty of Guadalupe Hidalgo, which formally ended the vicious and bloody Mexican War (which was the first American war to be photographed and covered by newspaper correspondents, and the first fought mainly beyond America's borders), resulted in Washington's eventual ownership and control of some 340 million further acres of western lands. It was a treaty that at last gave the United States clear access and title to the warmer shores of the Pacific Ocean—something that most of today's jet-bound Hollywood lawyers who claim their lives to be "bicoastal" probably seldom stop to consider.†

---

*To this day, many Americans like to remember that the purchase of Alaska was known as "Seward's Folly."

†In 1846, just before the start of the Mexican War, a brief revolt—the Bear Flag Revolt—had broken out inside California against Mexican rule, with Yankee settlers boldly declaring California to be an independent republic. But then came the war, American troops seized the Alta California capital of Monterey, installed the first of seven military governors, and rendered the revolution essentially moot. But the battling independence of the revolt's architects is revered, with their bear still the dominant image of California today, at the center of the state's official flag.

Some still regard the war as a manifestation of President James Polk's bitter ambition to control this huge swath of western territory (not least to deny it to the British, who were also eyeing it hungrily). But, whatever his presumed motives, the consequences are with us still: firm American title to the land that comprises today's New Mexico, Arizona, Utah, Nevada, and much of Colorado and Wyoming—and, most important for this story, the 100 million acres of incomparably beautiful and fertile land that had long been separated from the others and named "California" by its former masters. All this was handed over by Mexico on that February day. In return the Mexican government was given a U.S. Treasury check for more than $18 million and a series of promises that, like so many made in international treaties, came to be honored more in the breach than in the observance.*

California's previous rulers—first the Spanish, who had planted the flag of Castile in San Diego in 1769,[†] and then the Mexicans, who, once they themselves had wrestled free of Spanish rule in 1821, assumed control of what was then called the Department of Alta California—made pathetically little attempt to realize the potential of the terri-

---

*In the interests of historical tidiness it is worth noting that the fifth treaty, the Gadsden Purchase of 1854, won yet more land, while ostensibly seeking to iron out some inconsistencies in the 1848 treaty. The American railway companies were essentially behind the scheme to extract further concessions from America's then weak and cash-strapped neighbor, and eventually did win the handover of what is now the southern third of Arizona and a narrow strip of lower New Mexico. Whatever the merits and motives of the purchase, with Gadsden continental America was finally made whole.

†This is no place to enter into the argument over whether Sir Francis Drake landed his privateer in a cove near San Francisco—a subject that has exhausted more acres of fine forestry than most episodes of maritime lore. A small industry—populated by acolytes and writers and conventioneers, and furnished with Drake-named societies, any number of two-lane highways, not a few navigational features, and a scattering of fan clubs—currently exists in California and neighbor states in support of the notion that he did so, in 1579. That he did not obviously lay formal Elizabethan claim to the newfound land allows fervid imaginations to run riot over the possibilities spawned by the simple thought: What if he had?

tory. One after another visitors from other parts of Europe and from the American East Coast expressed bewildered astonishment at its woefully undeveloped condition, at the sadly unkempt ordinariness of a place that had the possibility to be, in more competent hands, truly extraordinary.

Richard Henry Dana, for example, the Harvard aesthete who briefly became a common seaman and worked his way aboard a sailing brig to California by way of Cape Horn in 1834, wrote scathingly of the inhabitants in the book that was to make his reputation, *Two Years Before the Mast.* The Californians,* he wrote, "are an idle and thriftless people, and can make nothing for themselves. The country abounds in grapes, yet they buy bad wine made in Boston." He added, with the scorn of a true-blue Brahmin who well knew his cobbling needs, that, rather than making shoes out of the hides so easily available on every farm and in every forest glade, these indolent and artless people imported them *from merchants back in Boston*—shoes that had been "as like as not made of their own hides, which have thus been carried twice around Cape Horn."

Not everyone shared this view. Walter Colton, the kindly Vermont clergyman who would edit the state's first English-language newspaper and whose book *Three Years in California* would become a classic soon after its publication in 1850, spoke fondly of the courteous and leisure-loving citizens of those early times, and was highly critical of the practical effects of the state's Americanization by those he saw as greedy and amoral Yankees. But his was a minority view. Most other early visitors—men like George Vancouver and George Simpson from Britain, Count Nikolai Rezanov of the Russian-American Fur Company, the French explorer Jean-François La Pérouse and his fellow countryman Jean-Louis Vigne (who, as his name suggests, grew wine grapes and urged the practice on Californians)—said much the same

*The Spanish residents of California to whom Dana was referring were actually known as Californios. They had unremittingly poor relations with their Mexican governors, whose rule was seen as capricious and irksome. Miniature rebellions erupted from time to time, which further sapped the locals' energies.

as Dana. Here was a place blessed by fair weather, pure streams, and fine-looking country; and yet the few thousand Californios who settled had made nothing of it. Content to be ruled by the primitive feudalism of the missions, or administered by the gasconading lickspittles of the Monterey bureaucracy, they were docile and placid and lacked all initiative, lived in gimcrack houses, drank bad wine, supped on rotten food, and had a look and feel about them that spoke of utter primitivism. Though to Walter Colton California was a noble paradise on the verge of ruin, to most others it seemed that, by settling in Mexican California, pre-American humankind had decided to step back in time. George Simpson, a grand panjandrum of the Hudson's Bay Company, thought they had done little more than adopt the ways of "the savages whom they have displaced."

Such criticism was largely based on racial disdain, of course. Most white Americans of the time held most Mexicans in contempt, and felt that since it was their Manifest Destiny, to use the phrase of the day, to settle themselves eventually upon all of the continent with which Providence had supplied them, the presence of this rabble of dark-skinned Spanish-speaking hobbledehoys was an inconvenience that should swiftly be brought to an end.

Which is what began to happen from almost the very moment California was formally inducted into the United States. And, as this change took place, there came a transformation in the character and outlook of Californians themselves—although the roots of that transformation, the first stirrings of the spirit with which the state would be so amply endowed at the time of the earthquake, were probably discernible as far back as 1841, almost a decade before full statehood.

For this was when the first rush of settlers from the American Midwest began to arrive. Their long overland treks first brought in scores, and then hundreds and thousands, of Americans of great energy and a restless spirit who were in search of the then-not-so-elusive phenomenon that has since come to be called the California Dream. New Englanders were already living in California's western seaports—having come around by way of the long sea route—and trappers from outside had been camping in the forests and deserts of the east for years; but by 1841, when the famous Bidwell Expedition set out in its immense wagon train from Independence, Missouri, word had gone out to all Americans that to the far west lay a land of perpetual springtime, and all who could come should come, and quickly.

John Bidwell, who died six years before the earthquake but who left a legacy of attitude and style that would direct the manner in which the city and the state dealt with the disaster, was the quintessential California Man—the fact that he was born into a farming family in western New York proving no bar to his becoming so. He lived in California for almost sixty useful and productive years, and in those six decades he saw his adopted state blossom into the greatest of all the American states.

With a caravan of six-horse Conestoga wagons and their lighter ox-drawn siblings, known later on as "prairie schooners," the Bidwell party pushed their way west across the prairies and via wide passes through the Rockies, right up to the wheel-breaking granite fortress of the Sierra Nevadas. Here they had to leave their wagons and journey on horseback or on foot, and with trains of mules that had to be persuasively lashed to make their way up and across these great peaks. Bidwell knew that he was almost certainly the first American-born man* ever to make this westbound journey.

(Jedediah Smith, among the most fabled of the American mountain men, a trapper and trader from Bainbridge, New York, was the first white man to cross the Sierra, in 1827. But he was going home, and was crossing in an easterly direction that would take him back into familiar territory. Smith had been the first American ever to cross into California from the east the year before; he had come by way of a much less mountainous—but more terribly hot and dry—southern route, through the hundreds of miles of the Mojave Desert. His reputation stems from this barely believable epic: With Indians slaughtering his men, with water running out in the scorching heat of cactus country, with long periods in the mountains when he had to endure

*The soldier-turned-trapper Benjamin Bonneville, who had crossed the mountains on his way to Oregon in 1833, had been born in France.

"my horses freezing, my men discouraged, and our utmost exertion necessary to keep from freezing to death," he eventually made it, crossing into California and out again, all by land, and without a map of any kind to guide him. Had he left more writings behind, or had he lived longer, he would probably have enjoyed a reputation that seems less like a tale embroidered by fable and more like the kind of substantial life story of a Daniel Boone or a Davy Crockett. But it was not to be. He was eventually killed in his beloved Southwest, where a party of Comanches stabbed him to death at a watering hole near the Cimarron River in what is now New Mexico. He was on his way to trade in Santa Fe and was just a few days shy of his thirty-third birthday.)

Bidwell was more fortunate: He and his thirty-strong party found the proper westbound passes, endured the terrible mountain weather, and eventually descended into a valley, winning his first sighting of the Mexican department that would soon be America's newest acquisition. He was instantly aware of its riches. "It was apparently," he wrote, "still just as new as when Columbus discovered America, and roaming over it were countless thousands of wild horses, of elk, and of antelope."

He made some unsuccessful forays, got himself into trouble with the local Mexican authorities, and then decided to seek sanctuary in a curious European settlement of which he had long heard rumors. Some years before the Mexicans had granted a tract of 48,000 acres in the valley of the Sacramento River to a group who had come all the way from Austria. They had set up a tiny and self-sufficient colony and called it New Helvetia, and when Bidwell arrived it was still in the process of being built. The colony's leader was an Austrian-Swiss pioneer named Johan Suter, a man long fascinated by the American West who had sailed the long way around to California from New York by way of Honolulu and the Russian-Alaskan capital of Sitka. Once there he had changed his name to the one that is well known to all Americans today, but more for the wooden race of the mill he owned than for the details of the life he led: John Sutter.

Bidwell supposed he would get to New Helvetia in two days, but it took him eight. Winter, he wrote in a lengthy essay in the *Century* magazine, had come in earnest,

> and winter in California then, as now, meant rain. I had three companions. It was wet when we started, and much of the time we traveled through a pouring rain. Streams were out of their banks; gulches were swimming; plains were inundated; indeed, most of the country was overflowed. There were no roads, merely paths, trodden only by Indians and wild game. We were compelled to follow the paths, even when they were under water, for the moment our animals stepped to one side down they went into the mire. Most of the way was through the region now lying between Lathrop and Sacramento. We got out of provisions and were about three days without food. Game was plentiful, but hard to shoot in the rain. Besides, it was impossible to keep our old flint-lock guns dry, and especially the powder dry in the pans. On the eighth day we came to Sutter's settlement; the fort had not then been begun. Sutter received us with open arms.

The bustling settlement, which at times had as many as a thousand residents, went on to become a model of prosperity. Bidwell and Sutter would become the firmest of friends as they set about turning Northern California into a place of gardens and croplands like no other region then known on the American continent. There were vineyards and orchards and wheatfields around Sutter's sawmill; there were a smithy, a distillery, and hundreds of acres of grazing pastures for herds of longhorn cattle.

John Bidwell adopted Sutter's ideas for a while, and raised cattle and wheat on a 22,000-acre estate that he managed to buy close to the present-day city of Chico.* But soon the promise of horticulture seemed more thrilling than the pursuit of agriculture—gardening was more alluring than cultivating, as it has remained for the truck farmers of the Central Valley ever since. And so, on the wildflower-filled mead-

---

*Nearby is the small town of Paradise, one of about twenty that have been so optimistically named across America. When I made an effort to visit them all during one summer in the eighties, I was dismayed to find each had been spoiled to some degree or other by tourism, greed, development, or a simple lack of care. Except, that is, for the tiny town of Paradise, Kansas, close to the geographical center of the United States, which is quite lovely still, and happens to be home to four families named Angel.

ows of his Chico land, Bidwell started a magnolia nursery; he planted large groves of peach trees, orchards of pears and apples, and small armies of different varieties of cherries. He experimented—at first with low-risk crops like oats and grapes; and then, as his confidence grew, with figs, quinces, walnuts, raisins, Egyptian corn, almonds, melons, sorghum, and olives. He imported threshing machines and built a cannery.

His energy and ambition were prodigious, virtues matched only by his piety (he became a devout Presbyterian) and his zealous abhorrence of strong drink. And it was in this one regard that John Bidwell did, eventually, come to depart from what would soon be the California norm. For once gold was found at his friend John Sutter's millrace in 1848, the trickle of immigrant pioneers grew to an overwhelming flood, and such niceties as Christian charity and respect for the ladies and abstinence from strong drink vanished, almost overnight, from California society. The well-mannered order that men like Sutter and Bidwell had brought to replace the idle apathy of Californians was itself suddenly replaced by a tidal wave of abandon and turpitude.

When James Marshall spotted those flakes of gold late in the afternoon of January 24, 1848,* the population of California was not much more than 18,000—a fivefold increase from ten years before, to be sure, but an increase that was by no means suggestive of the gold-struck migrant invasion that was about to begin. Came the Gold Rush, however, everything precipitously and dramatically changed. It was by the purest coincidence that just nine days after the find, the territory on which the gold was discovered changed hands, and what had been Mexican became, indubitably and eternally, American.

Yet when thousands upon thousands abandoned their roles as

---

*Marshall was not the first to find gold. Six years before, while taking a lunch break in a meadow some thirty-five miles north of what is now Los Angeles, a man named Francisco Lopez y Arballo found nineteen ounces of gold pellets sticking to the roots of the local onion plants. He and his colleagues collected them and sent them off to the Philadelphia Mint for assay: They were valued at more than $40,000. But, despite churchly fervor and pious incantations, no more gold was found—until six years later, 400 miles to the north.

wretches and wage slaves in faraway factories and headed off to California, they found a country that, though technically American, still had no constitution, no settled system of law, no firm notion of statehood or any timetable for it, no reliable system of justice—not even a fixed eastern boundary.* This was America raw and unprepared, and it was about to undergo as profound a change of nature as it is possible to imagine.

When newfound nuggets of placer gold were brought out for display, everyone was amazed. Walter Colton wrote of the mood of unbridled elation that gripped almost all who saw the metal and understood its promise:

> The excitement produced was intense; and many were soon busy in their hasty preparations for a departure to the mines. The family who had kept house for me caught the moving infection. Husband and wife were both packing up; the blacksmith dropped his hammer, the carpenter his plane, the mason his trowel, the farmer his sickle, the baker his loaf and the tapster his bottle. All were off for the mines, some on horses, some on carts, and some on crutches, and one went in a litter. An American woman, who had recently established a boarding-house here, pulled up stakes and was off before her lodgers had even time to pay their bills. Debtors ran, of course. I have only a community of women left, and a gang of prisoners, with here and there a soldier, who will give his captain the slip at the first chance. I don't blame the fellow a whit; seven dollars a month, while others are making two or three hundred a day! That is too much for human nature to stand.

From all America, and from all across the world, they raced to the foothills of the Sierra. Buoyed by entirely accurate reports that gold

---

*Some delegates to California's first constitutional convention, which met in September 1849 in the then capital of Monterey, wanted the state to extend itself across all of what is Nevada, and to incorporate much of Utah and Arizona, too—making it the biggest state in the Union, and too big, thought Washington, to control. The idea was rejected by the Monterey delegates themselves, avoiding what could have been an unpleasantly divisive argument with the national capital.

In the early, crazy days of the California Gold Rush, most miners conducted placer mining, as in this later photograph—looking for the "bloom" of gold flakes in the river sediment they caught in flat pans. Later, massive hoses were used to break up rocks and flush out the gold, at immense environmental cost.

---

was to be found in vast abundance—in streambeds, in deposits of gravel, on the sandbars in estuaries and around lakes, in the potholes in rocks—a tidal wave of humanity, most of them young, single, and rudely energetic men, began to surge its relentless way westward. The men came from the East by the two ways then possible: by sea or by land.* If by sea, they could choose (on the basis of either anecdote or blind and hurried ambition) to get to the gold by one of two routes: either the five-month, 15,000-mile journey by way of Cape Horn, as Richard Dana had done, or through the marginally less stressful Strait

*One scientist, named Rufus Porter, planned a steam-powered dirigible service that would speed prospectors west at 100 mph. He floated a company, but never an airship.

of Magellan; or via the shorter route of two months, which went by way of a ship to the Isthmus of Panama, across that narrow and malaria-infested neck of land by foot and mule train, and thence by a steamer that would be waiting on the Pacific side of Panama, and on up to its West Coast destination. In 1849 more than 500 sailing ships and the occasional steamer left ports like Boston and Gloucester and New York and Baltimore for the goldfields: The vessels were hurriedly provisioned, scantily crewed, and poorly maintained, and the journey was in most cases the most hazardous part of the adventure—a large number did not survive it.

For those braver or more foolhardy men who elected to travel overland,* four routes soon opened up—along a handful of previously vaguely defined trappers' paths that within weeks had become scarified and rutted wagon ways that could soon be seen for miles, lacerations that would scar for years to come the hitherto unmarked landscapes of the West. These were the trails—from Memphis to California across Texas, from St. Louis to California across Utah, from Hannibal along the valley of the Platte River toward Oregon, and from Chicago up across the peaks of Montana and via the empty rain-shadowed wastelands to the tiny army encampment at Fort Walla Walla—from which the roadbeds for all of America's future surface links would be formed. The routes of the telegraph cables, the tracks of the Pony Express, the Wells Fargo stagecoach lines, the transcontinental railroads, the two-lane and then four-lane highways—Route 66 most legendary among them—and then the roads of the interstate system of today—all of them first followed the routes that those Gold Rush migrants had taken, once John Marshall's find had cried its havoc and the newcomers had slipped their traces and begun.

They came in absurd numbers, and in many cases with either an abysmal lack of preparation, or entirely the wrong kind of preparation. All that a miner really needed, according to an early how-to guide for

---

*Alonzo Delano, the New York humorist who went out west and wrote some of the most acutely observed essays on the Forty-Niners, observed drily that anyone traveling overland to California positively deserved to find a fortune.

Forty-Niners, was "a good but light Pick, a round pointed shovel, a light Crow Bar, a Pan or light cradle, a short knife and a horn spoon." But a writer named Archer Butler Hulbert, who seems to have been less infected than most by the mythic hyperbole of so many Gold Rush accounts, spoke of

> wagon trains lurching their way across the country loaded instead with everything that a man's wife or a boy's mother could think of ... sheet-iron stoves, feather-beds, pillows, pillow-slips, blankets, quilts and comforters, pots and kettles, dishes, cups, saucers, knives and forks ... some had trunks full of white shirts and plug hats ... one man was hauling a great walnut bedstead.

Most of what the migrants tried to take was about as much use, said Hulbert, "as two tails to a dog."

And, in any case, little of this ever arrived: Hundreds of the more hopelessly overloaded wagons broke down, or bogged down, in Death Valley, or on the Great Salt Lake, high on the mountain passes, or in the awful loneliness of the Humboldt Sink. There they were abandoned to rust and woodworm, monuments to the folly of these particular pioneers, so many of whom were impelled merely by avarice, rather than by the real rewards of settlement.

Gold, and the greed for it, played havoc with the morals of the California-bound miners. At this point in America's history towns in the East and in the Midwest were fully sophisticated, with citizens behaving toward one another with mature civility. But those who initially went to the West were overtaken by the barbarism of the frontier with astonishing speed—think *Lord of the Flies* or *Heart of Darkness.* There was murder, mayhem, robbery, alcoholism, depression, and suicide, and all of it on a positively Homeric scale that still has cultural anthropologists enraptured. The presence of thousands of unattached and desperate men, all gathering in California with a single ambition—to break the chains of their past and to grow rich—made each man a risk to every other. There was so little brotherhood, so much individual assertion of right and claim. It was, quite literally during those Gold Rush years, *sauve qui peut,* and the devil take the hindmost.

And yet many of the men who streamed west were far from being simpleminded members of the *lumpenproletariat;* far from it, indeed. These, after all, were men who had in large measure the good sense and the gumption and the moxie, call it what you will, *to get up and move:* They were a crew who as often as not were intelligent (an unexpectedly high proportion had university degrees), besides being more obviously motivated and determined. They were men who wanted keenly to put whatever past they had firmly and forgettably behind them and to start anew. The fighting and the tedium with which it was interspersed prompted many of the brighter souls to creative endeavor. Walter Colton's newspapers, for instance, were able to write about the establishment of musical ensembles, the offering of lecture series, the formation of debating clubs and amateur dramatic societies that staged ambitious new plays and operas.

So if the journey, and the rivalry, and the crime, and the temptations of the bottle and the brothel (for not inconsiderable numbers of women did go west, and not all of them angels either) did not do for them, then those who survived, even if they did not all make their fortunes, provided the new territory with a substrate of cultural energy, intellectual achievement, and classless ambition that eventually helped turn California into by far the most *interesting* state in the Union. And in time, of course, into the wealthiest as well.

Although the goldfields had played themselves out after only three years (and after metal worth $2 billion had been extracted), many still seemed bent, once they could draw breath long enough to see how magical were their surroundings, on settling there for good and making California home. "If a man comes to California and stays two years," wrote one early settler, "he will never want to leave." "I don't intend to call California home," said the wife of one Lewis Gunn, a Philadelphian who had come out to mine in 1849, failed, opened a drugstore and a small newspaper in Sonora, and persuaded his wife and children to make the endless journey out to him, around the Horn. She was horrified by what she saw—drunken women staggering through town on election day, lurid pictures of naked girls tacked up in the bars, shoot-outs staged almost every night. But she persevered, stayed put, brought her children up as she thought proper, and endlessly looked forward to a time when the state might adopt more sober ways. She remained there, the first of thousands of long-staying Californians, for the rest of her life, dying only in the earthquake year of 1906, when she was ninety-five years old.

Mrs. Gunn lived her long life in San Diego, in the south of the state, but the vast majority of those people who migrated during the Gold Rush went to the north, with most living around San Francisco Bay, or along the postroad that led east to Sacramento and the Sierra, where the ultimate reason for their being there could be found. The six southern counties of the new state had just 20,000 people, of whom Mrs. Gunn was one. The tiny pueblo that was Los Angeles had a mere 1,600 inhabitants in 1865, most of them ranchers, and the total value of their property—and this at a time when the area declared that it wished to be known simply as "Queen of the Cow Counties"—was just $800,000. By contrast the city of San Francisco, which got its name and its first American mayor in 1847 (a year before the treaty and three before statehood), already had a population of 40,000, and that a scant three years after James Marshall discovered his first flecks of gold. Provided that these people stayed—as they did, no matter what happened to the gold that first drew them—there could be no doubt by the mid-1850s that the state, and its bounty, were there to stay, too.

The Civil War, which affected California little during the fighting itself,* spurred still further immigration once the dire days of Reconstruction got under way. This immense and open territory, with its fair weather and its huge skies and fields in which, it seemed, anything that might grow in the South would in fact grow twice as well, was so much more agreeable than the wrecked landscape of Dixie. There was a huge demand for labor in the West at this time, too—and so began a flood

---

*California, once established, had been admitted to the Union on September 9, 1850, as free state, meaning in essence that slavery was banned. The first civilian governor, a Tennessee Democrat named Peter Burnett, had resigned a year later because of "certain personal prejudices"—among which was his view that blacks should be barred from entering the state at all.

of new people from Alabama and Mississippi, Arkansas and Louisiana, and soon kept up amply with demand: The state's population rose from 380,000 in 1860 to 560,000 in 1870.

The transcontinental railway was completed in 1869. The cheap and willing labor needed to build it was provided in large measure from China, which led to the creation of a scattering of big-city Chinatowns (the *Oxford English Dictionary*'s first citation for this term is 1857, quoting a newspaper in California, where the phenomenon was born). Trans-Pacific immigration—with the concomitant pressure to ban it, which was on occasion a sorry feature of California's history during the nineteenth century—thus became a prominent feature of the state's story, and led in large measure to the extraordinary demographic diversity that had already begun to characterize its great cities at the turn of the century.

By now the state was in good administrative order, with an abundant supply of cities, roads,* railways, schools, hospitals, prisons, universities, and all the other accoutrements of settled western life. It had shifted its capital more than once—from the Mexican capital of Monterey it had been moved in 1851 to Vallejo (close to Mount Diablo, from where the state's land survey lines had first been drawn), then two years later to Benicia and finally in 1854 to Sacramento, where the first railway terminus was built to link California with Chicago and New York, where the state capitol buildings were completed in 1869, and where, somewhat inconveniently, the capital remains today.

Governors came and went—most of them Democrats (though one of them, in 1860, was a Democrat who ran on a ticket favoring the admission of Kansas on a proslavery constitution—racist passions sometimes ran high in early California). The first Republican governor was Leland Stanford, in 1861, who saw the introduction of the telegraph to his state and the start of the building of the railway, the Central Pacific; Leland Stanford, Collis Huntington, Mark Hopkins, and Charles Crocker, soon to be known as California's legendary "Big Four" busi-

*The first official state highway was the Lake Tahoe Wagon Road, taken under the wing of the newly created Bureau of Highways just before the turn of the century.

ness tycoons, were the founders and owners. The first California oranges reached New York in 1885. The state's population passed a million a year later. Yosemite became a national park. There was an outbreak of bubonic plague in San Francisco. Anti-Chinese riots, anti-Oriental legislation, vigilantism—the ills of society all seemed to be manifest in the extreme in California, as were the blessings.

And then came the 1902 election, and a rather dull ear-and-nose doctor named George Pardee, a Republican who had been the mayor of Oakland, was elected to the governorship. He is little remembered today—just a dam, a reservoir, and a recreation area stand as memorials to him—and the inaugural address he offered in January 1903 speaks only of the banal quotidian demands of his position, with no great reform required, no major overhaul called for, simply a need for a steady hand on the tiller of state. He stood up in the newly finished rotunda at the capitol and addressed the politicians before him:

> We take office under conditions which are most flattering. California is blessed to-day with a material prosperity for which her citizens may well thank the bountiful mercies of the God who rules the destinies of nations and of men. Almost every interest is thriving as it has not thrived before in many years; our homes are homes of peace and plenty; work and employment abound, and the rewards of industry and enterprise were never greater. Wealth is increasing; and the proportion of the increase which is represented by the $350,000,000 deposited in our savings banks indicates that prosperity is widely distributed among all our people and is not the exclusive enjoyment of a few. Upon such industrial conditions as these we may well congratulate ourselves; and, if we are wise, we shall carefully refrain from any course which might produce a change for the worse.

He expressed some concern about the state's trees and its waters (John Muir, the archdruidical founder of the Sierra Club, was ministering away among the redwoods at just this time); he asked for more money for the 2,400 students then being educated at Berkeley—taxes that would be "most cheerfully paid," he predicted; he called upon the citizens to desist from their eager habit of obtaining injunctions from the courts for every imaginable complaint; and he recognized that

> California, probably on account of her geographical position and her fame as a land of wealth and easy conditions of life, which serve as an attraction to the restless and idle, has an overplus of inmates in her penitentiaries and reformatories.

He sought, in consequence, more funds for prisons and for those institutions then known as Hospitals for the Feeble-Minded. He urged expansions of the ports of San Francisco and Oakland. And he reminded his listeners that international expositions were to be held during his tenure in the cities of Portland (as a memorial to the Lewis and Clark Expedition) and St. Louis (similarly reminding the country of the Louisiana Purchase—which President Jefferson had sent Lewis and Clark to explore and to map). It would be prudent, he said, for California to be represented at both, the better to advertise its wares and its charms to those few unfortunates who might be still unaware.

In short, the state was in exceptionally good shape, and Governor Pardee had merely to tinker, to tune, and to tread water. The statistics of the time underpin the notion of comfort and prosperity. There were a million and a half citizens in a state that, just fifty years before, had comprised an indolent handful of Mexicans clinging to a thin strip of land along the fogbound coast. Come the new century and the state's mines were still producing precious metals of considerable worth; exports of wheat, barley, wool, milk, butter, cheese were being loaded onto ships in the great ports of the Bay and down at Long Beach; freight trains on the three great lines that now linked California with the Mississippi Valley were hauling away cars loaded with oranges and lemons, prunes, sugar, wines, brandies, beans, raisins, and oil. And Stanford University had just been founded: $20 million of private money had been slapped down for the foundation and endowment—a gift "unparallelled in magnitude," said the *Daily Telegraph* across in London, "in the history of mankind."

All was, in other words, set fair for California, and for its principal city of San Francisco, as Governor Pardee and his team commenced their gentle program of governance and supervision. The first two of his four years in office went as smoothly as might be expected, with a host of unanticipated challenges but not a single problem that even approached the level of a crisis.

And then came the spring of 1906—when, almost halfway through the governor's elected term, everything suddenly went spectacularly and memorably wrong.

SIX

How the West Was Made

> The West I liked best. The people are stronger, fresher, saner than the rest. They are ready to be taught. The surroundings of nature have instilled in them a love of the beautiful, which but needs development and direction. The East I found a feeble reflex of Europe; in fact, I may say that I was in America for a month before I saw an American.
>
> OSCAR WILDE, *quoted in the* St. Louis Daily Globe, *February 26, 1882*

ON GOING WEST, EVERYTHING CHANGES ON THE FAR side of Amarillo. The word means "yellow," presumably from the dun-colored dust of this parched part of Texas—and the town was set up first as a railway-construction camp. Now it is a thoroughly up-to-date place, with tall buildings that house banks and the headquarters of large cattle ranches and small private oil companies, and there are lines of strip malls ringing the town just as the Conestoga wagons used to ring the campfires.

The town ends suddenly, with neither sprawl nor suburb to extend its reach. And, as it does so, the road ahead empties and the horizon becomes quite flat, with just the faintest line of settled cloud in the distance hinting at a chain of mountains somewhere far away. The gas stations are fewer here, the radio signals fade away into static—except for the evangelical preachers, who rant on endlessly into the otherwise unpopulated ether. Ed Ruscha, an Oklahoman contemporary artist who became fascinated by the lonely majesty of this part of America, wrote in an essay that he had discovered here the importance of gas stations. They are like trees, he said, but really only *because they are there,* and there is not much else besides—nothing but the bleached jawbones of cattle, old wind pumps creaking with rust, broken-backed barns and barbed-wire fences that provide a barrier against the bundles of tumbleweed that bounce casually past in the ceaseless hot wind.

The landscape is dominated by sagebrush and chaparral; it is a place of mesas and buttes, of canyons and arroyos, of lodgepole pine and saguaro cactus and rattlesnakes and golden eagles and museums of wagon trains—and all underpinned by a geology, moreover, that is nothing at all like the geology back in the East. For a start, it is all so very much on display: The ribs of the land show through, landscape laid out as textbook.

I drove across the slowly ascending flatlands of far-west Texas, up and over the low hills of what is called the Front Range—the gradient of the highway ticked upward a few miles outside the junction town of Tucumcari. Then, near Albuquerque and Santa Fe, the road passed through into the southern Rocky Mountains, with the Sangre de Cristos to the right, the Sacramentos off to the left, and the Black Range and the San Andreas Mountains directly ahead. The flanks of the hills were covered with pine trees, and on the late-January day I was there, fresh snow glistened from their summits. After that I endured some hours of driving over the open deserts and sharp-edged ranks of peaks in the region that is formally known as the Basin and Range Province, 500 miles of wide-open land that has been favored for its emptiness and romantic beauty by writers and filmmakers for years past.

I was heading for the small town of Winslow, Arizona, where a terrific meteorite collision had occurred 50,000 years ago, leaving behind a 600-foot-deep hole punched into the desert, 4,000 feet in diameter. I had learned about it as a small boy: My geography teacher in Dorset had given us lectures about the very rare allotropes of quartz known as coesite and stishovite, formed when quartz is suddenly subjected to enormous pressures, as it would be when hit with an immensely heavy

Somewhere deep below the meteor impact crater near Winslow, Arizona, is the immense body of iron and nickel that hurtled in from outer space. It has never been found; but the crater's owners, thwarted as miners, now run the site as a highly profitable tourist attraction.

---

and fast-moving body of iron and nickel. Both of these newly named (after their discoverers) quartz minerals can be found at the bottom of the Winslow crater, I remembered him saying. When I got there I found something else, something quite unanticipated, and in truth something rather trivial—but something that added immeasurably to my feelings about western American geology.

Up on the lip of the crater, I listened as the guide talked about its history: of its discovery in the 1870s by one of General Custer's scouts; of the visit made twenty years later by the great geologist Grove Karl Gilbert,* who fretted publicly over whether it was a volcano or an impact crater of the kind seen on the surface of the moon; and then of

---

*See the prologue, page 18.

its acquisition by a Philadelphia mining engineer named Barringer, who fancied that he might one day be able to find the buried meteorite and make himself a tidy fortune. In the end he never did find it, but, said the guide, his family still owned the crater. The firm was based in Pennsylvania and was called the Barringer Meteor Company.

And then I remembered that I had once known a family named Barringer, a fairly well-off and very elderly couple who lived in one of those rather comfortable horse-country neighborhoods close to the Delaware border. My cell phone had good signal, and so I called them. Old Mr. Barringer answered. He was rather deaf and shouted to ask who might be calling. I told him, and explained that I was telephoning from the very lip of the Meteor Crater in eastern Arizona. There was a pause.

"Bless my soul!" he exclaimed happily. "Our family owns that, you know!" And so I put the phone into speaker mode, and the guide and I listened attentively as Mr. Barringer told us how delighted he was that we were all there and how he hoped that the staff were being as pleasant as possible. Then he thanked everyone for paying up their $20 admission fee, "as it keeps us all in good champagne, and for that we are most grateful."

A meteor could have fallen anywhere in America, of course; but it seemed somehow appropriate that the biggest of all should have collided with the earth in one of the most beautiful, unspoiled, and most geologically complex regions of America. And the westerly path that I followed over the next few weeks took me to other, similarly emblematic places, all created by the forces of geology. From the Meteor Crater I went on to the Painted Desert, the Petrified Forest, the Grand Canyon, and the Canyon de Chelly in Arizona, to Zion and Bryce Canyons in Utah, to Death Valley and the Salton Sea and the southern Sierra Nevadas in California. All are stupendous examples of the geological mayhem that is the American West; all spectacles of ancient topographical miracle work that manage to draw millions from around the world.

The story of the geological exploration of the American West, which resulted in the discovery of nearly all these marvels, begins in the spring of 1860, when the Geological Survey of California was founded. Untold hundreds of exploring scientists—for who would not want to explore here, where the geology is so spectacularly *on view,* so seldom covered by such inconveniences as soils and forests?—amassed a wealth of geological information the likes of which few countries have ever known.

The diligence and derring-do and sheer romance of the geologists who came through here still haunts the territory. And there is a nice symmetry in the fact that their years of pioneering work—which in most cases had the official blessing of the American government—had its origins in the discovery of gold at Sutter's Mill, California, in 1848. When President Polk announced the importance of the find to the world, he also demanded that a serious and extensive geological survey of the region be undertaken—and, by doing so, he kindled a sudden interest in geology in dozens of other states, where governors and legislative chiefs started to demand that their local scientists investigate what wealth might be in the rocks beneath their feet.

Up to this point agriculture, rather than industry, was the dominant force in the American economy. Such a mining industry as existed was in poor shape, and in the late 1850s coal production was actually going into decline. But as soon as gold mines started to be sunk in the Californian mountains, everything changed: A realization spread like brushfire that other worthwhile minerals might also be lurking underground, and the mining industry swiftly began to go into overdrive clear across the country.

It took little more than a decade for the situation to become so changed. In 1848, the year of Sutter's Mill, agriculture still reigned supreme, industry was primitive, mining in the doldrums. But by 1859—the same year that a new gold strike was made in Colorado, the year that the great silver deposit known as the Comstock Lode was found in Nevada, the year that the first oil well was drilled in Pennsylvania, and just when work on the four great transcontinental railway routes was starting—government figures showed that at last the value of products made in American factories had overtaken the value of produce from the farms and fields. And with America's transmutation into a fast-growing industrial power, there came a sudden and anxious need for minerals. Such natural resources as the country possessed were all being swallowed up by the furnaces and the foundries, and in 1866 a frankly worried government in Washington decreed that the proper development of the nation's geological and mineral wealth had to be of "the highest concern of the American people." It was imperative, the White House said, that the untapped resources of the country be found and wrenched from their hiding places belowground.

Congress agreed. And so in March 1867 it was decided that the entirety of the West—unsettled, poorly charted, and inhabited, when at all, by understandably unhappy nations of indigenous peoples—had to be properly and methodically explored and mapped. Four great surveying organizations were promptly inaugurated—the Great Western Surveys, history has come to call them—and for the next decade enormous parties of soldiers and scientists, most of them surveyors but many of them geologists, went off under the auspices of one or another of these government-financed surveys, and investigated every nook and cranny of the immense landscape. Eventually they threw up all* the stunning details of the fantastical worlds that lay in the far beyond, in that immense unpeopled wasteland that lay to the west of longitude 102, which the U.S. government had decreed as marking the formal edge of American settlement.†

Each of them had a grandiose name. *The Geologic and Geographic Survey of the Territories* was the first of the four to be formed, and it was led by one of geology's least-favored pioneers: the illegitimate son of a Massachusetts drunk, an explorer named Ferdinand Vandeveer Hayden. Few have kind words to say about this curious man: The only biography is entitled *Strange Genius;* one writer denounces him as a vindictive, insecure, manipulative, and self-promoting plagiarist, and even

*Or nearly all: In the summer of 2004 a hitherto almost unknown site of an ancient American Indian civilization was unveiled in eastern Utah, with hints that there are more finds yet to come.

†Amarillo, which has come to symbolize the edge of American settlement and the formal beginning of the West, sits almost precisely astride the 102° meridian.

the biographical dictionaries, with their tendency toward detached kindliness, remark on Hayden's ruthless ambition, impatience, and combative style.

But Hayden was both a brilliant field geologist and a great popularizer of science—and it is through his efforts and those of his survey colleagues (two of whom went on to help found the National Geographic Society, which was gestating all the while this surveying of the West was unrolling) that the unexplored parts of Nebraska, Kansas, the Dakotas, Montana, Idaho, Utah, Wyoming, and Colorado were properly mapped and the more obvious aspects of their geology recorded. He had a budget of $75,000 from U.S. Treasury funds—more than enough to hire the best surveyors, scientists, and photographers. The legendary landscape photographer William Henry Jackson was a member of the Hayden team, and his stunning images of the mountains and extraordinary scenery around the headwaters of a river in what is now northwestern Wyoming set the capstone on what was Hayden's greatest achievement: the creation of America's first national park, at Yellowstone. An "unprepossessing man of no outstanding achievements" he may have been, according to some unkind biographers; and a man who became notorious for feathering his own nest and regarding the West as his personal empire; and a man whose personal life was checkered enough to leave him dying of syphilis. But Hayden, with his superbly organized and eloquent 500-page report on what he had found, persuaded President Ulysses S. Grant to create Yellowstone National Park, and few greater memorials to the explorer's art can there ever be.*

But Hayden had a rival. At the same time as he was pushing out westward from the Missouri River, the man who would in due course

---

*By chance it happens that Yellowstone, and in particular the behavior of its famous geysers—Old Faithful being the best known—have been shown recently to enjoy a curiously unexpected connection with the earthquakes that occur along the edge of the North American Plate, miles away to the west. I will try to explain the connection and what is known about its causes in a later chapter; but for now this link serves to remind us that Ferdinand Hayden's own connections to this story are rather more intimate than one might initially suppose.

become the first director of the U.S. Geological Survey was leading a group of explorers eastward, out into the unknown from California. The expedition was called the *Geologic and Geographic Survey of the Fortieth Parallel,* and its leader was Clarence King, perhaps one of the truest heroes of the American school of the Old Geology.

King was evidently a strong-willed, imaginative, and impetuous young man. After completing his four-year chemistry course at Yale in only two years, and then going on his own accord to hear Louis Agassiz lecture at Harvard, he decided to ride his horse the entire way to California. He joined the California Geological Survey (where he explored and then named Mount Whitney, the highest peak in the Lower Forty-Eight, after the survey's then director) and promptly became so enthused by the life of western exploration that as soon as was prudent he rode all the way back east to Washington, D.C., with a scheme.

He confronted the secretary of war and insisted that the government finance a survey, just like Hayden's, that would progress eastward across the continent along the line of the 40th parallel, following a swath of territory that lay along the course of the main Central Pacific railroad route. His proposed journey would start where the parallel crosses the Sierra beside Lassen Peak, continue through Nevada in the great desert south of Winnemucca and Elko, run across Utah fifty miles south of the Great Salt Lake, pass on through Provo and across the Green River, and cross the Front Range, where those mountains intersect with the parallel close to Boulder, Colorado. The expedition would then drop down from the high peaks and safely conclude in the settled farmlands of the Great Plains.

The government agreed, offering him a three-year budget of $100,000 for his troubles. King swiftly assembled a team of thirty-five, most of whom were polished European geologists or dapper American scientists of elegance and taste—men who found it mildly amusing to be asked to do such things as carry rifles and put on desert boots. As a result the King survey was at first regarded more for its élan than for its science—though as their results began to flow in, the science began to win respect as well.

The team—together with a large contingent of soldiery designed to protect the scientists against the anticipated anger of disconsolate Indians—embarked on a ship to Panama, then on another up to San Francisco, and finally crossed the Sierra to the starting point in Nevada's Virginia Mountains, just north of present-day Reno. It pushed relentlessly and steadily eastward, triangulating and hammering as it went, producing maps of both the topography and the geology with an accuracy and attention to detail that is impressive still.

But the event for which Clarence King remains best known came as the expedition was winding down, when the team entered northwestern Colorado and heard rumors of an astonishing find of diamonds and rubies and other precious stones around a mesa near a settlement called Browns Park, nowadays between the villages of Hiawatha and Dinosaur. To a geologist like King, this sounded most odd: Despite their meticulous surveying, they had found no diamond-bearing earths. There was a possibility, King concluded, that the story was a hoax.

As indeed it turned out to be. The party found the mesa, and in short order found an extraordinary scattering of gems littering the ground. But the stones were to be found only in disturbed areas, with footprints all around; wherever the soil was undisturbed, digging threw up nothing more than the occasional pebble. Very little geological detective work was required to find out that a pair of clever ne'er-do-wells had in fact salted the mesa with precious stones. They had set up a massive fraud, and convinced a staggering array of believers to invest money in a mining venture, to go after treasure that simply didn't exist.

The size of the swindle was remarkable. Investors from as far away as London had sunk money into some twenty-five companies, and to the tune of $250 million—especially after the pair had somehow managed to borrow and then (as the investment money began to flood in) to buy bags of uncut jewels that they used first to salt the mesa, and then to excavate and show to their growing army of backers. They had persuaded Charles Tiffany, the New York City jeweler, to vouch for their authenticity—though Tiffany later rather weakly confessed that he knew very little about stones other than about how to market them.

Clarence King, who appears to have been impelled by a sterling sense of propriety—and despite appeals from the villains' associates that he, too, could now make a quick profit from what he knew—went straight to the companies' lawyers and told them of the fraud—prompting the shell-shocked investors to close down the firms. The two confidence tricksters fled; one died as a coffin maker, penniless, the other from shotgun wounds after a fight near his pig farm in Kentucky. The investors spent years licking their wounds and cursing their gullibility.

But Clarence King's reputation soared. "God and Clarence King," it was said in the San Francisco newspapers, had saved the day. He became, in part because of his soaring reputation as a pillar of moral rectitude, a close friend of the powerful, men like Teddy Roosevelt, John Hay, and Henry Adams; and when the American government decided, somewhat late in the game, to set up its own national Geological Survey, it seemed entirely appropriate that he be installed as its first director.

Few seemed much to care when another, somewhat less acceptable aspect of King's character was revealed. It turned out that King had an unappeasable passion for dark-skinned women, particularly enjoying the delights of the Shawnee, the Comanche, and the Cheyenne while surveying among them. He went on to marry a black nursemaid named Ada Copeland, changing his own name to James Todd for this purpose. And though he fathered five children with Miss Copeland, Clarence King never breathed a word about his circumstances, or about the existence of James Todd, to his closest friends or to his employers in the federal government.

When Henry Adams idolized Clarence King as "the most remarkable man of our time," he probably had little idea just how remarkable he really was. All that Adams and the legions of King's admirers knew was that the King survey, when it was measured by the number of volumes of reports and the classic books and maps that were to be its legacy, had been a triumph, and that its leader had in addition performed brilliant service by exposing one of the greatest geological frauds in western American history.

With the surveying along the line of a parallel now accomplished, it must have seemed that for symmetry's sake a meridian should surely be explored as well. Hence the formation in 1872 of what is generally known as the Wheeler survey, with its formal, government-designated title *The Geologic and Geographic Survey West of the 100th Meridian.* It would still be a while before Washington would designate 102°, the Amarillo line, as the formal edge of western settlement. (Back in 1872 hardly anyone lived west of that line, which would pass from Bismarck, North Dakota, all the way down to the Texas town of Abilene.) George Wheeler, a swaggering young lieutenant in the U.S. Army, had a reputation for being able to look after himself in the field and for producing impeccable maps. He was thus given the task of surveying all of this enormous emptiness. It was to all intents and purposes a military expedition—Wheeler saw his role as the leader of one further instrument of the American conquest of the West, and he counted the Indians as though they were wildlife, in the hope that in short order they would all be exterminated. He was not a pleasant man.

But he was a highly competent one, and when the parties under his command ended their seven years of work, there were no fewer than 164 new maps of western America, beautiful, accurate, and of lasting value. He had mapped 327,000 square miles—with the challenge of mapping Death Valley chief among his cartographic successes—and he had spent rather more than 600,000 of the Treasury's dollars doing so. And he had taken along with him—to dilute the overwhelming military appearance of his party—a group of civilians, three of whom were so distinguished and influential that they, like the pair who rode with Ferdinand Hayden, also went on to be founders of the National Geographic Society.

One of this trio was Grove Karl Gilbert, yet again—the Meteor Crater visitor and San Francisco Earthquake witness whose name keeps cropping up in accounts of the American West. He was mild-mannered, quiet, shabby (because he liked to repair his own clothes), and is still widely regarded as the greatest of all American geologists.* His most lasting memorial, perhaps, is that it was he who named the Basin and Range Province, and recognized how very different it was, structurally, from the ranges and basins that were so characteristic of the mountains of Appalachia, near where he grew up.

But neither Gilbert nor Wheeler nor Hayden nor even the redoubtable Clarence King has won quite the lasting general repute that has since been enjoyed by the leader of the fourth great federal survey—the magnetically appealing one-armed former soldier who is renowned for first navigating the entire length of the Colorado River through the Grand Canyon: John Wesley Powell. This was a man who had lost his arm to a bone-shattering minié ball at the Battle of Shiloh: It did not to any measurable degree inhibit him from undertaking one of the most heroic explorations in American history.

This best-known achievement of the figure whom the writer Wallace Stegner once described as "a one-armed little man with a bristly beard, a homemade education and an intense concentration of purpose," actually came some while before the creation of the federally funded survey that he would eventually lead, and that was to be called the *Geologic and Geographic Survey of the Rocky Mountain Area, 1870–78*. Immediately after leaving the army (he continued in service after losing his right arm, seeing action at the siege of Vicksburg and the Battle of Nashville), he had gone off to Illinois to teach geology, to establish a small museum of natural history, and to collect specimens.

While pondering where best to get hold of specimens for his museum he conceived the idea that he might profitably explore the uncharted tracts of the West. And so in the summers of 1867 and 1868 he organized two collecting expeditions to the Rockies, returning with wagonloads of specimens and, so far as the geography of the place was concerned, a sorely whetted appetite. In particular he was fasci-

*He was also the subject of many laudatory (and usually pretty dreadful) poems, including one by an Australian fellow geologist declaring that Gilbert's "many faults were mighty ones."

Grove Karl Gilbert, one of the revered
figures of early American geology,
and a witness, inter alia,
to the San Francisco Earthquake.

nated by the enormous rivers that he had encountered on his travels: Where did these mighty rivers go? Where did they reach the sea? Might it be possible for boats, or even for ships, to navigate their waters?

So Powell went back out to the West in 1869, only this time with a party of eleven men, four boats, and a modest subvention from the newly formed Smithsonian Institution in Washington. He had a plan that was both simple and ambitious, but that had less to do with fossil hunting than with conducting exploration simply for the sake of it. He would put his boats into the waters of the especially fascinating-looking Green River in southwestern Wyoming; he would navigate his way down it as best he could to where he knew that it joined up with the Colorado; and he would then navigate his way down the Colorado, regardless of whatever hazards might be in his way, for as far as it might take him.

Despite his amputated right arm, John Wesley Powell—shown here with a Paiute guide—managed to lead a party of explorers through the entire length of the hitherto unexplored Grand Canyon, canoeing down the Colorado, a river famously "too thick to drink and too thin to plough."

He had little more than a vague idea, based on a few half-told stories, that there was a canyon somewhere down there: Such reports as existed were hazy and dismissive. So he told his colleagues that he was going into what he called "the Great Unknown," and, as the party slid their boats into the waters beside a railway crossing close to the settlement of Rock Springs, they must have been more than a little apprehensive. They had no real idea of what they would find.

The canyon had been seen by very few people. The Spaniards, who had first named the Colorado* in 1540 and who had ventured a hundred miles upstream from where it debouched into the sea, had sent out a gold-hunting expedition across the desert that same year and reached the south rim of the canyon; probably a man named García López de Cárdenas was the first European ever to see it. More than two and a half centuries would pass before another Spanish party spotted it and encountered some of the more accessible of the Havasupai Indian settlements that are so well known and protected today. In the 1830s a fair number of hunters and trappers began to stumble across it: Jedediah Smith, already noted as having been first to cross the Sierra, was one; a soldier named Joseph Ives and the geologist John Newberry descended into its depths, being the first white men ever to stand at the very bottom of this terrifying declivity. "The region," reported Ives, "is altogether valueless. Ours has been the first, and will doubtless be the last, party of whites to visit this profitless locality."

Hardly. A further geological survey was undertaken in 1859—and then, ten years later, the diminutive one-armed major, with his team, his boats, a small amount of army-gifted rations, and a wagonload of nerve, came to Wyoming to shoot their way through the rapids, force a way down the canyon, and establish the region's reputation as one of the greatest natural wonders on the planet.

After three months, 900 miles, and uncountable adventures, Powell and his diminished party—five of his men had deserted, and one of the boats and all of the food aboard had been lost—emerged into the still, wide waters of the lower Colorado, on the far side of the canyon. When word spread, this thirty-five-year-old became a national hero. To the public, he was the epitome of brio and dash and courage and vision. His book, *Exploration of the Colorado River and Its Gorges,* became a bestseller. His collections—for he managed to collect assiduously and intelligently, no matter how harsh the conditions

*The river has had three names: Río de Buena Guía ("River of Good Guidance"), Río del Tizón ("River of Half-burned Wood"), and finally Río Colorado ("Red-colored River").

of his travels—were sought after by researchers across the country. And his reputation within the scientific community entered the stratosphere. Almost overnight he became the natural candidate to lead the fourth and the greatest of all the surveys the government had planned—that of the Rocky Mountains, which then occupied the following eight years of Powell's life.

The maps he produced and oversaw during the survey were peerless for their time; his reports are classics still; and his tenure as the second director of the Geological Survey—once he had completed the survey he succeeded Clarence King in a post that made him unchallengeably the most powerful scientist in America at the time—was marked by innovation and discovery on a breathtaking scale.

Powell was, however, a keen environmentalist. His appreciation for the great outdoors and for the peoples who were indigenous to it, and his belief in the need to explore rather than to exploit, in the benefits of preserving rather than plundering (beliefs that were personified by such figures as the great Scotsman John Muir, who went on to found the Sierra Club, and such painters of the wilderness aesthetic as Thomas Moran), won him more enemies than friends. His eager support for the preservation of Indian culture, his abiding preference for sensible and sustainable development, and his obsession with the value of water in the West proved too much for many of the settlers, foresters, and miners who were heading in that direction simply to exploit the country. Increasingly Powell came under vitriolic attack from the unconcerned pioneers, and he retreated into a shell, languishing in a world of science and philosophy that was more comforting than controversial. He died a disappointed man at the age of only sixty-five. However, his profile in bronze presides at the South Rim of the Grand Canyon today—watching serenely over that mile-deep chasm of raw cordilleran geology, hoping that his successors take good care of all that he first explored.

The immediate legacy of the four Great Western Surveys—the realization that western America was profoundly different, geologically and topographically, from the rest of the country—was the birth of what was then considered almost a brand-new science. This came to be known as cordilleran geology, the study of the rocks that make up the extraordinary complexity of the American West. Specifically named societies, journals, conferences, university departments, and offices by the score were created to fill the needs of this new discipline; and when researchers gaily described themselves, without much fear of challenge, as "cordilleran geoscientists," everyone knew what that meant.

But, for all the romance and intellect brought to bear on the region, until the birth of the New Geology, cordilleran geology resulted in precious little fundamental understanding of what was going on there. The initiation of a real understanding—a post-tectonic-plates-New-Geology kind of understanding—came with one classic academic paper that was written and published in 1970.* It was a paper that would have given no clue to any lay reader, considering the recondite language of its title: "Ultramafics and Orogeny, with Models of the U.S. Cordillera and the Tethys." But its publication produced a tipping point in the comprehension of western American geology—and it made the career and reputation, many would say, of the one figure who today is most prominently associated with the arrival of the New Geology in the West. The author was Eldridge Moores, who, even envious colleagues will acknowledge, remains the best-known member of the geology department—he is now an emeritus professor—at the University of California's campus in the Central Valley, at Davis.

THE BASIC QUESTION that would niggle away at anyone working on the rocks and structures of the western states of America was simply: *Why?* Why this? Why here? How come?

It is something that probably any westbound traveler would ask as well. The plains over which he has been moving for days have been hot

---

*Inevitably, since geologists can be a fractious crowd, there are other contenders, each vocally supported. Their papers, all produced at around the same time—for it was around 1970 when plate tectonic realizations fully dawned in the American West—are noted in the bibliography.

and dry and endless and flat—and then suddenly, as if out of nowhere, a mountain range! It is climbed, surmounted—but then beyond it on the horizon there appears another, and beyond that a stretch of a desert, then another range, and then a network of canyons and waterfalls, and a farther set of monster mountains, and yet more and more scenery that is so spectacular and so utterly different from anything that lay behind, or anything that was even imaginable in the America already seen and known, that any wanderer with any sense of curiosity would be bound to ask that selfsame question, too: just *Why?* What force could possibly have compelled the world to do as it has done out here—just why is everything such a confused mess, so unsettled, and so manically disarranged? Why this? Why here? How come?

Eldridge Moores was one of the first to come up with a credible-sounding answer. It is an answer that hinges on a very special sequence of rocks that is found in many places in western America and in a number of other geologically fascinating (and, as it would later be revealed, structurally similar) parts of the world, and that invariably incorporates three distinct types of material.

There is always in each sequence one great thickness of very dark-colored volcanic rocks that includes lavas like basalt or their coarser-grained kin called gabbro; there are always a number of beds of deepwater shales and often a particular type of sedimentary rock known as chert, frequently complete with the fossils of plankton and slightly more preservable sea animals called radiolarians; and there is most essentially and most invariably a sequence of hard and greenish-colored speckled rocks that have markings in and on them that vaguely resemble the skin of a snake. Because of this resemblance this one particular rock has been called *serpentine,* a name it was first given in the fifteenth century.* The three-rock sequences that then incor-

---

*The *OED*'s first citation is from 1426: "my best covered cup of silver and gilt . . . with one serpentyn in the bottom." Serpentine, so important in the story of the state's making, has now been formally adopted as the state rock of California.

porate this serpentine—the sequence usually runs, from the bottom upward, serpentine-gabbro/lava-sediments (including chert)—are currently known by a combination of the two Greek words *ophis,* for snake, and *lithos* for rock, which gives us the highly distinctive snake-like rock, *ophiolite.*

The fact that such sequences of rock were very common to mountainous regions, especially the European Alps, was first recognized by the great nineteenth-century German geologist Eduard Suess. Another Alpine geologist named Gustav Steinmann then pinned down the three essential components, and for most of the first half of the twentieth century the sequence was known as "Steinmann's Trinity." Calling them ophiolites came about in the 1930s, once it was realized that the sequence—serpentine-gabbro/lava-sediments—was to be found in glorious display in geologically contorted areas all around the world, with ophiolites discovered in the Alps, in Oman, in the Troodos Mountains of Cyprus, as well as in the magnificently disarranged mountain chains of western North America and, most specifically of all, in California.

It is this unfamiliar word—though *ophiolite* is now a word known to all practitioners of the New Geology—that lies at the heart of what Eldridge Moores realized. It refers to a concept that is tricky to explain, but that, once understood, answers all the basic questions relating to the makeup of the American West and, most particularly, to the structural peculiarities there that led to all the San Francisco earthquakes, culminating in the disastrous event of 1906.

Professor Moores remembers the moment of his realization only too well. It was December 20, 1969, and he was in Pacific Grove, California, at the Asilomar Conference Center. He was listening, fascinated, halfway through a session of the second of the annual Penrose Conferences that the Geological Society of America now holds to ruminate on the most important new developments in earth science.*

---

*One might almost say that the Penrose Conferences, which were established in 1969, and to which any geologist in the world can seek an invitation, came about specifically to discuss the new whole-earth geology that had been born at much

At this legendary gathering "the full import of the plate tectonic revolution burst on the participants like a dam failure," he later wrote. Paper after paper was being read that overlaid the new theories on top of virtually every major process of geology that had shaped the planet—the location of volcanoes, the folding of mountains, the distribution of earthquakes, the shape of the continents, the history of the oceans. Everything was being answered by this devastatingly simple notion: that plates floated about on top of the plastic mantle and collided with one another, scraped alongside one another, broke into pieces, or welled up under the influence of the immense heat from below. The "marvelous dance of the plates" is how one of the conferees put it, with the rapture of the collective *Eureka!* It was, reflected Moores, "one of the most exciting moments of my life," and everyone else at this most remarkable gathering of geologists appears to have felt the same. Asilomar was a turning point in science like few had ever known.

His own moment came as he was listening to the conference convener, Bill Dickinson, presenting his summary. Moores had drifted off message for a moment, thinking about a discussion the previous evening about just where the world's ophiolite sequences were, when, "in a blinding flash of insight, it came to me." What came to Eldridge Moores would make him famous, in two distinct worlds. He would become well known throughout the geological community, one of the revered figures in the story of how a new science gave new answers to a very old question: How did the world as we know it come about?

And he became famous much more widely among the lay community thanks to a book that was written about him by his old friend and Princeton colleague, the similarly legendary John McPhee. The book, *Assembling California,* became a bestseller and remains a classic today.

---

the same time. The first conference held in Tucson a year before had more limited ambitions—it discussed copper—but what has come to be known as the Asilomar Penrose was designed as something else: a truly worldwide conference to support the fast-growing idea that geology had truly changed, and would never be the same again.

Eldridge Moores, one of the great discoverers of the processes that led to the making of the American West, is shown here in a suitably heroic pose, with a sequence of ophiolites, the key to the mystery, spread out before him.

Few are the academic geologists who have ever enjoyed the kind of fame that is currently enjoyed by Eldridge Moores. And that he is so well known is in part because of the sinuously elegant prose by which McPhee describes him—but also because, although the concept of ophiolites may seem at first a little cumbersome, the explanation of what Moores realized—how ophiolites fit into the story of plate tectonics and how they answered the three questions *Why this? Why here? How come?*—is, thankfully, very much less so.

For, basically, he grasped that each of the ophiolite sequences that are found today in the mountains of the West are the remnants of ocean floors. They are remnants of oceans floors that have been bounced, to put it most crudely, up on top of the continent. They have been bounced up in wave after wave as the tectonic plates—and that in relatively recent times means the North American, the Farallon, and Pacific Plates or their various antecedents—have collided with one another over the eons past.

Moores made his deduction simply by thinking about the nature of the three kinds of rock that made up an ophiolite—wondering why such strange and exotic rock sequences were sitting up on the mountaintops in the middle of a continent. After all, at the *base* of each sequence are the serpentines and coarsely crystalline igneous rocks that we know underlie all of the world's oceans. In the *middle* of each sequence are the "green rocks," the minerals of which have been subjected to a moderate amount of pressure and reasonably but not terrifyingly high temperatures, and so can be said to have been altered by a series of moderate geophysical events—they have not, for example, been hurled down to the edge of the mantle and tortured by the extraordinary pressure and heat that is found there. And at the *top* of each sequence are the lavas and then the chert-rich sediments, deposits evidently made by deep seas and that include the sieved-out remains of seawater itself.

What Eldridge Moores deduced from this complex mélange of rocks is a sequence of events that is exactly consonant with what is now known about plate tectonics. He was able to declare that the geological confusion so readily apparent in the Rockies, the Sierra, and all the other mountain chains of the West is in fact no more and no less than the result, time and again, of neighboring plates bumping into each other. What then happens depends on local circumstances. Material on one plate dives underneath the material of the other plate: Some of this melts as it does so and then (because it is so light and volatile) forces its way back up toward the atmosphere; some of it rides up and is scraped off onto the upper side of the other plate; and some of it slides past the other plate and is scraped off onto it in a similar manner.

The basic narrative is exceptionally simple: The plates bang into each other, and, where they do so, large amounts of material are dislodged and stay put, being heated and compressed by the enormous forces of the collision. Then the collision abates, and the plates reorganize and regroup for another assault—and the foreign material that is left behind, so mixed and contorted and folded, is already, or in time becomes, a mountain chain—ready to have another mountain chain piled on alongside it when the plates collide once more, and yet another when it happens again, and again and again and almost endlessly (since there is so much time in geology) yet again.

Each of these foreign rock slabs or slices or sequences—some of them hundreds of miles long and dozens of miles wide, and many now spearing up into the cold and snow-rich air of the upper atmosphere—is now known as a *terrane,* a word confusing because of its homonymy with the word *terrain,* and which means landscape. But a terrane is a piece of continent or ocean floor that is foreign—it is, to use the term more favored by today's geologists, *exotic.* It is a piece of the geological somewhere that is now settled in a place scores of miles—maybe hundreds or even thousands of miles—from the geological somewhere else where it was made, or from where it once existed. The West of America is full of such terranes, and full of the ophiolite sequences that embrace them or that they in turn embrace. Most of these bodies or sequences trend north and south in great elongated ranges of mountain peaks that fill the windshield when you are driving west, and that etch the land below when you are flying across America toward the Pacific Ocean.

And they trend north and south because they are the result of collisions that occurred millions of years ago, either between three north-south-trending plate edges—those of the Pacific and Farallon and North American Plates, whose edges very roughly parallel the Oregon and California coastlines that exist today—or between their antecedents, that formed other oceans and other continents, such as Pangaea and Rodinia and Kenorland and all the other bodies and seas that composed the earth in times that, unlike mankind's own history, were truly ancient. These bodies moved headlong into one another, banging, diving below, arching up, retreating, and banging into one another again, with one of the plates heading from the west toward the east, the other moving slowly but with immense and unstoppable power from the east toward the west. And all of these bodies and plates were compelled to move by the convection currents in the earth below, currents that are churning ceaselessly and that are making the plates execute these unending mazurkas and tarantellas up above on the mantle top.

Nowhere is this more visibly true today than along the 750 miles of the San Andreas Fault, at the very western edge of the North American Plate, where it meets up with its neighbor plate that underlies the Pacific. The San Andreas Fault is young by the standards of the West, and very young by the standards of geology. But it is a very great deal older than America, older than California, and older than San Francisco, and it has been making its own kind of mischief for many millions of years past, and will continue to do so for many millions more to come.

SEVEN

The Mischief Maker

San Andreas Fault
moved its fingers
through the ground
Earth divided
Plates collided
such an awful sound . . .

NATALIE MERCHANT, *from "San Andreas Fault," 1995*

The law against sodomy goes back fourteen hundred years to the Emperor Justinian, who felt that there should be such a law because, as everyone knew, sodomy was the principal cause of earthquakes. "Sodomy" gets them. For elderly, good-hearted audiences, I paraphrase; the word is not used. College groups get a fuller discussion of Justinian and his peculiar law, complete with quotations from Procopius. California audiences living on or near the San Andreas Fault laugh the loudest—and the most nervously. No wonder.

GORE VIDAL, Matters of Fact and Fiction, *1977*

AT A QUARTER PAST TEN ON THE MORNING OF SEPTEMBER 28, 2004, a woman named Christy Gieseke was taking a shower on her horse farm in San Miguel, in the vineyard country of central California. Suddenly the house began to shake violently. Ms. Gieseke ran out of the bathroom, noticed that the living-room chandelier was shimmying dramatically on its chains, decided to run, and eventually made it, unscathed, into the open air. Her dog, a usually well-tempered Rhodesian ridgeback, was panicking, running wildly across the yard, barking madly. But her horses, which she might have supposed would be bucking in panic, were by contrast unusually and strangely quiet—stunned, perhaps, by this inexplicably frightening set of circumstances.

After only a few more seconds, the shuddering, and the barrage of curious creaking and cracking sounds from the very earth itself, came to an abrupt halt. The event was, at least for the time being, quite over. The horses began to graze peacefully once again, pawing the ground with their hooves by way of seeming remonstrance; the chandelier's swaying slowed, then stopped completely; and whatever growling sound had welled up from the ground faded to mere white noise, and then was gone altogether.

It was at this point that Christy Gieseke realized that she had dashed into her garden quite naked but for a dishtowel she had grabbed in the kitchen on the way through. She ran back into the house, found her clothes—and hoped that would be the end of it. Earthquakes tend to make people nervous—and to behave spontaneously, even in a place as accustomed to the earth's violent movements as San Miguel, California, and the rather more notorious neighboring towns of Paso Robles and Parkfield.

This, as the locals like to say, is earthquake country. San Miguel, hilly, pretty, moneyed—horse ranches, wineries, and exemplary weather—suffers from a ceaseless onslaught of small seismic events, to which many of the locals have become almost entirely inured. Once in a while a larger earthquake strikes. The last of major significance hit just before Christmas in 2003; it killed two people in Paso Robles, toppled a church tower there, and did grave damage to the 200-year-old Mission San Miguel nearby.*

But let us concentrate on the event of September 28. Ms. Gieseke and her horses; a vineyard owner named Harry Miller, who observed

*The mission's builders created their church of out of adobe bricks, which are typically sun-dried mud mixed with straw. Damaged bricks examined after the quake were found to have ox blood, horse manure, and dead birds stirred into the mix as well.

"trees shaking like brooms, and dust coming everywhere," as 300 cases of wine stored in his cellar were upset and jets of water were sent spouting thirty feet into the air from his tank; and the 9,000 Californians living between Sacramento and Santa Ana who are officially recorded as having felt the event—all were affected by what is now as properly and fully annotated a disturbance of the earth as has ever been known.

Its first shock—the mainshock, as the originating shock of an event is known—struck at 24 seconds past 5:15 P.M., Coordinated Universal Time, or 24 seconds past 10:15 A.M. Pacific Daylight Time. The shock produced a strong shaking that lasted for almost exactly 10 seconds, and it was followed by a series of some 161 aftershocks. The event's epicenter—the point on the earth above which the shock-inducing rock rupture occurred—was seven miles southeast of the village of Parkfield. The hypocenter—the subterranean point where the rupture took place—was a little less than five miles down.

Once the rock had ruptured the shocks then traveled, and at a fantastic speed, in a northwesterly direction, disturbing people—Mr. Miller and Ms. Gieseke among them—in the myriad ways that a shock as impressive as this one can. The formal classifying number of the event (or the *eq,* as such happenings are generally known in the seismological community) was NC51147892, with the NC being the internationally recognized two-letter code for the Northern California Regional Seismic Network, based at the U.S. Geological Survey headquarters at Menlo Park, at the upper end of Silicon Valley. The regional moment magnitude of the quake, which is what is usually calculated and released to the press, was 6.0.* No one was hurt by it, nor was there any but the most mildly inconvenient damage.

In normal circumstances, and in most places, this would have been a merely moderately significant event. But the circumstances, and the place, were anything but normal—and as it happens, the event of the September 28, 2004, was probably more measured a seismic happen-

*The way in which earthquakes are classified is more than a little complex: An explanation more properly belongs in the appendix.

ing than had ever been recorded in the history of this planet. And that is because the event took place very close to this otherwise memorably forgettable little Californian settlement named Parkfield.

Parkfield, California—a sleepy, dusty farming town that nestles among rolling hills fifteen miles to the northeast of Christy Gieseke's home—enjoys a peculiar reputation in world seismic lore. It may not suffer most from a superabundance of seismic activity—that record, for what it is worth, appears to be held by a faraway and rather tumbledown Californian village called Petrolia, which has earthquakes

Drilling equipment in a rancher's field outside Parkfield, California. Measuring devices being placed at the base of the drill hole are expected to give vital information about what exactly happens at the very edge of the earthquake-triggering San Andreas Fault.

almost every other hour—but it is the place that suffers most from a superabundance of seismic science. For it has had so very many measurable earthquakes (dozens of very small ones every week, and larger tremors that are substantial enough to be felt by reasonably aware humans several times a year, and those of a magnitude to drive women from their showers every two or three decades) that in the last years of the twentieth century a program was begun that made sure that every Parkfield hillside and valley and suspicious-looking outcrop or knoll or riverbed was covered with an array of earthquake-detecting instruments. These have become more sophisticated, delicate, and expensive as the years and science have worn on.

Anyone who makes it to this tiny town rapidly becomes aware that it is a much-measured place. Beside the roads, up on the hilltops, lurking in groves of trees, and in strange boxes with menacing locks and orange notices that warn of the not-so-dire federal penalties awaiting those who interfere, are the devices that monitor, hour by hour and, in some cases, millisecond by millisecond, what happens below the earth around Parkfield. A photocopied guide handed to visitors relates the kind of thing: "on your right, look for a 4' × 4' × 4' structure . . . this contains a seismometer," "on the south side of the road there is a piece of PVC pipe sticking out of the ground with the letters JPL-GPS—this belong to the Jet Propulsion Lab," "on the left side is a USGS creepmeter . . . do not touch the thin metal wire."

The data from the machines is broadcast to thousands out in the seismically fascinated world. Some is sent via tiny satellite aerials to Colorado, some goes to a university near San Diego, still more to the Geological Survey's regional headquarters at Menlo Park, while other parcels of information are flashed to monitors in Pennsylvania, North Carolina, Oxford, London, and Brisbane. Parkfield may be a town very little known in most of the rest of lay California, but to members of the geological priesthood with a keen interest in how the world is believed to work, it is the center of the seismic universe.

Though Parkfield is no Grand Canyon, the local residents know they are onto a good thing. So the town's café, across the dusty street from a hotel made of logs, sports a water tower painted with the slogan

BE HERE WHEN IT HAPPENS, and the hotel tries to tempt passersby with SLEEP HERE WHEN IT HAPPENS. On the menu is a steak called the Big One, as well as a rather more modest version called the Magnitude Six; and the desserts are called Aftershocks. A sign nailed to a tree offers eggs produced by an evidently surprised chicken shown being shaken awake; and there are a handful of memorials and notices and walking trails that remind people of the dramatic moments to which this town is prone.

At the southern entrance to Parkfield is a white-painted iron highway bridge that takes the road across the oak-shaded and occasional stream known as the Cholame Creek. Anyone crossing the bridge will notice that it is spectacularly bent: The metal rails and supports all veer off to the right by well over a foot and a half. Moreover, there are patches of freshish asphalt on the pavement, where repairs have been made by CalTrans, the authority that fixes damaged state roads, and whose crews doubtless sigh whenever they hear of fresh earthquake activity up at Parkfield. The event that pitched Ms. Gieseke from her shower moved large chunks of concrete off their foundations and opened up a new gap several inches wide. For a time the bridge was closed, yet again, while the asphalters did their stuff. By measuring the bridge—and people have been doing this for many years, especially since it was particularly badly damaged by an earthquake in 1966—it can be shown that the abutments have moved more than five feet since the structure was built in 1936: The westernmost abutment, the side farthest away from Parkfield, has moved to the north, while the side closer to town has slid down toward the south.

The little iron bridge and the half-dry creek serve as all-too-visible reminders of just why it is that Parkfield occupies the place that it does in the seismic canon. For the creek marks the very edge of the North American tectonic plate, and is the center of the narrow zone that shows where the next-door Pacific tectonic plate butts up against it, and lurches and slides northward along it with predictable unpredictability.

SOME CHAPTERS BACK we began a long journey clear across the full expanse of the North American Plate. It was a journey that started in a valley in the center of Iceland, where a line of cliffs marks the plate's far eastern edge. Here, in the middle of this very ordinary-looking Parkfield town bridge, is the point that marks its extreme end—the plate's western edge, where it rubs against its Pacific neighbor. Central Iceland is divided from central California by 6,000 miles—a distance that marks, at this place in the world, the width of the entirety of one of the greatest and most important geological entities of all.

THE ABUTMENT of the bridge that stands closer to Parkfield is on the massive North American Plate; the abutment on the other end stands on the Pacific Plate, the equally massive and ever-moving part of the world that underlies such faraway places as Hawaii and Tonga, Pitcairn Island, Easter Island, and the South Island of New Zealand. Another part of the world begins right here. And the line that divides the two plates, the place that geologists like to call the plate abutment, or the plate boundary, coincides with the visible surface manifestation of what is perhaps the best-known geological fault that can be seen anywhere.

The line is some 750 miles long. It curves sinuously, in the shape of an elongated boomerang, from where it makes its landfall on an unstable clifftop up in Humboldt County, in the wet and windy north of California, to where it dives into the heart of the earth in a hot and muddy field beside a hissing geothermal plant, down by the fences and well-armed border guards at the Mexican frontier. The town of Parkfield lies almost halfway along the line; the buckled bridge is buckled because although it, too, is sited *on* the fault, it cannot be said to lie *along* it, but rather to lie *across* it, since the road in and out of Parkfield needed to cross from one side to the other. And it is axiomatic that any structure that is built across a fault like this is placed under enormous stress whenever the fault moves—something that this one does, uniquely, all the time.

Shifting, sliding, vibration, destruction—everything about this 750-mile-long line suggests that we are dealing with a living, breathing, ever-evolving giant that slumbers lightly under the earth's surface and stirs, dangerously and often, according to its own whims and its own rules. At almost every single place where the line can be traced there is evidence of movement, damage, some kind of mysteriously infuriating life. The bending of the Parkfield bridge is one instance, as is the crumbling instability of the sea cliffs up in Humboldt County where the fault first comes ashore; the regurgitating steam that hisses around that muddy southern field is another, as is every earthquake that is measured every day in the Cholame Valley, like the one that drove Christy Gieseke from her shower in San Miguel, like the one that knocked down the nearby mission some years before, and like hundreds of others besides.

And, while there have been countless other shakings and vibrations randomly scattered across the American West, once a series of systematic studies of California's seismicity had been made a century or so ago, a very obvious pattern soon emerged along this line. There seemed to be much more movement and nuisance of one kind or another along it than anywhere else in the state.

The activity was first noted as being concentrated in a valley south of San Francisco, which lay right on top of the line. This valley had been named the Valle de San Andreas by its Spanish discoverer, because he found it on November 30, 1774, the feast day of the apostle most English-speakers (like most Russians) know as Saint Andrew. Once it was realized that the valley lay on (and indeed was caused by) the very fault line where all this activity was concentrated, the fault itself was given the name of the valley: It became the San Andreas Fault.*

---

*The fault was first comprehensively delineated in the official Report of the California State Earthquake Commission, established to inquire into the 1906 events. Since the author of the report was a somewhat bumptious geologist named Andrew Lawson, a mischievous canard circulated suggesting that he named the fault after himself.

The San Andreas Fault is a feature so well known to geologists around the world that it is generally referred to by its initials, SAF. More important for this account, movement along the San Andreas Fault was the fundamental event that all but destroyed San Francisco in 1906. It is also the phenomenon that, on some unpredictable day in the future near or distant, will surely destroy any city built by those improvident enough to site it nearby.

THE STORY OF THE BIRTH of the San Andreas Fault is far from simple, and interpretations change as new information is uncovered. It can be summarized, however, and in essence it appears that the mechanisms that gave rise to the fault can conveniently be said to have begun about 150 million years ago, out on the edges of a Pacific Ocean that was very much larger than it is today.

A submarine chart of the ocean today shows a line of relatively shallow water that extends southward for several thousand miles from a point close to the west coast of Mexico, via the Galápagos Islands and Easter Island and the Juan Fernandez Islands,* all the way down to the Antarctic. The island groups that lie along it are volcanoes, some active and some not; these, and the submarine ridge of which they are the visible peaks, are the manifestations (just as the Mid-Atlantic Ridge is a manifestation on the farther side of the world) of a part of the earth that is splitting open, where plates are moving away from each other, and where new material is being oozed out onto the planetary surface (in this case most of it invisibly, beneath the sea). The ridge is called the East Pacific Rise—not least because it is the dominant feature of the eastern half of the Pacific Ocean.

But 150 million years ago it was not in the east: It was much more in the middle of what in those days was the very much wider Pacific. And just like the Mid-Atlantic Ridge, this then Mid-Pacific Ridge was also a spreading zone, a place where two tectonic plates were moving

*One of these is Isla Robinson Crusoe, since a local shipwrecking story is thought to have inspired Daniel Defoe.

steadily away from each other. The plate on the western side of the ridge was the Pacific Plate that we know today, and it was moving northwestward. The plate on the eastern side of the ridge, which geologists have named the Farallon Plate, was moving southeastward. To complicate matters further, the North American Plate, which was being forced by pressure from the upwelling of magma at its eastern edge, at the Mid-Atlantic Ridge, was shifting steadily westward, in the direction of these two spreading plates, the Pacific and the Farallon.

The consequence of movements like this one—the collision of the North American and Farallon Plates—is the familiar one of subduction, the phenomenon that is known in every corner of the world where there are active and very violent volcanoes, from Java and Sumatra to Japan, from Kamchatka and New Zealand to Alaska—places where a light continental plate hits a heavy oceanic plate square on, and the heavier plate is subducted beneath its continental collision partner. Wherever such a collision takes place, volcanoes and earthquakes are created and break out in dangerous abundance. In this case the heavy and oceanic Farallon Plate subducted below the light and continental North American Plate—causing many of the geological features, Eldridge Moores's ophiolite sequence in particular, that make up the American West.

The North American Plate's westward movement continued until, sometime around 30 million years ago, it hit a snag: It ran into the selfsame East Pacific Rise. The rise then began subducting below the North American Plate as well—and this created the kind of tectonic confusion that is much better seen on a pool table. (Physicists with a penchant for divining the way that a number of colliding forces all moving in different directions can impinge on and interact with one another work out this kind of thing mathematically; in *The Hustler,* which some say it parallels, it was more a matter of intuition, and gin.)

Once this was done, or under way, a portion of the North American Plate started to collide directly with the Pacific Plate—most of the Farallon had by then vanished and was wallowing deep below what would in time be Southern California. As it did so, the relative motion of the plates changed—just as the directions of cannonading billiard balls do. The two plates did not hit each other square on, one of them moving west to east, the other east to west: The North American Plate ran out of steam and essentially stopped in its tracks, while the Pacific Plate began to move northward. It began to slide up and along the outer edge of the North American Plate. It began to *slip* along its *strike,* as the geologists who first discovered the phenomenon declared; it began to slip along the line that marked the edge of its outcrop. And it began to do this a little less than 30 million years ago. Where the plates scraped against each other, the land, up on the surface, became crazed and fractured with a pattern of faults.

At first these faults, marking the strike-slip movement going on below, were some distance away from their present-day track, and ran in different directions from it as well. Near the Californian seaside town of Santa Cruz there is a fault called the Gregori-Hosgri that represents one of the early sliding-plate tracks; and another called the San Gabriel Fault a little way north of Los Angeles also shows, in what can be thought of as a fossilized way, where the plates used to slide alongside each other. And both of these go off in a very different directions from the more recent fault, which displays the more recent relative plate movements.

About 10 million years after the process had first begun—about 20 million years ago, in other words—the relative motion of the two plates settled down, running essentially along the line of one principal fault. And though even today in most places the plate-against-plate contact line cannot quite be pinned down to a matter of inches—usually it is more a zone than a line, and is some hundreds of feet or, in places, even miles wide—today's center point, the zone where the maximum annual slip between the two plates is noticed, is the track of that most infamous darling of seismicity, the San Andreas Fault.

It should be noted that this plate-on-plate strike-slip zone extends between two "triple junctions"—places where the two principal plates meet up with two small relict pieces of the old Farallon Plate that did not get themselves subducted. (These parts were not subducted because, in essence, they were too far north or too far south of the main westward thrust of the North American Plate to be affected.) So the

San Andreas Fault — Northern Section

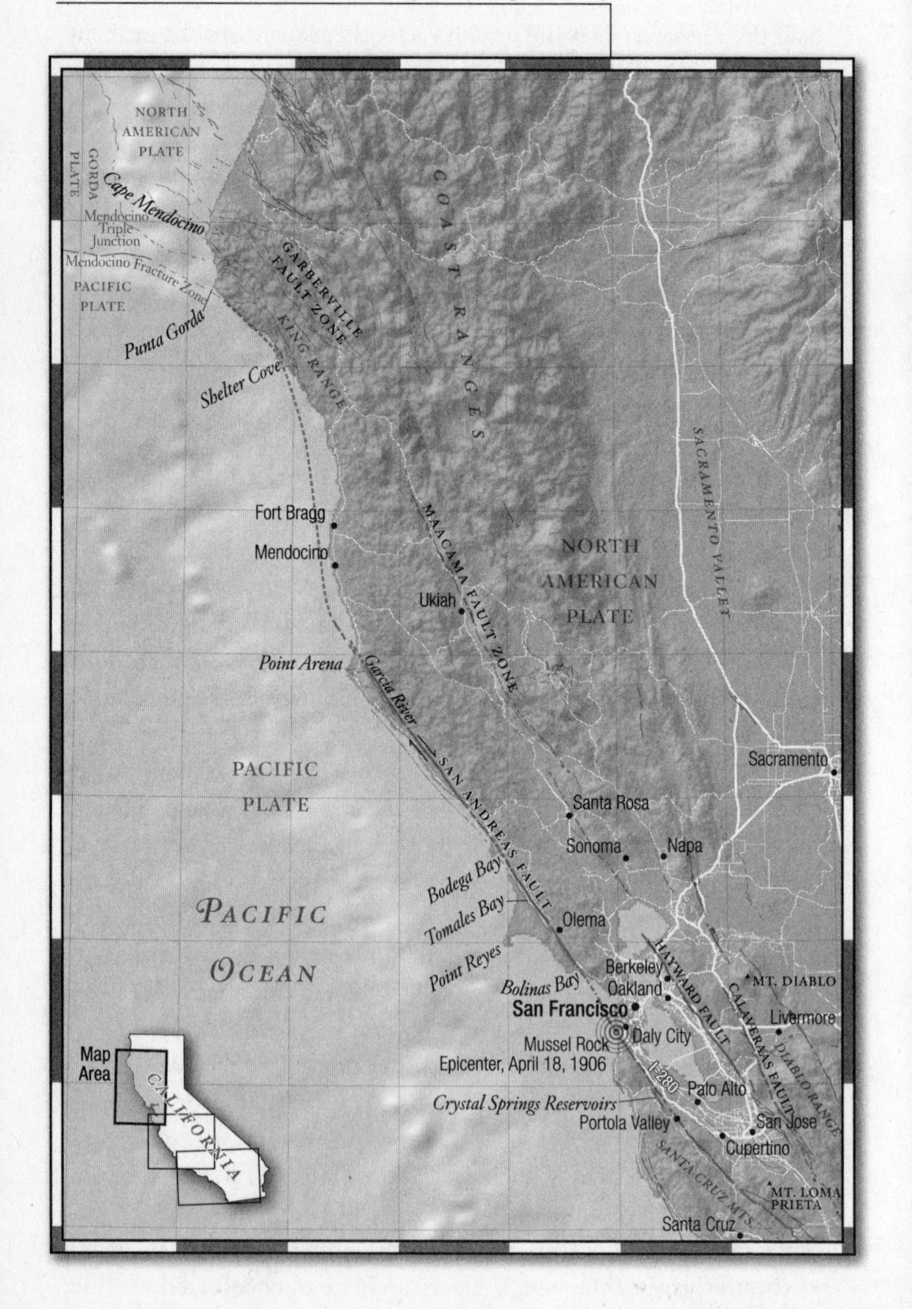

fault zone technically runs from what is called the Mendocino Triple Junction, off California's Cape Mendocino—where the Farallon Plate's little relict piece is called the Juan de Fuca Plate*—down to the Rivera Triple Junction, off Mazatlán and Puerto Vallarta on the Mexican west coast—where the tiny relict piece of the Farallon is called the Rivera Plate.

These two little plates—the Juan de Fuca up north and the Rivera down south—are still subducting, as their predecessor once did. And, sure enough, where they do subduct, there are, as always in such situations, volcanoes. In the north, as a consequence of the beautifully named Cascadia Subduction Zone, there is a fleet of active volcanoes, with Mount St. Helens being the most recently notorious; and in the south there are volcanoes such as Mexico's Colima and Paricutín, the former old and still active, the latter young and, in spite of spectacular eruptions in the 1940s and 1950s, now apparently quite defunct.

Between these two triple junctions, then, runs the San Andreas Fault—this 750-mile-long zone marked by its near ceaseless activity and occasionally by very lethal seismic outbursts. The land on each side of the fault is moving, all the time—though not everywhere along the fault's length at quite the same rate. Overall it is moving at a speed that in real terms seems very slow: Up at a place called Telegraph Creek, close to Cape Mendocino and the northern triple junction, the Pacific Plate is moving northward at about an inch and a half every year. In the terms that geologists understand, however, an inch and a half a year is something approaching raceway speed. According to the USGS, a velocity like this makes the San Andreas a fault like very few others, anywhere in the world. So, by most geologists' lights, it is very fast, very interesting, and very, very dangerous.

Its effects have been noted for a very long while and noted, more-

---

*To complicate matters still further the southern part of the Juan de Fuca Plate, at the point where it meets the Pacific and North American Plates, is called the Gorda Plate. Where it connects to the Juan de Fuca there is a spreading zone, making the geology of Northern California and southern Oregon—especially the mysterious and fascinating Klamath Mountains—weirdly complex.

The very dangerous and geologically active western edge of the North American Plate, with the San Andreas Fault dividing it from the Pacific Plate (which is the landmass on the left of this photograph). The section of the 750-mile fault shown here is in the Carrizo Plain, some 200 miles southeast of San Francisco. Compare with the plate's quieter eastern edge, shown on page 68.

over, in places far removed from where the fault makes itself topographically obvious, such as in the San Francisco Valley where it was first named. Its real extent was noticed in 1906, for instance, just after the San Francisco Earthquake, when a man named F. E. Matthes was sent up north to Humboldt County to help the California State Earthquake Commission see if what had happened in the Bay Area had spread into the north of the state. He found that it had—and moreover he then mapped what he imagined to be the trace of this selfsame fault, finding that all along its never-before-noticed path there was a pattern of instantly recognizable breaks, shears, landslips, and a host of other peculiar and damaging phenomena.

All of this evidence convinced Matthes that this was not an ordinary fault line, but one that had been occasioned by two sets of rocks *sliding past each other.* To him it was a revelation. Back in 1906 most geologists assumed that faults operated only vertically, with huge forces throwing rocks either sharply upward or downward. But here, 200 miles north of San Francisco, there was evidence of the fault that had caused all the trouble—and it was a new kind of fault, a strike-slip fault as it came to be known, and one that was all too obviously an extension of the San Andreas. It was making itself dramatically felt up among the mountains, the clouds, and the mists of far northern California, farther away from the most gravely affected areas than most scientists of the time could ever have imagined.

This was the first time that geologists fully realized the extent of the fault's spread. It was the first time that the whole of California was seen to be playing host to so threatening and so massive a danger.

CLOSE TO SAN FRANCISCO the fault, and the drama it causes when it moves, is very much more obvious than it is in the tortured coastline and hills of the state's far north. A town called Olema, forty miles north of the city (*ole* is "coyote" in the language of the coast Miwok Indians), is generally reckoned to be the place on the fault that moved most dramatically during the 1906 earthquake—and such is the local certainty of this that, quite wrongly, local residents like to think of Olema as having been the event's epicenter. There is a small clothes shop that calls itself the Epicenter. In it and other local stores you can buy caps adorned with the words OLEMA and EPICENTER. I have one. I wear it. But what it implies is not true.

For a general truth should be pointed out here. An earthquake's *epicenter* is not necessarily the same as the place of maximum ground displacement. The epicenter is the point on the earth's surface directly above that point where the seismic energy of the quake begins to radiate outward, the earthquake's originating point. But the parts of the earth where displacement and damage occur depend on a number of details—most important of all the nature of the underpinning rock, the kind of soil, and the slope of the land. And in Olema—where the fault zone is half a mile wide, and filled with soft soils, crumbling limestones, rotting granites, and generally pulverized rocks, much of it lying on well-watered slopes—the sudden shock of a sideways-moving fault rupture did indeed shift the ground in a dramatic and dangerous way. But that does not mean the earthquake was centered there. It means merely that it had an impressive effect there—with few casualties and less publicity, however, since it was a place where few people lived.

Grove Karl Gilbert, the distinguished geologist who, it will be recalled, was roused by the quake as he lay in his bed in Berkeley, was among the first of the state commission members to go to investigate the situation in the Olema Valley. The devastation he found was most impressive. Buildings made of redwood and oak were crumbled as though fashioned from balsa chopsticks, huge water towers had been tossed around like cruets, long fissures spread across fields. The local dairy farmers told him fanciful stories, including one that was widely reported in the papers: A cow supposedly fell into one fissure, which swallowed her up as it closed shut behind her; she then suffered the postmortem indignity of having her tail eaten off by packs of marauding and hungry dogs.

However, it was measurement—or one measurement in particular—for which Gilbert is perhaps best remembered. And he measured anything he could lay his eyes on, particularly anything that seemed to have been displaced across the trace of the fault. He measured fences, stands of eucalyptus trees, roads and farm lanes and tracks—and, from the displacements he found, he came up with a litany of earth movements that has been exceeded only by a very few other earthquakes in history. One farmer's fence snapped, and its posts were shifted thirteen feet apart, declared Mr. Gilbert. A line of eucalyptus trees near Bolinas Lagoon was broken and moved, also by thirteen feet. A road southwest of Point Reyes Station was displaced by twenty feet. The rough gravel pavement of the Sir Francis Drake Highway was broken clean across, and one part of the road was moved—Gilbert photographed it, memorably—just over twenty feet relative to the other.

Although the small Northern California village of Olema saw the greatest displacement of the earth—this roadway was shifted some twenty-one feet by the right-lateral motion of the San Andreas Fault—it has lost its much-prized position as the epicenter of the 1906 event. That distinction now goes to a point under the sea off Daly City, forty miles south.

Not up or down but from side to side: If you stand and look at where the centerline of the road might be, it suddenly ends, and then reappears twenty feet off to the right, with the highway then continuing and vanishing over the horizon as though nothing had happened.

To the right. The effect of movement on a *right-lateral fault:* This is how the fault is officially known, as a *right-lateral strike-slip* fault. Stand anywhere and look across the fault—and the land on its far side will have been moved to the right. Hills, streams, roads, lines of trees—all of them, if they are on the distant side of the fault, will be to the right of those that lie on the side near the observer.

Today, at the Point Reyes Visitors Center, the fault is plotted through the meadows with a line of blue posts stuck into the ground. In one place the posts show the fault spearing underneath the barn that once belonged to a local farmer named W. D. Skinner. He had told Gilbert many things about the event (including the story about the cow) and showed him how his barn had been torn apart, his neat rows of raspberry bushes offset by fourteen feet, his fences ruined and scattered. And indeed there still is a fence on the property dating from long before 1906. At one point the fence posts seem suddenly to vanish—until you look and spot them once again, half hidden in a patch of woodland. The fence had been built with one continuous line of hastily carved redwood posts, with pine or redwood crosspieces between. But at the moment of the earthquake this fence line was snapped and sundered by the force of the event, with the western end of the fence moving north. And it moved a lot—to such an extent that the two ruptured ends that are visible today have been left no less than twenty-one feet apart.

This figure is most often cited, and impressively so, as illustrating the maximum amount by which the San Andreas Fault, and by association all of California as well, had moved on that extraordinary April morning.

But still, Olema was not the epicenter.

EARTHQUAKES EMIT WAVES that ring through the solid earth just as sound rings through a bell made of brass. Moreover, they emit a number of different types of waves that, most fortunately and most crucially, move through the earth at very different speeds. It is the difference in speed between the two fastest-spreading kinds of waves (which are known as body waves because they travel through the entire body of the planet) that allows us to determine generally where each earthquake has its point of origin.

All that is required is a minimum of three recording stations, equipped with seismometers, that can observe the earthquake, measure the arrival of those two waves, and record the time between them. The first of the two body waves to arrive—and so the faster of the two—is a pressure wave, a P-wave, one that presses the rock and releases it, presses and releases, as if a Slinky toy were being stretched and then given a hard shove along its main axis. The second, slower body wave is the shear wave, the S-wave, which ripples horizontally through the rock strata—just like the sideways ripple that can be made to course through a Slinky.

(When terrified observers speak of the ground rippling toward them, or rising and falling in great fast-moving wavelike motions, they are almost certainly seeing the next family of waves—surface waves—that propagate more slowly because they only involve the outer surface of the earth. These most destructive waves—which are very easy to see on a seismograph because they are very large and have relatively large amplitudes—are further subdivided into what are known as Rayleigh Waves and Love Waves; the behavior of these, though they can be devastating to buildings and lethal to people, generally has less relevance to determining the location of earthquakes.)

Providing that both the P- and the S-waves travel through the same kind of rocks on their way to the observation station—which of course they would; but it is worth pointing out that waves run at very different speeds depending on whether they pass through granite, say, or shale—then there is always the same differential in their velocities. The farther the origination point from where the earthquake is felt and measured, the greater that differential. An earthquake originating in Olema might show, on the seismographs in the USGS machines in Palo Alto, a differential between the two waves' arrival times of just ten seconds; that same earthquake, noted by the machines in Columbia University's famous Lamont-Doherty Earth Observatory in New York City, might show a differential of eight minutes. Each single second of separation in the waves' arrival times equates to about five miles of distance. This is similar to the rule-of-thumb approach that is used to compute how far away a storm might be: One sees the lightning flash and then counts the number of seconds until the thunder is heard—that number is very roughly how many miles separate you from the storm. So from Palo Alto the quake would be 50 miles distant, and from New York, 2,400 miles.*

So, taken singly, any receiving station can tell, simply by noting the difference in the arrival times of the two waves from an earthquake, just how far away the earthquake's center is. But, armed with at least three receiving stations, one can then triangulate these data and find out not just how far away but also exactly where the earthquake is. The exercise is simplicity itself: All one does is draw a series of circles representing the distance (calculated from the time differential) that the earthquake appears to be from each one of the three stations. If the time differential indicated the earthquake was 50 miles from Palo Alto, then draw on the chart a circle with a 50-mile radius around Palo Alto. If the time differential suggested the same quake had happened 2,400 miles away from the Lamont station in New York, then draw a second 2,400-mile-radius circle around New York; and finally if a Calgary observatory measured the quake as occuring 1,200 miles from its seismometer, draw this third circle with a 1,200-mile radius. The three circles intersect at just one point—and this point where they meet is the epicenter—the place on the earth's surface that is directly above the originating rupture. And the exact point of that rupture, and the depth of it below the surface, is something that can itself be deter-

*A diagram of a typical seismographic record of a large earthquake annotated to show the subtle and not-so-subtle distinctions between the various waves used to chart the event's distance, depth and magnitude, can be found in the appendix.

mined by performing the same P-wave, S-wave, and time exercise all over again, but this time in three dimensions. That will give the hypocenter, the true originating point of the event. All, in other words, extraordinarily easy, for anyone with three seismographs and three accurate stopwatches.

Which is what a Berkeley seismologist named Bruce Bolt had—although actually he had many more seismographic records than these—when, in 1968, he recomputed the data responsible for the widely believed notion that the 1906 earthquake had its epicenter in the Marin County countryside near Olema. The 1906 event was recorded not merely by three observatories but by ninety-six seismographs around the world, and just about every single one of them had an accurate recording clock, each tuned to what was then Greenwich Mean Time. Moreover, a prodigious number of the local clocks—ordinary timepieces on living-room mantels, long-case clocks in parlors, clocks on church towers, clocks displayed on public buildings—were stopped dead by the force of the event, further evidence for the time of arrival of at least the strongest jolt of the quake. (It does take some technical expertise, however, to work out whether a clock from a historic earthquake has been stopped by a P-wave or by the subsequent S-wave, or perhaps even later, when a brick might have fallen on it. Generally, though, the shear waves are stronger, and so calculations on stopped clocks often assume that the stoppage was due to the arrival of the S-wave.)

However questionable some of the data might have been, Bolt had a very great deal of them, enough to calculate what is now generally accepted to be the true epicenter of the 1906 event. And so, according to the paper that he published in 1968, the recalculated epicenter turned out to be in the sea, a little more than a mile offshore, southwest of the Golden Gate Bridge and northwest of a particularly crumbled section of the coast near the sprawling suburban community of Daly City.

It has long been recognized that the track of the San Andreas Fault passes into the ocean a few miles south of Olema, close to that expanse of sand and shingle where so many San Franciscans go to swim known as Stinson Beach. The coastline is indented eastward from there on south, forming a funnel toward the Golden Gate and the main shipping entrance to San Francisco Bay. The fault, however, remains unpersuaded by this deflection, and continues to roll on southeastward in its usual die-straight line. It scythes on through the sea to the immediate west of where the Golden Gate Bridge now stands, and reconnects with the coastline five miles farther on, in a particularly wretched part of the suburban mass of Daly City called Mussel Rock, where—under the fault's malign influence—houses seem always to be sliding into the ocean, landslips are an all-too-frequent occurrence, and the coast is a veritable construction site of seawalls, patched roads, and fractured gas lines. The benighted residents of Mussel Rock have paid a stinging price for the luxury of being able to watch the sun set over the Pacific each evening.

These days the community of Mussel Rock is the site of a ceaseless and wearying three-way battle going on between those residents who like to live there and gaze dreamily out at the foggy blue sea; those who say that the fault below them is moving so fast and so unpredictably that nature will not allow for indolent dreaming, or at least not for long, and that no one should be allowed to live there; and city officials who claim that the fault's energies can all be sapped or countered by man's cunning energies, and that Mussel Beach can be made safe and permanent. It can be made as enduring, some say, as that similarly situated rock in the Mediterranean, Gibraltar—and, in any case, Daly City people are perfectly at liberty to live just where they want to. They may not get good insurance, but they can live there if they feel they must. The tussle continues.

And then there was the row that broke out over Daly City's reputation. Just as soon as Mr. Bolt proved that the epicenter lay offshore a mile from Mussel Beach, a local historian, inspired by the news, designed a heavy brass plaque that memorialized what now appeared to be Daly City's historic importance in the event. His plan was to erect this monument—which he had bought and paid for—on a nearby beach.

But the mayor of Daly City, an elected official who at first blush appears to be a man with rather little sympathy for history, said no. He didn't, he said, want people to associate his city with an earthquake. "We don't need to put a blemish on Daly City's shine," he declared. "There's no reason for it." So the plaque could not be erected anywhere in Daly City.

The historian tried three further times, even appearing before the city council armed with the support of a host of the region's geologists and seismologists. But the councillors remained adamantly intransigent, too, with the result that Daly City will not now be publicly associated in this way with the events of 1906.

Except that it will be associated with the event—thanks to a much more acceptable story that is already well known locally. Refugees from the earthquake's firestorms fled south across the city line and camped on plots beside what were then the meadows of one John Daly's dairy farm. Rather than return and risk another tragedy, many of them then set up home there. Daly City is, in other words, a place that owes its very existence to the 1906 earthquake, is in its very essence a memorial to the tragedy of 1906, and probably does not want to be reminded any further, thank you very much. Besides, the crumbling of the cliffs at Mussel Rock provides a daily reminder of the presence of the fault, which streaks under the entire length of the community on its way south.

THE SAN ANDREAS FAULT can be handily divided, like Gaul, into three parts—though, unlike in Gaul, they are three unequal parts. The division is made by geologists, since in each of these three parts the fault behaves very differently.

First, to the north, there is a 275-mile-long sector that stretches from Cape Mendocino, past Olema and San Francisco and Daly City, to a point just south of the old Spanish mission at the dust-dry ranching town of San Juan Bautista. Then there is a rather longer southern section, which runs for about 350 miles from the beginning of the Sierra Madres southwest of Bakersfield to the fault's southern terminus at the Salton Sea, close to the Mexican border. And in between these two sectors is the shorter, 125-mile-long midriff of the fault centered around the small town of Parkfield. This middle section is very different in one hugely important way from the sections that lie to the north and the south of it.

Some things about the fault remain essentially the same. Throughout its length the San Andreas is a right-lateral strike-slip fault: that basic characteristic, with the North American Plate on the eastern side and remaining resolutely in place, the Pacific Plate on the western side and shifting northward at an inch and a half a year, is unvarying. All the movement along the fault is sideways, and it all goes in the same way. (Having said that, there are a few places that have the fault going up or down, in the way that faults were once thought to do. But these places are few. They reflect local aberrations in rock structure, and do not affect the overall thesis, which is dominated by the now all-too-familiar scenario of two plates sliding past each other.)

However, there is one signal difference. In the middle section of the system, in and around Parkfield, the fault is for some inexplicable reason always *moving*. It is shifting, all of the time, little by little, millimeter by millimeter, day by day. But in the rest of the system—in the northern section centered around San Francisco and in the southern section centered on places like Palmdale, out in the desert to the east of the eastern suburbs of Los Angeles—the fault on a day-to-day basis is not moving at all. The north and the south are locked solid. And yet the middle part of the fault is moving—a circumstance that places the locked northern and southern parts of the fault under an ever-increasing amount of stress.

The best analogy for this is to think of a line of railway freight cars standing in a marshaling yard, each car connected to the next by a firm iron clasp. The cars in the yard are all on a slope that makes them want to move downhill headfirst, as one. But the wheels of the cars at the head of the line are rusted solid, as are those of the cars at the distant end. Only the cars in the middle have well-oiled wheels—and it is these, under the force of gravity, that move down the slope under its influence. They move, but their movement is checked all the while by the sheer inertia of the rust-blocked cars in front and behind. Enormous stresses thus build up in the iron clasps that bind the cars together, with those at the front being pressed tightly together, those clasps at the rear being stretched beyond endurance. Some of them may even bend a little, their inherent elasticity storing huge amounts of energy as they wait for the day that they either break or spring back to their original shapes.

And then, every so often, the compounded weight of the cars in the midsection becomes simply too much for the rust holding fast the wheels of the locked cars—and, under this terrible mounting pressure, the rust gives way and the cars shoot forward, jerking anyone foolish enough to be standing on them off their feet and upsetting any bales of cargo piled on top or inside. The pressure from the midsection of the cars suddenly eases, because all of the cars have now moved as much as they are going to—for a while. But then, day by day, the middle cars begin to shift once more—and after a while, with the intolerable stress again built up both ahead of and behind them, one of these rusted sections gives way, suddenly, dramatically, and violently.

There is a shudder as the pressure is relieved. The energy that has been building up, the potential energy that has been growing, waiting for the weakest link in this particular chain—in this case, the rust in the locked wheels—to give, is suddenly released and converts itself in an instant into kinetic energy, which tumbles cargo and throws people around. This would be very dangerous were it ever to happen in a railroad marshaling yard. It is also, in its essentials, precisely what is happening in these sections of the San Andreas Fault.

The forces caused by the unstoppable, irresistible movement of the plates deep below are building up and up and up. Because of the movement, barely recognizable or measurable shifts in the landscape are occurring more and more. Rocks close to the fault, in what is called the near field, are all the time bending, straining, distorting. As they do so, they absorb huge quantities of energy—and this process continues until that moment when, suddenly, the weakest link in the affected rocks—the friction that prevents them from moving under all those accumulating pressures—ruptures. This happens because at last sufficient forces have accumulated to overcome its resistance.

In a matter of microseconds two events occur. First, the rocks that have been bent and strained and distorted rebound to their original relative positions. And second, the plates are suddenly released to jerk forward to their long-awaited new positions. As these two things happen, both plates below and surface rocks above release all the energy that has been stored in them in the years during which the stress had been a-building. This long-stored potential energy transmutes itself immediately and demonstrably into energy—kinetic and thermal and sonic—and, as the transmutation occurs, with drama and turmoil that is directly proportional to the amount of energy that is transmuted, buildings are torn apart, heat is created, enormous sounds rumble across the land. It becomes, in short, an earthquake.

This the famous theory of elastic rebound. It was first adduced by the celebrated geologist Harry Fielding Reid, who had been appointed by Andrew Lawson to his 1906 Earthquake Commission, and has remained the dominant theory of the cause of earthquakes along this kind of fault ever since. In places like California, where the earth is clearly under the influence of an ever-moving fault, it is in constant tension, much like a tightly wound mainspring in an old pocket watch. The tighter it is wound, the more energy it stores; should it be wound too tightly, then some weakness within it, or its anchor point, will eventually give way and the innards of the watch will collapse dramatically. This, in its own limited way, is an earthquake, too. Although Reid's theory has been tinkered with over the century since he introduced it (as faults are now known to be much more complicated entities than is to be supposed from their cartoonlike representations in textbooks), what he declared in 1906 continues to be central to seismic study. The earth is elastic: Tighten it and it changes; overtighten it and it breaks.

Which brings us, if circuitously, back to the current fascination with Parkfield.

I HAVE ALREADY mentioned that the San Andreas Fault at Parkfield is constantly moving, and as a result there is a ceaseless blizzard of very tiny earthquakes, and every so often a rather larger one. This is

San Andreas Fault— Central Section

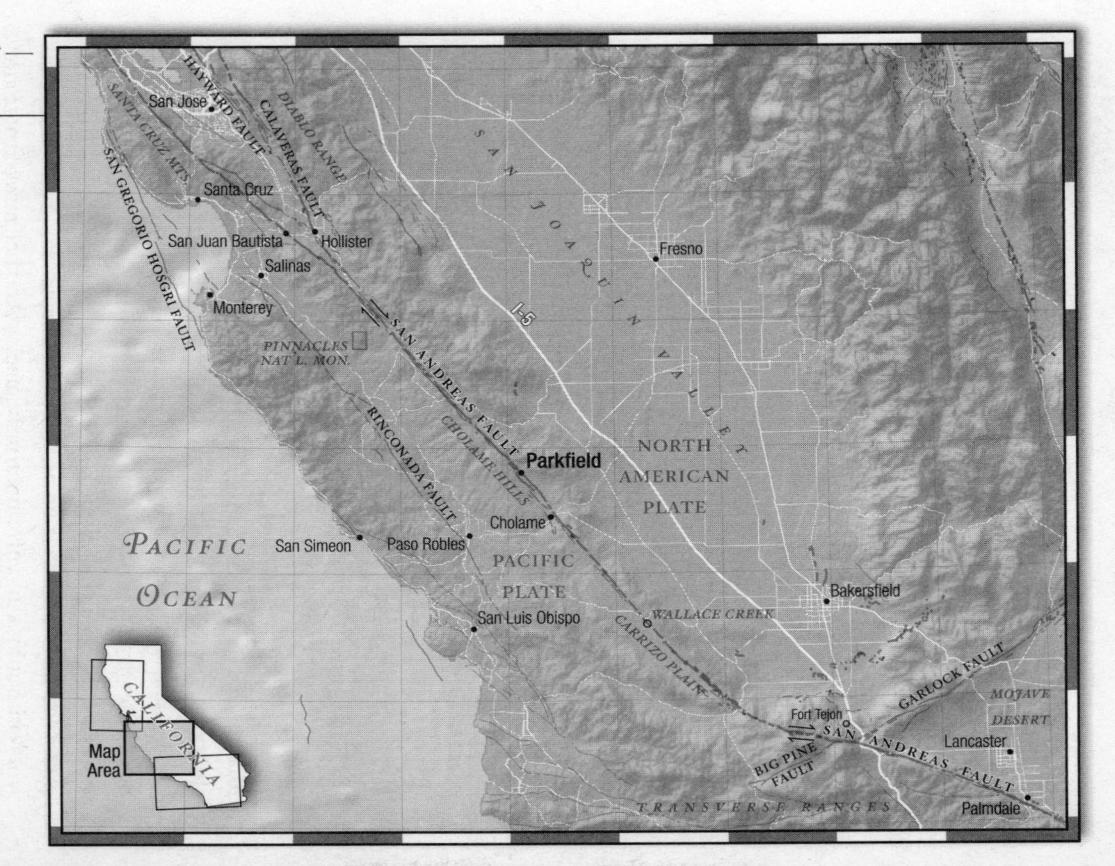

true—except when one looks at the situation very closely indeed. For like many other kinds of movement, Parkfield isn't so much moving as moving in a very rapid series of very small jerks. There is a similarity, for instance, in the different ways an automobile moves when in a straight line and when going around a bend. In a straight line its tires move constantly, seamlessly; when taking the car around a curve the tires move in a series of thousands of very tiny skids, each one requiring traction that can make the tire hot and wear away its rubber. An airplane, too, skids its way around the air when it is turning; when looked at closely, in a wind tunnel, there is nothing smooth at all in the way an airfoil negotiates a turn—it merely looks smooth and seamless.

The fault behaves in Parkfield in much the same way. Deep down, the plate is moving here just as it is everywhere else; and the rocks on top of it in Parkfield appear to be moving fluidly past one another—unlocked—in a way that they do not move up north in Olema or down south in Palmdale, where they are locked. Except that, on close inspection, the supposedly fluid movement is not always quite so fluid as it looks. To be sure, part of the movement is fluid, and is known as aseismic creep—it just stretches and bends and slowly moves along under pressure. But, as far as researchers can tell, this is a very small component of what goes on.

Most of the time the fault in these parts is actually locking and unlocking itself, but at so fast a rate that it looks as though it is in continuous motion. Recently it has been deduced that the process of locking and unlocking that goes on scores of times each year underneath the ground in and around Parkfield is mechanically and physically exactly the same as the process that a researcher might have to wait years to see farther north or farther south along the fault (and where the unlockings produce earthquakes). And because of this Parkfield is probably the best place in the world to study, on a small scale and in real time, exactly what happens to the San Andreas Fault—and, by extension, to any other fault—just before, during, and immediately after it locks, unlocks, and finally relocks itself.

They are doing this by drilling a hole—a modern-day journey, say some of the project's more enthusiastic boosters, to the center of the earth.

USING A LARGE oil rig, and employing a team of tool pushers and roughnecks hired directly from the oil industry, geologists are forcing a hole two and a half miles down into the fault itself. Scores of instruments have been placed at the bottom of the hole, and an array of branching tubes has been drilled out from the main shaft—and it is with these that the scientists are examining what happens at the depth at which all of the fault's movements are known to originate.

The experiment is called the San Andreas Fault Observatory at Depth (SAFOD), and it was formally begun in 2002 with a pilot hole that took the instruments down a mile and a half, stopping just shy of the active fault zone itself. Then, in the summer of 2004, a new drill team put up its $20 million derrick and its mighty stands of pipe and drill bits and the great square turning nut known as the kelly, and started to drill the main tube.

The National Science Foundation and the U.S. Congress had earlier in the year approved the $250 million to be spent on a hugely ambitious program known as EarthScope, of which the Parkfield pipe was a major part. (Parkfield accounts for only a tenth of the budget: $100 million is going toward a great number of GPS devices that will measure movements all along the boundary between the tectonic plates in western America; and $70 million will go toward a huge network of seismometers that will be installed around the West to paint a seismic picture of the continent with an unprecedented degree of accuracy. SAFOD costs only $25 million, and yet it is the project that of the three has managed most handsomely to capture the romance—and the danger—of this kind of research.)

EarthScope has had a profound effect on modern geology and played a major role in the making of the brand-new science into which the Old Geology has now indisputably evolved. At last the U.S. Geological Survey and Stanford University (and more than a hundred other universities and institutions in America and around the world that are committed to studying SAFOD's eventual results) have at their disposal a hugely costly and very-large-scale experiment that, in its own way, seems likely to present as good a picture of a fundamental process of the planet's working as the moon landers and bathyscaphes and cyclotrons and linear accelerators had already presented to the practitioners of their respective disciplines. Such a thing has never happened before—except for the 1969 moon landing and the astronauts' collection of moon rock. The SAFOD experiment is undeniably Big Science, and the geologists involved are happily amazed at their good fortune, finally able to hold their heads high in the company of all those nuclear physicists and genome researchers and oceanographers who have taken Big Science for granted for so long.*

The first part of the SAFOD drilling project, which was executed with the Nabors Industries No. 633 rig positioned on a cattle ranch that stands firmly on the Pacific Plate, was completed at the beginning of October 2004. The end of the hole was then left a tantalizing 1,600 feet above the football-field-size cube that seismologists have decided incorporates the most active part of the fault, where they want their various probes to be planted. In the second phase the shaft is to be extended to reach directly into this cube-shaped zone. (At first it was drilled entirely vertically but later bent eastward so that it runs through the fault and right out into the North American Plate.) Next, branching tubes will be bored with special directional drilling equipment, reaching to other parts of the cube and beyond it. Sensitive equipment in protective pods will be lowered into the end of each of the subterranean shafts, where all manner of physical data from within the deep will be measured: the stress, the pressure of fluids, the tem-

---

*One other very Big Science experiment that once had geologists excited was the attempt to use a deep-sea research vessel to drill through the seabed to the so-called Mohorovicic discontinuity, the Moho, which marks the division between the earth's crust and mantle. It proved costly and difficult, and was abandoned in 1966. The Japanese drilled close to the Nojima Fault on Awaji Island in 1995, but they stopped also, after less than half a mile down.

perature, the heat flow, the chemistry of the fluids, the rocks, and any gases—and, of course, the seismicity.

The big worry among local residents is that, somehow, the drilling program itself will trigger an earthquake. There is a widely held concern—held, at least, among those who are somewhat skeptical of science—that interfering with the delicate balance of the plates will start something that might prove unstoppable. Was it not possible, some vaguely exercised Californians asked rhetorically, that the drilling rig might set something off? The USGS said there was no possibility. Look at Texas, someone said. There can be no more comprehensively drilled place than the oilfields of Texas—and yet the state is seismically almost stone dead.

Leaving aside the questions begged by such a statement—Texas is basically very much more geologically stable than California—it turns out that the skeptics did have some cause for concern. For it so happened that the SAFOD drill site was at its most active, and the drill bit was probing most closely to the fault, when, on the morning of September 28, 2004, a magnitude 6.0 earthquake hit Parkfield square on.

It was the largest earthquake to strike the town for thirty-eight years. The USGS and the drill team insisted there was no connection. Locals were not so sure.

THE COUNTRYSIDE SOUTH of Parkfield rapidly becomes very much drier, sandier, and unpopulated. Only eight inches of rain fall each year on the Carrizo Plain—a high, hot place of salt flats and soda lakes, with kangaroo rats and leopard lizards and other improbably named fauna—but with almost no trees and little vegetation to obscure what the fault does on its way through. The result is that the San Andreas is laid bare as it pushes down along the desert, as if pinned to a dissecting table. Almost all books related to the fault sport an aerial photograph of its signature across the Carrizo Plain on the jacket, so clear and obvious is its passage.

But, leaving aside the scimitar-like line it cuts across the scenery, it is the right-lateral displacement of the rivers running across the fault that presents it in an even more dramatic light. Such oddities show the San Andreas not merely as a static entity but as something that shifts: *E pur si muove.*

To the east of the plain rise the low hills of the Temblor Range, and from their upperworks course innumerable rivulets—usually dry—known locally as washes. On those rare occasions when rain does fall, these washes fill up with water and form raging streams, which rush downhill, carving deep canyons from the loose and friable soil as they do so. The dry ravines that are left behind are known as arroyos, and, were the hillsides perfectly geologically stable, one would regard such things as merely pretty, the serried ranks of parallel valleys proceeding from summit to valley floor, peculiar only in their number and symmetry.

But the edge of the hills marks the line of the fault. The Carrizo Plain happens to be on the Pacific Plate, and the Temblor Range is on the North American Plate. Every so often the Pacific Plate jerks itself northward, and when this happens it takes with it the paths of the arroyos, just where they begin to run across the plain. Where this has happened the effect is startling: The ravine starts off running downhill, then suddenly turns ninety degrees and begins to flow along a straight line parallel both to the edge of the range and to the valley floor—or, to put it another way, parallel to the line marking the joined edges of the two plates. It goes on like this for a few hundred feet—and then with equal suddenness it turns back ninety degrees, flowing in its original direction once more.

And every stream does the same—in echelon, every single stream exhibits the double jog, each in the same direction, each serving as a marvelous reminder of the fact that the fault moves and of the direction in which it does so.

This is not to say that the movement along the fault is horizontal only—which is what a cursory look at the offset streams of the Carrizo Plain might imply. There is a vertical component, too—the Temblor Range, for example, has been caused by vertical uplift along the fault. (The reason it exists at all is that the fault does not exactly coincide here with the direction of plate motion: It is slightly skewed to one side, and as a consequence of this skewing the material bunches up, like a pleat in a carpet, as the plates move alongside each other.) But in this part of the San Andreas the horizontal component far outweighs the vertical, by a factor of between 10 and 20 to 1.

One of the most dramatically offset of these Carrizo Plain streams—a stream that has become the poster child for all San Andreas phenomena, so clear-cut is its display—has been named Wallace Creek, honoring the field geologist who has perhaps been most prominently associated with the fault, Robert Wallace of the U.S. Geological Survey.

It was Robert Wallace who reversed a notion among some geologists that had become fashionable during the mid-twentieth century: This was, that since the newly invented machine called the seismograph gave all the necessary answers about earthquakes, there was little need for geologists to go out into the field to study them. For years around the middle of the century, quakes were primarily in the purview of white-coated technicians, people sitting in air-conditioned laboratories working their bloodless mathematical wizardry to determine the nature of the spasms that occasionally afflicted the earth. Robert Wallace changed all that: Armed with a hammer, a compass, a series of very good maps, a tent, and endless cans of beef stew*—which he liked to eat cold—he went back out among the rocks and worked for years to delineate the fault as it charged southward from the Carrizo Plain.

He mapped the places where the fault exhibited what has come to be known as the Big Bend—the long and lazy turn it executes midway along its length, which gives the San Andreas its approximate overall shape of a boomerang. To create this bend, the line first turns through thirty-five degrees to the left—to the southeast, that is—then lazes back again toward the right, until finally, sixty miles of zigzagging contortion later, it settles straight, back on its original track.

The turn is made as it traverses an unusual set of mountains called

---

*He also took a violin, on which he would play serenades to the coyotes.

San Andreas Fault — Southern Section

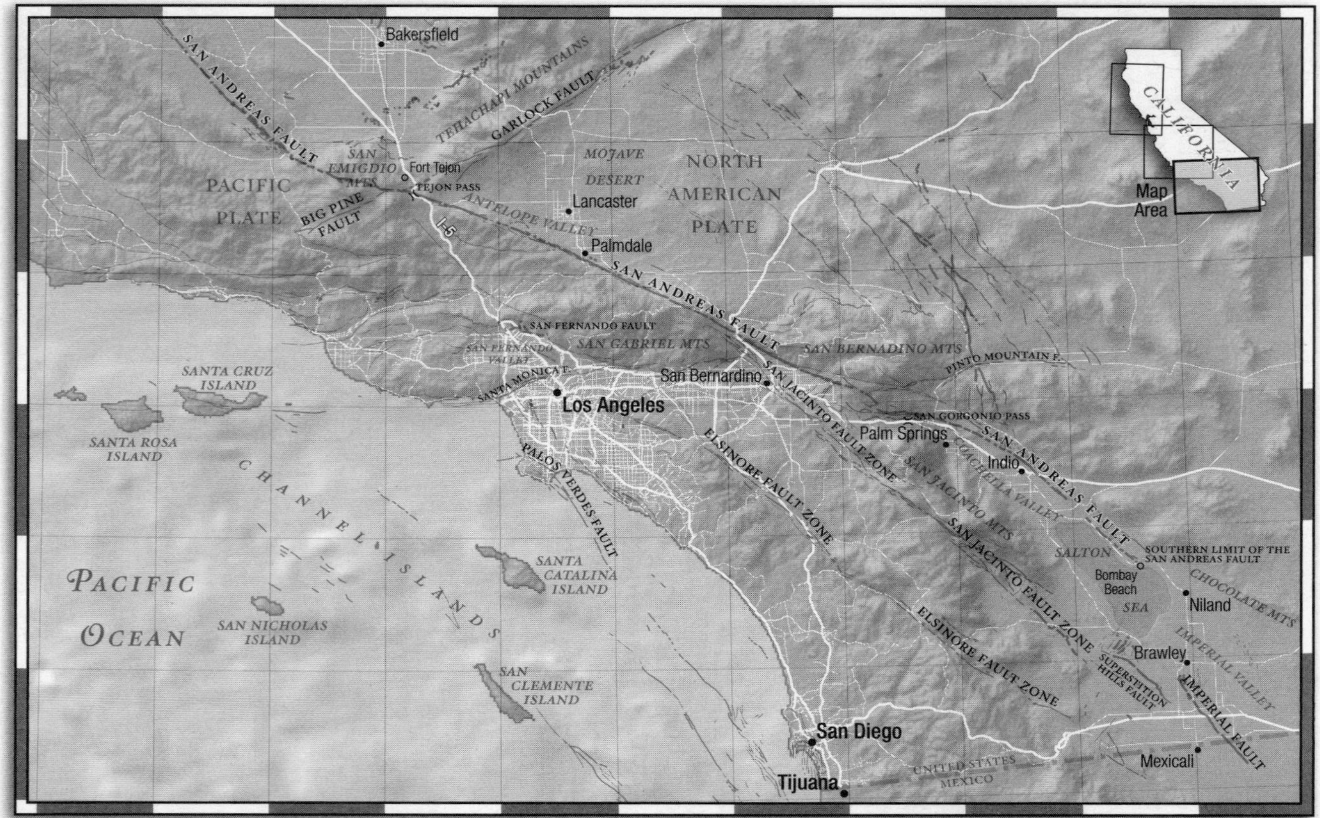

Bakersfield
SAN ANDREAS FAULT
TEHACHAPI MOUNTAINS
GARLOCK FAULT
SAN EMIGDIO MTS
Fort Tejon
TEJON PASS
PACIFIC PLATE
BIG PINE FAULT
MOJAVE DESERT
Lancaster
ANTELOPE VALLEY
Palmdale
NORTH AMERICAN PLATE
I-5
SAN ANDREAS FAULT
SAN FERNANDO FAULT
SAN GABRIEL MTS
SAN FERNANDO VALLEY
SAN BERNADINO MTS
PINTO MOUNTAIN F.
SANTA MONICA F.
San Bernardino
Los Angeles
SAN GORGONIO PASS
SAN JACINTO FAULT ZONE
Palm Springs
SAN ANDREAS FAULT
Indio
COACHELLA VALLEY
SAN JACINTO MTS
ELSINORE FAULT ZONE
PALOS VERDES FAULT
SANTA CRUZ ISLAND
SANTA ROSA ISLAND
CHANNEL ISLANDS
SANTA CATALINA ISLAND
SAN NICHOLAS ISLAND
SAN CLEMENTE ISLAND
PACIFIC OCEAN
SALTON SEA
SOUTHERN LIMIT OF THE SAN ANDREAS FAULT
Bombay Beach
Niland
CHOCOLATE MTS
SAN JACINTO FAULT ZONE
ELSINORE FAULT ZONE
SUPERSTITION HILLS FAULT
Brawley
IMPERIAL VALLEY
IMPERIAL FAULT
San Diego
Tijuana
UNITED STATES
MEXICO
Mexicali
CALIFORNIA
Map Area

the Transverse Ranges—unusual because, unlike all other Californian mountain ranges, these mountains slice across the state from west to east, and do not parallel the coastlines and the faults, as do the Sierra and the coast ranges, as well as all the lesser ranges like the Temblors and the San Gabriels.

The fault crosses this range at a remote mountain cwm called the Tejon Pass—so named because the Spanish lieutenant who discovered it at the beginning of the nineteenth century found a dead badger, *el tejón,* at the mouth of the canyon—and then passes on into the achingly dry wilderness of the Mojave Desert, close to the city of Palmdale. Robert Wallace was one of the first to work out just how far the fault seemed to have slid in these parts over its millions of years of life. As a brash young graduate student he declared that at Palmdale* the two sides had shifted 75 miles apart from each other—a figure that, in those more conservative, pre-tectonic-plate-theory days, was so improbably large that his colleagues chortled in disbelief. They should not have bothered, since Wallace was, of course, almost right—at least so far as orders of magnitude were concerned. The two sides of the San Andreas Fault have moved at least 250 miles apart since it was born—sometimes slowly and steadily, sometimes through the savage interruptions of earthquakes.

California's other truly great earthquake occurred in these parts, close to the Tejon Pass, in the spring of 1857. It was an event that was at least as big, and may well have been even larger, than the quake that was to wreck San Francisco nearly half a century later.

It occurred on April 9, and it had a supposed summary magnitude

---

*It is in Palmdale that highly contorted strata resulting from the fault's movement are displayed to advantage in one of the most famous and often photographed road cuttings in the world. They are on the east side of California Route 14, and, because there are three top-secret airbases also visible from the site, the local police take a rather dim view of photographers stopping there. An innocent geologist—especially a foreigner—can get into a good deal of trouble, and offering the excuse of merely trying to photograph the San Andreas Fault doesn't cut much ice these days.

The ceaseless grinding of the two plates where they meet at this roadside cutting near Palmdale, California, causes spectacular folding and distortion.

---

of 8.25—just a notional amount less than that later ascribed to San Francisco. Its intensity (which is not the same as magnitude, the distinction being explained in the appendix) is thought to have been in the region of VII, also rather less than the VIII and IX that were experienced in San Francisco. Some calculations insist that it was a greater event than San Francisco—that more energy was released and that the displacement of the ground was far more extensive. But there is so little by way of historical record—much less than that of the New Madrid Earthquake of 1811, mentioned before—that it has to be regarded as an event of less historic importance. It cannot be classed as one of the most significant earthquakes of all time—which 1906 most certainly was, because it ruined so much and killed so many. But it was nonetheless very, very impressive.

Its epicenter was actually nowhere near Tejon Pass. It has since been shown to have been back in Parkfield. But since virtually no one lived in Parkfield in 1857, and since Tejon had an army base, Fort Tejon, this bleak mountain pass has ever since been put forward to enjoy or to suffer the notoriety.

The base had been created three years before, ostensibly as a means of controlling—the army's word—the resident Indian population. It was little more than a handful of adobe huts huddled among the valley oak trees, beside a stream. The only thing that was unusual about the fort was its small and temporary population of camels, which had been brought to Texas from Egypt and Turkey because some military planner thought they might be useful; they proved worse than useless, however. The Texans got rid of them and sent them to California, and the army in California had no real idea what to do with them either, and they soon vanished. Some still associate Fort Tejon with something called the U.S. Army Camel Corps: There was never, in fact, any such thing.

Just after eight on the Friday morning of January 9, 1857, an almighty earthquake rolled down onto Tejon from the north. Witnesses speak of huge wavelike shakings of the earth; and, though some speak of up to three full minutes of shaking, an unprecedented duration, most agree that it was just some forty or fifty seconds' worth of nightmarish movement that wrecked all the army huts, tore most of the trees from the earth, and killed a woman at the nearby Reed's Ranch. The local Kern River ran backward; fish were thrown hundreds of yards from where they swam in Tulare Lake; long zigzag cracks appeared in the ground at San Bernardino; massive ridges, five feet high and fifteen feet across, rose and started to snake through fields; artesian wells suddenly failed; the Los Angeles River was hurled out of its bed and began, if only briefly, to flow along another channel; and up on the Carrizo Plain the fault jerked so dramatically that many of the rivers coursing down from the Temblors were thrown off course by as much as thirty feet in a matter of microseconds.

The event was felt all across Southern California. It was not felt at all north of Parkfield, perhaps because of the more lubricated nature of the ever-moving fault up there. Had it struck in modern times, it would have caused dreadful damage forty miles away in Los Angeles. But, as it was, only two people (the rancher, and one other man in a village plaza) were killed; and the 4,000 people who lived in the sprawling village that was Los Angeles got little more than a jostling.

Matters were somewhat more serious in Santa Barbara, then a pretty coastal hamlet of 2,500 thirty miles west of Tejon. Villagers, who were accustomed then as now to an idyllic setting with equable temperatures, balmy weather, and no more than the gentle plashing of Pacific waves, streamed out onto the streets, panicky and terrified at something they could hardly have imagined. In huge numbers they fell to their knees and struck beseeching attitudes (a mode of behavior that seems now to have been supplanted by a need to turn on CNN), then waited anxiously until the swelling vibrations fell away and the cascades of aftershocks abated. The local newspaper turned to doggerel to explain the dreadful majesty of the moment: "How awful is the thought of the wonders under ground / Of the mystic changes wrought in the silent, dark profound."

The implications of what was truly an awe-inspiring event at Fort Tejon go some way beyond the simple matter of the quake's enormous magnitude. The geometry of the San Andreas Fault's Big Bend has an effect on the local topography that is very complicated and still being properly worked out. But it boils down to one reality: that because the Pacific Plate is still pushing northward on this part of the fault—moving in the same direction, in other words, here as everywhere else on the 750 miles of its length—and because the Big Bend thrusts a prow-shaped bulge out into its path, the effect of the movement is here not simply a sliding-along-the-side affair; it is also a pushing-up-from-beneath kind of movement, a movement that tends to lift the prow of the northbound plate up somewhat and produce a range of hills in front of it.

Uplift like this was noticed up in Parkfield, where another section of the fault had become very slightly misaligned from the plate boundary—and the Temblor Range, as we have seen, was thrown up as the result of that. Here at Tejon the misalignment was evidently more spectacular: Rather than throwing up a low range of hills on just one side of the boundary, the plate in these parts seems to thrust itself head-on into the rock mass, right into the prow-shaped bulge of California, and has lifted up hills that would in due course become the Transverse Range, hills that look like a giant raft rising up onto the plate's advancing bow wave. The analogy is an imperfect one—the precise nature of what is going on below these mountains is still being analyzed—but in its essentials it holds.

There are other faults working here, too—very ancient faults with names such as the Clemens Well–Fenner–San Francisquito Fault and the San Gabriel Fault, which displaced the surface rocks for scores of miles, tens of millions of years ago; and shallow thrusting faults that never break surface today but that all the time are helping to accommodate the compression of today's plates as they ram themselves together, breaking and buckling like the crush zones in modern cars or the water-filled "impact attenuators" you see at dangerous road intersections.

And it would be idle to pretend that matters get any less complex as the fault spears ever onward, southeastward, passing as it does so along the zone that separates the San Gabriel Mountains from Antelope Valley (where there are countless supersecret defense establishments, making and testing costly warplanes and missiles), past the suburban sprawls of San Bernardino and Loma Linda and Redlands, then through the San Gorgonio Pass with its throbbing forests of wind generators, and along the Coachella Valley, where the elderly rich like to live out their final years in luxury and perfect weather, in communities like Palm Springs, Rancho Mirage, Indian Wells, and Palm Desert.

The fault underlies all of these places. When Andrew Lawson wrote his earthquake report in 1908, he supposed that the San Andreas in fact petered out here, but his confusion can be excused. For in this area the fault ceases to be a neat and tidy trolley track, or even a pair of tracks running in parallel; it is very tricky to recognize. It does not appear to cause much by way of local seismic problems, nor does it seem to determine the local lie of the land—which is, in any case, a mess of mountains and valleys spreading in all sorts of directions at once.

But in fact the fault, or the fault zone, continues on its merry way for at least another 150 miles—the first part of which is unutterably confusing, with a significance that must have been easy for a 1908 geologist to miss. Subsurface maps of this particular region, now that much of the fault geography has been worked out, show that the fault does indeed continue, but that it has the crazed aspect of a spider's web, or of a car's broken windshield. Newly identified sister faults hiss and sidle out from the main line, faults with names like the Pinto Mountain, the Garnet Hill, the Vincent, and the Arrowhead.

It was close by this point that the San Andreas may have had its first publicly acknowledged workout. No small amount of mystery attends the event, but it took place in the summer of 1769, at the very beginning of Spanish settlement of California, when an expedition of militarily supported Franciscans, led by Gaspar de Portolá, was pushing north from San Diego, hunting for an overland route to Monterey. On July 28, while they were camped beside the Santa Ana River, they were interrupted by a severe earthquake that, the explorer noted, "lasted about as half as long as the Ave Maria."

A little more can be gathered from the diaries of two other members of the expedition: They were perplexed by the aftershocks that went on for a full week. Juan Crespi, one of the diarists, recorded feeling as many as a dozen shocks a day while his team was winding its way around the southern hills of Los Angeles and down into the San Fernando Valley.

From all this anecdotal evidence, and from comparisons that can be made with more recent earthquakes, it appears likely that what Gaspar de Portolá and his party felt was a series of magnitude 6 quakes that had resulted from movement on either the San Jacinto Fault or, more probably, the San Andreas Fault. If so, then the fault that has caused so much mayhem during all of California's existence showed itself to be capable of great fury to the very first explorers of the region. They were given a warning—one that they and all who visited subsequently chose pointedly to ignore.

AFTER ALL THE topographic confusions to the east of Los Angeles, the San Andreas splits itself into two parts near Palm Springs, then repairs itself again near the town of Indio and becomes one once more. And finally, following this, after all of the excitements back at the Parkfield drilling site, after the ample confusions of the Big Bend, and the terrible complexities that are apparent as its licks its way around to the east of Los Angeles, it arrives in the city of Indio itself.

There is little that is memorable about the most southerly town on the fault's long track; it is no more than a scalding hot little railway community, a town where they have festivals to celebrate the dates they grow and the tamales they cook, and precious little else. The fault steals through without remark and without much evidence of its passing. It realigns itself slightly, kicks back onto its customary course, and resumes its orderly journey down toward its southern end.

And then, quite without ceremony, it disappears. It vanishes away in a muddy little field, beside one of the more unusual physical phenomena that is to be found anywhere in the country.

The Salton Sea is an enormous brackish lake, thirty-odd miles long and fifteen wide, that was created in 1905 as a result of a very foolish and very avoidable accident. At the turn of the century a firm called the California Development Company built a series of large irrigation canals in far Southern California to divert water from the Colorado River, which was then very close to its outlet in the gulf known in Mexico as the Sea of Cortés. The basic idea was to help local farmers on the edge of a very-low-lying part of the Imperial Valley called the Salton Sink (a relic of an earlier inland sea, as it happens) to grow fields of well-watered asparagus and broccoli. The company would charge the farmers, the farmers would sell their products, and everyone, in the classically American way, would make money.

Except that, unhappily, on one memorably unfortunate day, the levees protecting the trunk canal broke, and the entire flow of the Colorado River poured down into Salton Sink—which was more than 220 feet below sea level, at almost the same negative altitude as Death Valley. The waters kept coming and coming for more than a year, until finally railway wagons loaded with boulders plugged the hole in the levee and the newly created Salton Sea stopped filling and began to do what it has been doing ever since—evaporating.

The lake today is an odd, vaguely unpleasant place, rich with a strange smell that somehow mixes heat with dankness, rimmed with broken-down towns made of rusting trailers, with a reputation for the widespread manufacture of that particular nasty and highly addictive drug crystal methamphetamine, known variously as ice, Tina, Tish, or crank. The locals are seemingly obsessed with constant stories of death (as in fishermen drowning during sudden storms, birds perishing in their hundreds of thousands, beaches made exclusively from the crushed bones of fish skeletons, various species of flora and fauna dying out as the level of salinity, which is already close to that of the Pacific Ocean, keeps on climbing in the hot and pitiless sunshine).

Down at its southeastern end, just south of a formidably unpleasant junkyard of a settlement called Bombay Beach—alive with seismometers and GPS sensors, since the fault runs right beneath it—lies a grassy field, a few yards from a levee that halts the occasional rainstorm-induced flooding of the Salton Sea, allowing farmers to grow a threadbare harvest of soybeans and alfalfa. Not far from the field is a large white building with chimneys that belch a continuous cloud of white water vapor. It is a geothermal energy plant, an electrical generating station that spins its generators with the steam that pours from the ground in unstoppable volumes, scalding hot.

The area is riddled with hot springs and geysers and blowholes—and in the field beside the levee there are scores upon scores of mud volcanoes. The Cerro Prieto Geothermal Area is centered here, and it attracts businessmen who believe they might profit from all this energy that is bubbling up—quite literally—from below. It attracts in addition a small number of the curious. The local chambers of commerce shake their heads in sad acceptance of the fact that, were the area less displeasing, the mud volcanoes of the Salton Sea could be as big an attraction as Old Faithful. But it is not a pleasing corner of the world: It is more than unlikely that the Salton Sea will ever come to rival Yellowstone, not by a long chalk.

This is the southern end of the San Andreas Fault. It vanishes here; it dives underground, heading southward deep below until it reaches the edge of what is left of the Farallon Plate, at the so-called Rivera

Triple Junction in the Mexican Sea, off Mazatlán. Its final fate is so very different from everywhere else on the 750 miles of its length that it gives rise to one final big puzzle.

Why the geysers, the bubbling mud, and the streams of superheated water? There aren't any of significance anywhere else on the fault—so why are there here?

These geological characteristics, of what is officially called the Brawley Seismic Zone, are much like the characteristics of the middle of Iceland, or parts of Hawaii, or northern New Zealand. They are the signature characteristics of a spreading zone—a place where tectonic plates are sliding away from each other, and where new material is welling up in between, all of it hot and volcanic and geothermally interesting.

But what, one might ask, is a spreading zone doing anywhere near a plate boundary where the dominant characteristics have to do with slipping and sliding and thrusting and (in the very far north) subduction?

The answer to this seems to lie in geometry—and in the simple fact that while the overall heading of the Pacific Plate itself appears to be in one direction, that of the San Andreas Fault is very slightly different and, moreover, different in different places.

Relative to the North American Plate, the Pacific Plate is heading in a direction that is 36 degrees to the west of north, and it is doing so at a rate of just a little less than an inch and a half a year. The San Andreas Fault, on the other hand, is generally moving rather more slowly, at about 1.3 inches every year. Moreover, and most important, it is moving on a heading of 41 degrees west of north, which is some 5 degrees away from the direction of the plate movement. This 5-degree discrepancy between plate and fault, and the difference in the relative speeds of plate and fault, brings with it a welter of geophysically created wrinkles—unexpected mountains, additional faults, unanticipated earthquakes—and a host of other features that are all a consequence of the imperfect nature of the world.

But in a couple of places south of Parkfield, the discrepancy is different from this mean. In the Big Bend area, for example, the fault moves in an almost due northerly direction—and so its movement compresses the coast of California and thrusts itself inward in a way that makes mountains rise out of the solid ground. In the area beside the Salton Sea, on the other hand, the fault moves outward in the direction of the sea on a heading of almost 45 degrees west of north. Rather than compressing anything here, the relative heading of plate and fault takes each away from the other—and it causes stretching—extension, in other words—and the kind of phenomena—geysers and mud volcanoes—that are associated with spreading.

North of Parkfield the fault and the plate march along a broadly similar path, which is why the slipping and sliding between the plates is so much simpler to record. It is also why, since all is so firmly locked in place, there is so much devastation when movement finally occurs.

As it did in San Francisco, of course, on the morning of April 18, 1906, and, though generally forgotten, four memorable times before as well: first in 1836, in 1838, then again in 1865, and finally and much more destructively in 1868.

The event that struck so savagely on October 21 in that particular year—when a neighbor fault called the Hayward unlocked itself and suddenly moved—had long been known as the Great San Francisco Earthquake. It was a name that survived until the event was obscured in its notoriety by what was to happen, as all had been well warned it one day would, nearly forty years later.

EIGHT

Chronicle: City of Mint and Smoke

Dance then wild guests of 'Frisco
Yellow, bronze, white and red!
Dance by the golden gateway—
Dance tho' he smite you dead!

VACHEL LINDSAY,
"The City that will not Repent,"
from Collected Poems, *1923*

IN A HALF-HIDDEN COVE IN THE NORTH OF WHAT WAS then called Alta California, in a bay six miles inshore from the Pacific Ocean, on the eastern side of a bluff that was protected from the cold sea frets by ranges of hills and sand dunes, there was once a patch of fertile grassland in which grew a particular profusion of a bright green herb. It was a kind of mint, its leaves highly aromatic and its tiny flowers attractive enough for a nosegay. Now properly called *Clinopodium douglasii,* it was given this name by the nineteenth-century Scots botanist David Douglas, the man who also gave us the Douglas fir and many types of primrose. He had been commissioned by the Royal Horticultural Society to collect plants to bring back to London, and he traveled widely along the Pacific Coast searching for specimens; but he met an early and inapposite death in Hawaii after falling into a hole and being gored by a bull that had fallen into it first.

In 1776, when a party of Spaniards journeyed north from their local capital of Monterey and came across this fertile and sheltered spot, they thought of the pretty green plant as little short of miraculous. They took their cue from the local Miwok Indians, who seemed to use the plant for all sorts of cures, and by all accounts were delightfully healthy as a result. The Indians infused it into teas and decocted it into tisanes, they prepared poultices, munched it as a breath freshener, rubbed it on their skins to ward off wild animals, or wore it in their hair for cosmetic effect. The Spanish settlers followed suit (aside from the skin rubbing, which they found distasteful) and named the fragile little plant yerba buena, the "good herb." And more than that: They named the well-sheltered bay beside these herb-rich meadows Yerba Buena, too, which might well have been the name it still enjoys today but for those vicissitudes of Californian history that ensured that it ended up as San Francisco.

Before settling civilians there, the Spaniards had already dis-

The wild mint known as yerba buena, much favored by the local Miwok Indians, grew in such abundance beside San Francisco Bay that its name was the first given to the settlement.

patched to the neighborhood the two essentials of their rule, the soldiery and the clergy, and constructed the kind of dwelling houses they thought appropriate to their respective needs. A fort was built on the southern side of the main entrance to the harbor,* and the detachment of military men who had constructed it was ordered to remain there on sentry duty. Three miles southeast of the fort (and, as it happened, three miles southwest of Yerba Buena, too), in a valley filled with fruiting manzanita trees, a modest adobe mission house was thrown up, a Father Palou was appointed abbot, and a ceremony was held with, as the cleric noted, rifle fire taking the place of organ music and gunsmoke "supplying the want of incense."

Ever since the Franciscans had taken over the Spaniards' missionary work in California from the Jesuits (the latter order was banned in Spain), there had been pressure to name something, somewhere, after their patron saint back in Assisi. None of the previously founded California missions—San Diego, San Juan Capistrano, San Luis Obispo, and San José most prominent among them—had apparently proved worthy. But in 1776, both by chance and design, there were two happy coincidences. The fort that had been set down beside the Bay was formally founded on September 17, which happens to be the day that Catholics commemorate the impression of Saint Francis's wounds; and the nearby mission was consecrated a month later, on October 9, shortly after Saint Francis's feast day. There was ample reason, then, to name both structures after the much-esteemed saint—one the Presidio San Francisco and the other Misión San Francisco, although the latter was more familiarly given the name Dolores, in commemoration of the suffering of the Virgin Mary. Both buildings still stand today, with the white adobe structure of the Mission Dolores the oldest surviving structure in the city that eventually spread and grew around it.

---

*It was a harbor they came upon very late in the day: Scores of seamen sailing the Pacific Coast had managed to pass by the entrance without ever turning inside, and the *Clinopodium*-rich meadowlands were first discovered by a land-based expedition, and not by a sailor at all.

Yerba Buena, though only two and a half miles from the mission, was to remain uninhabited for nearly sixty years. But in 1835, by which time Alta California had become Mexican territory, an English sea captain named William Richardson broke its isolation, and became the first man to build a home there. Right beside the herb meadow, far enough from the shoreline to guarantee he would keep dry, he drove four long redwood posts into the ground and covered them with a foresail. He then moved in with some supplies, and ran this lonely and wretched outpost as a trading station, for the sole convenience of the owners of a pair of shallow-draft schooners that plied the upper reaches of what was fast coming to be known as San Francisco Bay. These little ships collected the hides and the tallow produced by the farms on the Sacramento and San Joaquin Rivers and brought it all back to the seagoing transport ships whose masters liked to lay up in Yerba Buena Cove. For the management of this station, the solitary Captain Richardson was paid a modest fee.

He had to wait only a year for fixed company. The shelter afforded by the cove quickly proved tempting for what was to become a harborful of other mariners and traders. It fast became a popular stopping-off place, and the hide trade expanded rapidly. The Russians called in often, and so did British vessels on their way to the Sandwich Islands—now Hawaii. There were French sloops, too, and American survey vessels making the first tentative explorations of a shore that, though currently Mexican, would before long be entirely their own.

Eventually one of these visiting Americans decided to join Richardson, and stay. He was named Jacob Leese and came from Ohio. Richard Dana remembered him later as a wild fellow who liked to shoot at wine bottles he had suspended from the ends of the top-gallant stunsail booms (just hanging the bottles would be quite a feat).* Leese negotiated with the Mexican *alcalde* for land and was granted a

*Dana, who had first come around the Horn from Boston to Yerba Buena in 1835, had forecast, with remarkable prescience, a glittering future for the region. "If California ever becomes a prosperous country," he wrote in his famous *Two Years Before the Mast,* "this bay will be the centre of its prosperity."

lot directly across from Richardson's tent; he imported boards and building materials. By Independence Day, 1836, he was able to stage a block party—the section of land between his own house and his neighbor's theoretically constituting a street—and invited the mission settlers, Mexican infantrymen from the Presidio and farmers from the Sacramento River Valley, to join him in celebrating America's sixtieth birthday. It was the very first party to be given by the first settlers—and from that day the tiny settlement never looked back, becoming renowned as a place for saturnalia, jollity, and drink. (And on April 15, 1838, Leese, who had married a sister of a Mexican general, became a father: His daughter, Rosalie Leese, thus became the first child to be born in the steadily expanding settlement.)

Ten years later, when the Mexican War broke out, President Polk

Fanciful though the imagined topography may be, the crowding of barques and brigantines in the harbor and the cluster of newly built houses and stores suggests a great city in the making: a view of San Francisco from the south in the mid-1840s, just before the madness of the Gold Rush.

ordered a Captain John Montgomery of the sloop-of-war USS *Portsmouth* to sail into the bay and seize what had by now grown from a settlement into a well-established trading village of fifty houses with a population of about 200. On July 9, 1846, Montgomery raised the American flag on the field the locals called its plaza and installed an American military governor in the Presidio. Six months later one of his officers, a lieutenant named Washington Bartlett, who had been appointed chief magistrate of the town, posted this notice in the January 30, 1847, issue of the *California Star:*

> **An Ordinance**
>
> Whereas, the local name of Yerba Buena, as applied to the settlement or town of San Francisco, is unknown beyond the district; and has been applied from the local name of the cove, on which the town is built; Therefore, to prevent confusion and mistakes in public documents, and that the town may have the advantage of the name given on the public map,
>
> IT IS HEREBY ORDAINED that the name of San Francisco shall hereafter be used in all official communications and public documents, or records appertaining to the town.
>
> *Wash'n. A. Bartlett*
> *Chief Magistrate*

The city grew at an exuberant, almost irrational rate. Will Rogers once remarked that it was "the city that was never a town." At the time of its renaming it already had two newspapers, and a census conducted by one of them found that in only the first six months of American governance it had more than doubled in size. The paper noted the presence among its 459 inhabitants of a minister, three doctors, and three lawyers, together with two surveyors, a teacher, eleven farmers, seven bakers, six blacksmiths, a brewer, six brickmakers, seven butchers, two cabinetmakers, three innkeepers, four tailors, eleven merchants, a morocco-luggage maker, a weaver, a watchmaker, and no fewer than twenty-six carpenters. Less scrupulously counted were a number of "Indians, Sandwich Islanders and negroes." The Hawaiians, said the paper, generally acted as pilots and navigators, since they had the uncanny ability to grope their way around the skerries of San Francisco Bay, no matter how thick the frequent fog. And men outnumbered women in the earliest settlement by almost three to one—a condition that would dog the city for years and lead to its later reputation as a place where most of the inhabitants would misspend their days in one kind of abandon or another.

The temptation for quite another kind of abandon began in the spring of 1848, with the event that more than any other would come to define early San Francisco. This, of course, was when gold was first found at Sutter's Mill, and, though it would later result in the explosive expansion of the town, the immediate initial effect was one of true abandonment—for, as the *Annals of San Francisco* put it, "Gold was the irresistible magnet that drew human souls to the place where it lay, rudely snapping asunder the feebler ties of affection and duty." In an instant almost all of the 500 men who were then living in town at the time dropped everything and headed out to the goldfields, leaving the city, and their women, to fend for themselves. "Day after day the bay was covered with launches filled with the inhabitants and their goods, hastening up the Sacramento ... leaving San Francisco like a place where the plague reigns, forsaken by its old inhabitants, a melancholy solitude."

In the summer of 1848 both of the town's newspapers suspended publication—but only briefly, because by early autumn came the first harbingers of the social revolution that was to come: The local government agreed to accept gold dust, to fix a price for it ($16 an ounce), to establish a mint and to manufacture coinage. Almost overnight everything changed. From the goldfields 100 miles away to the east poured gold dust worth hundreds of thousands of dollars—almost a million dollars in the first eight weeks alone—and all of it to finance the purchase of the goods that the miners needed, the shovels and trowels, the picks and knives, the bags of flour, the bottles of cheap whiskey, the tents, the boxes of eggs, and the vials of laudanum with which they would negate the effects of their otherwise untreatable ailments.

Prices for all of these goods swiftly began to skyrocket—a 500 percent rise in the price of beef, a fourfold increase in the price of flour, apothecaries making certain that a single droplet of laudanum would go for as much as $40. Merchants found they could charge more or less what they wanted: There were men out there desperate to find the wealth that they knew was waiting for them in the hills, and who would pay almost anything for the wherewithal to find it. Corruption and violence, cheating and envy, started to become a feature of everyday life. The papers began printing again, though only fitfully at first. The town's single school, which had closed briefly for want of gold-hungry teachers, reopened in October when more dedicated replacements arrived. The docks began to bustle with activity, first with crates of imported goods bound for the diggings and then with people, who suddenly and in prodigious numbers began to pour into the city from almost every corner of the hemisphere.

Earlier in the year a messenger had taken a tea caddy filled with gold dust to Washington, showing it to the country's leadership to convince them of the worth of the strike in the Sierra. The sight of it prompted President Polk to make formal mention to Congress of the discoveries—even though most of the country was already fully aware of them, and had packed its bags and organized its westbound travel accordingly. So now, with the president of the United States acting as cheerleader, the influx began—and the briefly moribund town promptly sprang back into lusty and exuberant life, assuming in a matter of weeks the role of gateway, outfitter, and *comprador* for thousands who were lining up on faraway quaysides, readying themselves for the one-way journey to Northern California to find their fortunes.

One sees in those few short months at the start of the gold fever a city that was made hurriedly, almost as though being assembled from a kit. First a town council was appointed; then justices of the peace; and then a settler with some experience of planning was asked to create a grid of streets along which the shanties might be put up in some kind of order, with duckboards laid down outside the houses to keep down some of the dust, or the mud, and keep it off the boots and the skirt hems of the tidier citizens. A bank was built—Naglee and Sinton's Exchange and Deposit Office. Small and primitive churches sprang up, hastily built, and small wooden halls were set aside for arriving Freemasons and Oddfellows, too. A concert was given in the local schoolhouse. A New York newspaper opened a correspondent's office. The first Chinese settlers found their way from Canton and introduced a small slew of service shops—a laundry, a café or two, a hardware store. A ship arrived from Panama with a U.S. postmaster aboard, a man federally charged with providing post offices and deciding on local mail routes (and he also brought with him the first regular mail that had been sent from New York to this newly won corner of the country).

Next, a brigadier general arrived to set up a proper military establishment. A harbormaster was appointed—a figure of increasing importance because of the armada of ships that was piling up, quite literally, in the roads immediately offshore. More than 200 sailing vessels were lying off the quayside in July 1849, almost all of them abandoned by the crews who had sailed with them, their sailors gone a-digging up in the placer streams. By the end of that same year the number had risen to 600, with most of the ships moored or abandoned. Grainy pictures of the time show behind the town's buildings a huge forest of masts reaching as far as it is possible to see.

Moves were then made to elect a proper state government, and the city thought about hiring police to keep some kind of control on the free-spending, ill-mannered miners who returned from the mountains with gold dust in their bags and mischief on their minds. Gambling houses opened, saloons proliferated in vast numbers, and, inevitably, whorehouses—a scattering at first—started up in business. These, which would eventually number in the hundreds, were staffed by professional women from all over the Americas, who answered the call to come to California every bit as eagerly as did the men.

San Francisco had its beginnings as a city "raised from the ground as if by magic," with its buildings "hatched like chickens by artificial heat." As many as a hundred new buildings a month went up in 1849, and the demand for them was so great that a small shop might rent for $3,000 a month and a modest tent in a good location might go for

$40,000 a year. A more substantial building on what was formerly the plaza but which in honor of the first visiting U.S. Navy vessel had been renamed Portsmouth Square,* was rented in 1850 for $75,000; its owner might have considered himself fortunate to get $200 in annual rent just one year before.

Most of the arrivals stayed in flimsy canvas tents, forests of which went up on the slopes of those hills deemed too steep for the wooden frame buildings being built on the shore. At night they presented an extraordinary sight, lit from within by oil lanterns—a sailor moored in the harbor reported that the hillsides looked like "an amphitheater of fire." A New York manufacturer named Sydam came to town offering canvas houses that weighed 125 pounds and could comfortably—his word—sleep twenty, with twelve in hammocks and eight on the floor. But the rains, the cold Pacific winds, and the gritty miasma of breeze-borne sand (for sand dunes lay everywhere to the west of the little settlement, stretching six miles to the sea) made tent life far less of an idyll than Sydam advertised. Moreover, he can hardly have anticipated the effect on his invention of the feral donkeys that wandered down from the hills and did their bit to add to the universal misery—by chewing at the tent canvas, biting down guy ropes, and in one case breaking into a tent that was occupied by a snoring drunk and munching away half of the man's hair.

The town was filthy in those early days, and known primarily for rats, fleas, and piles of empty liquor bottles. Cholera outbreaks were dismayingly frequent, and in the early years it was common for the bodies of the dead to be abandoned by the shore, in the hope that the tide might carry them off into the open sea. There was little by way of indoor plumbing, and the water supply was halting, with what there was invariably polluted. Gaslights had been invented but not installed, and so the city at night was dark and dangerous, crowded and unhealthy—and yet regarded with tolerant fondness by all who looked back on those heady first Gold Rush years. Those who survived the

*As it remains today, in the heart of San Francisco's Chinatown.

very early San Francisco were armed with a pride that was quite unknown to the later immigrants.

Lateral thinking had more than a little effect on the housing shortage. It was of course perfectly reasonable to use some of the abandoned ships crowding the shoreline for housing or for prisons (one, the British China clipper *Euphemia,* was anchored off the main city wharf and used as a holding pen from 1850, especially for the state's insane). But some bright spark decided that rather than keep the vessels anchored off in the roads—which meant that the patrons had to be rowed back and forth in lighters—it would make sense to sail them in fast, ram them head-on into the muddy shoreline, and berth them there permanently. All they then had to do was hammer together a frontage that would offer the illusion that the ships were in fact properly made buildings.

The best known of these ship buildings was the Niantic Hotel, a former North Pacific whaler that its skipper decided to employ hauling Gold Rush immigrants up to San Francisco from the west side of the Isthmus of Panama. He made only one journey: The ship was becalmed en route for weeks, scurvy broke out, and the moment they eventually arrived all the passengers and his crew jumped ship for the mines. The skipper promptly sold the hulk to a firm called Gildmeister, Fremsey and Co.—a name hinting at both the town's growing prosperity and the geographical spread of the newcomers' national origins—which used it as a warehouse, and then built shacks on deck to turn it into a hotel. It eventually burned, as such ships often did, after which its hulk was rammed up on shore and used as the foundation for the hotel that would remain in place until it was demolished in 1872. Pictures from the time show the Niantic standing foursquare among other businesses—the Boggs Liquor Store, the Tract Society, Colonel Tibbs the Dentist, the Eagle Saloon, and Bubb, Grub and Co.—each one of them ships, some still recognizably nautical, with bowsprits and masts and rigging, others storefronted, as if they had been standing for years on any main street anywhere in America.

Throughout the 1850s the ever-expanding city began to be plagued by fire—hardly surprising, with the highly flammable combination of flimsy buildings, stiff winds, abundant fuel oils, and strong drink. The first blaze to cause a good deal of damage came on Christmas Eve 1849, after which there were three outbreaks in 1850 and two more in 1851, all of which consumed tents and frame buildings, and even the newfangled iron-framed (but "Rustproof!" said the makers) prefabricated houses that were being shipped out from New York at a freight cost of only $18 apiece. Given that the houses cost as little as $100 to buy, it seemed likely that these might provide an affordable safe alternative to the fragile tinderboxes of before. But no: It turned out that in a fire the metal holding up these new structures "got red hot, then white hot, then fell together like card-houses," and that their metal doors expanded in the heat and couldn't be opened. People were often found trapped inside—and all of a sudden fire casualties started to become a significant feature in the city death rolls.

After a fire in May 1851 wrecked eighteen city blocks, did damage worth $12 million, and killed scores, the city decided it should have a properly organized fire department. Hundreds of young men responded, and by 1852 fourteen volunteer companies began competing to be the fastest to respond and the quickest to douse. But the hope that combining a passion for flammability with testosterone might solve the city's fire problem turned out to be a vain one. A furious fight broke out among three of the companies that were competing to be first to one particular fire, and there were broken bones and one man hurt by gunfire. The event prompted the city to authorize a paid fire department instead. At the same time City Hall's bureaucrats began to write building codes to protect structures against the ravages of flame, hot gas, and high temperatures. As we shall see, such reforms did little to help, but they at least imposed some kind of order on what for a few heady months was little more than a Wild West shambles.

And the fires also played a part in giving San Francisco a particularly appropriate official seal. In 1859, after presiding wearily over a ceaseless rash of destructive blazes, City Hall commissioned a designer to draw a stylized phoenix rising from flames, and to place it front and center of a shield that had the more obvious and expected motifs of gold mines, sailing ships, and patriotic flags. The seal also carried the motto *Oro in Paz, Fierro in Guerra:* "Gold in Peace, Iron in War." But it is the phoenix that remains in the mind—a symbol that would be cited many more times, and not least, of course, in the spring of 1906.

For a while the new building codes helped to slow the pace of the city's expansion just a little. Contractors became more circumspect about how they built, structures had to be made with more caution, and the more prosperous citizens began to put up bigger and more ornate houses and offices. Generally the pace of the city's growth assumed a sedate character—not least because the finds of gold started to wind down, though the great silver strike among the blue clays of the Comstock Lode in Nevada in 1859 perked matters up again, transforming scores of city beggars into fat and urbane rich men.

The steadier pace of growth did not, however, change the habits of those living there, which were as dissolute as ever, with San Francisco enjoying or suffering a reputation as the wildest city in the country, one that aspired to be the Paris of the outer West, but in fact looked and felt much more like the docksides of Marseilles. The local newspapers noted that in the first six years of the 1850s no fewer than 1,400 murders had taken place, and only three of the murderers had been hanged. Vigilance committees sprang up to try to apply rough street-corner justice to the worsening situation, and hanged a handful of supposed criminals after hasty trials at drumhead courts. During the 1850s San Francisco's notoriety was fully and widely established; it was a den of iniquity, a lawless town where men in unrestricted mobs drank, gambled, and whored their way from street to street, unchecked by family, by conscience, or by law.

The area on the seaside of Portsmouth Square was without rival the most louche part of town; and because some visiting sailors thought its reputation for iniquity and chicanery bore more than a passing resemblance to the pirate-infested coast of North Africa, where the truest villains were the Berber cameleers, it quickly came to be known as the Barbary Coast. It was here that a uniquely San Francisco patois was born: The music halls, where male passersby were first lured by extravagantly painted women, were known as *melodeons;* if the customer could be persuaded—and strong drink helped—to abandon the singing for sex, he would be taken to a whore's *crib;* if he preferred to keep drinking, running up his bar bill to a level where it was impossible for him to pay, his debt would be passed on to a *crimp,* a loan shark who would sell his indebtedness to a sea captain, who would in the end come looking for the unfortunate and take him off to work onboard ship, as the only known way to reduce the balance. And if debt was not persuasive enough, then the captains could rely on the violent mickey-and-blackjack administering thugs of San Francisco, the so-called *shanghai men,* who for a small fee would deliver a hog-tied and unconscious bundle to the deck: When the poor man awoke he would find himself somewhere out in the cold Pacific, heading west for China, and working on a ship whether he liked it or not.

The word *hoodlum* comes from the San Francisco of the time as well. Some say it was an anti-Chinese cry of "huddle 'em," a signal for mobs of ne'er-do-wells to surround and harass an innocent "Celestial," as they were widely called; other etymologists claim that the word comes from a seldom-seen German term for "ruffian." Whatever the word's origins, the fact that it was born in San Francisco says much of the temper of the times. There were indeed rough men down by the new town's waterfront—hoodlums who would shanghai you, or who would guard a working girl's crib, or tip off the crimp with a name like Blinky Tom or Whale-Whiskers Kelley in the melodeon—all the kinds of nuanced nastiness that for much of the closing quarter of the nineteenth century infected the unsavory streets of the Barbary Coast.

The Bonanza Kings, as the new Comstock silver millionaires were known, made certain that San Francisco also offered the nouveaux riches an array of glittering pleasure palaces that were attuned to their own peculiar needs. So fine restaurants and music halls and bars and high-class brothels sprang up, and the streets of the new plutocracy were fast being paved with cobbles; handsome signs were being put up with their names—Montgomery, Clay, Sacramento, Commerce, Battery, and California among the first—and the luxury of piped water brought in from small reservoirs was being offered to all residents fortunate enough to live in the gleaming center of the new city.

AS THESE TRAPPINGS of major metropolitan status continued to gather, the city's wild reputation began to abate, and a kind of bohemian normality gradually settled on the place. By 1863 there were 115,000 people living within the city boundaries, and very slowly the proportion among them of women—which had been as low as one in seven just ten years before—was edging up to its customary level. Large hotels were thrown up—sixty in all by the time of the American centenary in 1876—some of them, like the Palace, with its courtyards and marble splendors, ranking with any in the world. There were something like 600 saloons, 40 bookshops, a dozen photography studios (in consequence there are today more black-and-white images of San Francisco* in archives around the world than of any other city of comparable size and standing). Omnibuses and hackney coaches took passengers around the town; some of the coaches were quite magnificent, their carriages silk lined, lacquered, and brightly painted, their horses enormous and covered with leathers tricked out with Comstock silver.

Fleets of huge smoke-belching ferryboats began regular services across the Bay to Oakland and Berkeley, stopping en route to discharge passengers at Yerba Buena Island. (This island, the only major relic of the city's original name, is still preserved today.) Eventually, given the importance of the cross-bay ferries, an imposing terminal building would go up: The Ferry Building, with a 200-foot clock tower modeled on Seville Cathedral, remains today, having survived every earthquake and fire that has afflicted the city since.

In time the small rowboats that took passengers around the shoreline and down the soon-to-be-filled-in creeks (whose landfill would

---

*Among the great San Francisco–based photographers was Arnold Genthe, the remarkable German best known for his sensitive and tender portraits of Chinatown; and Eadweard Muybridge—born Edward Muggeridge outside London, he pretentiously amended his name to achieve greater artistic credibility—who made some immense panoramic photographs of the entire city, and was the first to surmount the technical problems of photographing a horse at the gallop, thus settling a bet with Leland Stanford about whether a galloping horse lifts all its hooves off the ground at once. (The pictures proved that it does.)

make them the most earthquake-vulnerable parts of town) were replaced by small railways that brought people in from the growing suburbs. In 1870 the mechanisms that would in due course allow the creation of one of the city's most enduring icons, the cable car, were invented and duly patented (and very forcefully protected). The first of these cars began to run along Clay Street in 1873, using A. S. Hallidie's patented grips, levers, sheaves, and pulleys; it successfully hauled carloads of passengers up and down the rather steep (though not particularly high) hills, making light of gradients where no horse could possibly tread. Before long the cars were everywhere, their clanging and the steely singing of their cables beneath the roadway part of the city's magical soundscape to this day.

Soon there were an opera house, an art gallery, a synagogue, and various asylums for the troubled and afflicted, and in time there were public gardens galore. The open space where rallies were held, sympathetic to the Union cause in the Civil War, was left open after the passions of that conflict had subsided, and not unreasonably called Union Square. But it was not a pleasure garden, not one of the "lungs of the city" that planners elsewhere thought important.

The city had expressed a need for true public spaces as early as 1855, when one of the newspapers denounced the successor to the old Mexican plaza, Portsmouth Square, as a "barnyard for human and other cattle" that was "an eyesore and a disgrace." There were envious glances back east, where Frederick Law Olmsted was just then starting work on Central Park in New York, and Boston was seeing the beginning of its public gardens. But at first it was left to entrepreneurs, rather than to the city, to provide green spaces, and so before long there was Russ's Gardens, begun by a local jeweler; there was the Willows (next to the Willows Beer Hall) and the privately run City Gardens, which offered lawns lit by Chinese lanterns on the estate of a businessman named Shaw. All provided some relief for the swelling throngs on the city streets.

One of the best known and longest lived of these privately run gardens was Woodward's, started by a man who made a small fortune out of a hotel he opened in 1852 down by the seafront on Sacramento Street, which he called the What Cheer House. Robert Woodward was from Rhode Island, had come to California in 1849 by sailing around the Horn, and had started What Cheer as a workingmen's hotel that eventually had a thousand rooms, most of them going for as little as fifty cents a night, providing that the establishment's strict rules of temperance were observed.

He made a fortune thereby, brought out his family from Providence, and built himself a large house on Mission Street—which had itself been built on the old plank route between Yerba Buena and the Mission Dolores. He stocked its gardens with exotic plants and a collection of even more exotic animals. He opened an art gallery—filling it with impeccable copies of old masters—and started a lending library near his hotel. Eventually, being at heart a philanthropist of the deepest dye, he opened up his own gardens to the public. Inside the walls there were a boating pond, a lake with sea lions, enclosures with panthers and kangaroos, camels, tigers, opossums, and monkeys—every imaginable plant and animal and item of astonishment and delight to please the thousands who poured in daily to see it all. People flocked to fire-eaters from India, acrobats from Japan, an eight-foot-tall Chinese man, and a dancing bear called Split-Nose Jim. For the next twenty years Woodward's Gardens were San Francisco's equivalent of Copenhagen's Tivoli. It was only when the city created the even larger, more remarkable (and still surviving) expanse of meadows, gardens, and lakes known as Golden Gate Park that the citizenry permitted Woodward to close his creation in 1894, and to have the place leveled and turned over to the great commercial buildings that still occupy his tract of land today.

AND THEN THERE WAS *Tangrenbu*—Chinatown.

The Gold Rush had attracted men from everywhere. According to one of the drier accounts, there were "Indians, Spaniards of many provinces, Hawaiians, Japanese, Chinese, Malays, Tartars and Russians." There were people "from Chile and France and China, Vermont and Tennessee," wrote another. Or, as the *Annals of San Francisco* put it, there were

> the people of the many races of the Hindoo land; Russians with furs and sables; a stray, turbaned, stately Turk or two, and occasionally a half-naked shivering Indian; multitudes of the Spanish race from every country of the Americas, partly pure, partly crossed with red blood—Chileans, Peruvians and Mexicans, all with their different shades of the same swarthy complexion, black-eyed and well-featured, proud of their beards and moustaches, their grease, dirt and eternal gaudy serapes or darker cloaks.

There was little by way of tactful correctness about the authors of the *Annals*: The Spaniards in town were "dignified, polite and pompous," there were "fat, conceited and comfortable" Englishmen, and in large number there were Germans, Italians and Frenchmen who were known for being "gay, easy-principled, philosophical ... their faces covered with hair, and with strange habiliments on their person, among whom might be particularly remarked a number of thick-lipped, hook-nosed, ox-eyed, cunning, oily Jews." (One of those Jewish immigrants—a twenty-one-year-old German named Levi Strauss, with no known facial peculiarities—went on to invent and give to the world Levi's, an invention that was born of the sartorial demands—cheapness and durability—of the Gold Rush.)

The Chinese, however, were different, and, in the eyes of most of the Yankee San Franciscans of the time, they were infinitely the worse for being so. They were worse, remarked the *Annals,* because, unlike settlers from other races, they lacked qualities that could be "materially modified, and closely assimilated to those of the civilizing and dominant race. They were aloof, infuriatingly haughty, separate, stoically indifferent. They were regarded by the other San Franciscans as so strange and unsettlingly peculiar that they could never hope to become part of the social continuum of the town. They were outsiders from the very start.

That start came early on in the city's history, when, in 1848, two men and a woman arrived in the Bay from Canton on board the Pacific brig *Eagle,* and immediately made their way the hundred miles east to the goldfields. They promptly sent back word of inestimable riches, and before long the waters teemed with vessels churning through the Golden Gate to what they at first called *Jinshan*—"Gold Mountain"—bringing in scores and then hundreds of families from the coastal ports of south China, from every southern settlement between Shanghai and Hainan Island. Soon, in addition to the countless hundreds of Chinese who went out to the Sierra,* there were those who came and settled in the city itself. There were 800 listed as being there in 1850, 3,000 in 1851, and 10,000 the year after. The Taiping Rebellion was taking place in China: Thousands were dying, and millions were fleeing. Many went to this rich and secure American city where they could create for themselves a sanctuary, a place within Jinshan they would call *Tangrenbu*—the "Port of the Tang People."

At first they settled where they could—some making for themselves a pretty little fishing village at the mouth of Mission Creek—but eventually, as working-class Chinese are wont to do, they clustered together for companionship and succor. Their chosen home was eight or nine central city blocks, surrounded on one side by the plaza and its businessmen, and on the other by Nob Hill and its more patrician householders. This made them ideally positioned to trade with, and provide services for, both the merchants downtown and their families up in their mansions.

---

*More than 12,000 Chinese also worked uncomplainingly—though only for gold, not the distrusted paper money—on the most difficult sections of the Central Pacific Railroad's route east across the mountains to where they would meet up with the Union Pacific's rails heading west from Omaha. Their legendary courage in working on the fantastically dangerous Cape Horn cliff face near Colfax is memorialized in a plaque—as is their involvement with hundreds of Irishmen in laying the final ten miles of track through the Utah desert in twelve hours flat. They were paid at the rate of a dollar a day, slept in the open, and lived with exemplary frugality on fish, seaweed, and dried oysters.

In the 1870s there were said to be 45,000 Chinese crammed into their tiny oasis, or laager, or ghetto. A writer named B. E. Lloyd wrote that

> the side-walks are monopolized by them, with their little tables of fruits, nuts and cigars. The cobbler, tinner, chair-mender and jack-of-all-trades claim, by squatter right, a seat upon a box or a door-sill, where to ply their trades; the alleys, lanes and by-ways give forth dense clouds of smoke from the open fires, where cooking is performed, and the house-tops are white with drying garments, fluttering from the network of clothes-lines that are placed thereupon by enterprising laundrymen. Even across narrow streets lines are thrown upon which are placed to dry all manner of wearing apparel.

In 1857 a small newspaper known as the *Butte Record* of Oroville, California, became the first publication to recognize this crammed urban creation for what it was—a mention now cited by the *OED* as the earliest appearance of the word. This, the first in the world, was *Chinatown.*

The fact that the Chinese kept to themselves and made little effort to speak the language or to fit in with the cultural niceties of the majority, caused them to be regarded much as Jews were in other cities: first suspect, then loathed, then feared—and finally, and for all too long in San Francisco's history, ruthlessly and cruelly discriminated against.

It began with relative innocence, with casual references in the local newspapers to the settlements crammed with "natives of the Celestial Empire, and subjects of the uncle to the moon, with their long plaited queues or tails, very wide pantaloons bagging behind, and curiously formed head coverings."

Before long, though, some of the Yankee residents were complaining. Permit the Chinese to rent, said one, and before long they will "make the building uninhabitable for decent white folk.... They will divide the rooms into numerous diminutive compartments by unsightly partitions, and the smoke and rank odor from their open fires

Ross Alley, the "Street of Gamblers" in San Francisco's old Chinatown: One of the moody images created by Arnold Genthe, the German philologist who became fascinated by this concentration of Eastern life in the midst of this quintessentially western American city. The characters at the right of the photograph are part of the notice for a noodle take-out stand.

---

and opium pipes discolors the ceilings and walls and renders the whole building offensive."

It got worse. "The manners and habits of the Chinese are very repugnant to Americans in California," remarked the *Annals,* a magisterial book that often appears to be presciently attuned to the majority view. "Of different language, blood, religion and character, inferior in both mental and bodily characters, the Chinaman is looked upon by some as only a little superior to the negro . . . his person does not smell sweetly, his color is unusual, his penuriousness is extreme; his lying, knavery and natural cowardice are proverbial."

From remarks like this it was a few short steps to active discrimination. As early as 1852 one governor of California asked the legislature to ban the immigration of "coolies," but his proposal was ignored. Then began an insidious campaign against them. Three years after the failed ban, a law was passed prohibiting the testimony of any Chinese in a court case that involved white people. Race-based covenants were written into property deeds, forbidding the sale of houses outside Chinatown to Chinese, and effectively keeping the Chinese mired eternally in their nine-block encampment. Chinese patients were briefly stopped from using the San Francisco County Hospital. Taxes were imposed selectively against the Chinese by targeting laundrymen, fishermen, and vegetable hawkers, an ordinance was passed banning the use of bamboo laundry poles, and an order was promulgated insisting that no man in prison could wear his hair more than an inch long—which meant that the Manchu queue, the pigtail by which most Chinese men displayed their allegiance to the faraway Celestial Throne, became illegal, and prisoners were compelled to cut them off.

The Chinese fought back. The Chinese Six Companies was a group of locally raised benevolent societies that helped the poor and the homeless, tried to mediate in the endless wars among the Chinese tongs, and attempted to neutralize the violent thugs called *highbinders* who did their dirty work for them. They endeavored to have the discrimination ordinances struck down, and hired lawyers to battle the city in the courts. But in the end even their cleverness and acumen could not overcome the immense prejudice of the Yankee residents.

All manner of new regulations were brought in. The wearing or sporting of the queue was officially banned in 1876, for instance; and after dreadful anti-Chinese riots in the city in 1877 (with fires deliberately set in Chinatown, and young hoodlums, as christened here, cutting the firemen's hoses with razors to keep the fires burning), and under pressure from anti-Chinese labor unions claimed that the "yellow peril" was now endangering American jobs, the U.S. Congress passed the Chinese Exclusion Act of 1882, which forbade the immigration of coolie labor and prevented existing residents from bringing in their families. Shameful as this law, which had its origins in San Francisco, seems today, it actually remained on America's statute books until 1943.

(One might well say that San Francisco's Chinatown had the last laugh, however rudely its inhabitants were treated. Sun Yat-sen stayed there for a while, and it is believed by many locals that he wrote the outline for postrevolutionary China's constitution there. Regardless of whether there is any truth in this, he most certainly received word of the long-awaited overthrow of the Manchus when reading a newspaper on an eastbound express train near Denver. He was riding on rails linked to those laid down by his fellow countrymen, whose growing wealth, cosmopolitanism, and confidence, in part built by their Gold Rush experience, had hurried the revolution that would begin China's long and steady climb to the vast power it commands today. Hardly comforting at the time, but a sweet irony that some might now appreciate.)

THIS UNATTRACTIVE gallimaufry of attitudes and morals that made up pre-earthquake San Francisco was presided over, most appropriately, by a city government that was as corrupt as it was incompetent.

The city boss at the close of the century was a dapper little crook of lawyer named Abraham Ruef, essentially the local point man for the all-powerful Southern Pacific Railroad, which wielded immense influence over not just the city but all of California. Ruef handpicked as mayor a handsome, courtly, black-bearded Irish-German violinist, president of the city's musicians' union, named Eugene Schmitz.

Mayor Schmitz had many foes but was generally popular, and, though suspected of graft, never had any charges pinned on him. He was re-elected in 1905 to a third term, and it was later universally acknowledged that he rose amply to the occasion that was later provided for him by the earthquake—something that came as a complete surprise to his critics, and probably even to the mayor himself.

It was widely recognized that the degree of corruption in Edwardian San Francisco was remarkable, even by comparison with the relentless thievery of the Tammany Hall machine back in New York. Here, out in a corner of the country that could well live up to being called the Wild West, was a city where "French restaurant" often meant specifically a two-story building where meals were sold on the first floor and girls on the second. The entire operation was neatly licensed, with the liquor license fees going to the city; these were collected by officials so eager for graft that they invariably took bribes from both sides in any dispute. The city fathers were said by Abe Ruef to have been so bent and hungry for graft money that they would eat the gold paint off the City Hall walls. Ruef's own men took a handsome cut—"oiling the skids," they called it—for the provision of almost any service the city might offer. The city fathers also turned a blind eye to the Barbary Coast's legendary "cowyards," assembly-line brothels that provided sexual services on a titanic scale. The Hotel Nymphomania, for example, offered 150 cubicles on each of its three floors.

The water company, the gaslight providers, the local railways, and even the big transcontinental railroad companies all found it more convenient to see that city officials were kept happy, fat, and replete than to worry too much about the niceties of democratic needs and popular demands. Ruef himself was paid monthly retainers of up to $400 in unreceipted cash by all the corporations wishing to do business with the city.

Just before the earthquake hit, an inquiry into all this alleged extortion and venality was getting under way. One of the local newspaper editors, Fremont Older of the *Bulletin,* had gone to Washington in high dudgeon. He had petitioned President Roosevelt (this was back in the days when American presidents made themselves available to assist those in want), requesting that he help clean up the city. William Burns of the Secret Service—the founder of the famous Burns Agency and later a founder of the FBI—agreed to help, as did Rudolph Spreckels, son of the city's wealthy sugar baron Claus Spreckels. The city bosses heard the unsettling news of a possible inquiry on the morning of April 17; they also heard that the *Bulletin* was about to run an exposé. Given that they knew the Spreckels family was involved, they must have realized that this had be taken seriously. The quake would provide them with only a brief delay.

NO MATTER THAT THE city had great and glittering hotels, with the St. Francis just opened and the Fairmont just about to. No matter that one could dine in restaurants—Marchand's, Tait's, the Poodle Dog, Zinkand's, as well as all the French ones, where the food was said to be every bit as good as the sex—that, according to their proprietors, could rival those of Paris and New York; that a four-story amassment of well-built stores and restaurants on Montgomery Street known as the Monkey Block,* once the tallest structure west of the Mississippi, had an array of cafés and clubs that attracted writers like Robert Louis Stevenson, Ambrose Bierce, Jack London, Bret Harte, and the then-Samuel Clemens, scribbler of bawdy doggerel; that there were ten-story steel-framed skyscrapers† and grand municipal buildings with churchly domes and acres of gold foil; that there were three working opera houses and orchestras in abundance and stores filled with the finest goods imaginable. No matter that the brand-new City Hall had taken twenty-six years to build, had cost $6 million, and was by far the biggest building west of Chicago and the grandest civic structure west of the Mississippi. No matter that the celebrated Big Four railroad bosses—Collis Huntington, Leland Stanford, Mark Hopkins, and Charles Crocker—maintained immense palaces at the top of Nob Hill, that one of the Comstock Lode bonanza barons, James Flood, did the same, and that scores of lesser millionaires had similar ambitions. No matter that Adolph Sutro had constructed a complex of swimming

---

*Well built enough to survive 1906 in near-perfect shape, but demolished in 1958, to make way for a parking lot.

†There was also a domed sixteen-story building at the corner of Market and Third Streets, the tallest building in the American West. It was owned by Claus Spreckels, the sucrose magnate whose fondness for younger women is said by some to have given us the term "sugar daddy." It housed the offices of the *Call,* one of the more prominent daily newspapers of the time. Under the dome was a celebrated restaurant, with a much-envied view.

Eadweard Muybridge lugged thirteen cameras to the turreted roof of Mark Hopkins's uncompleted mansion on the 900-foot summit of Nob Hill, and twice, in 1877 and 1878, photographed the entire city. This is the second of his panoramas: Almost every building in sight would be destroyed by earthquake and fire thirty years later.

---

pools at the Land's End side of the Golden Gate that could accommodate 25,000 visitors a day. No matter that a citizen named Joshua Abraham Norton declared himself Emperor of the United States and Protector of Mexico in 1860, issued money, was followed by two stray dogs called Bummer and Lazarus, was known widely as the Mad Hatter, gave himself a uniform with epaulets and a sword, fined people

$25 if he heard them calling the city "Frisco," had a funeral that was attended by thousands and a headstone that proclaimed his imperial standing. No matter that luxury and decadent pleasure seeking was now the hallmark of the place, and that unimaginable wealth and sordid poverty and exclusion existed cheek by jowl in a more demonstrable way than anywhere else in America. No matter that one notably excessive writer named Will Irwin declared it to be "the gayest, most light-hearted, most pleasure-loving city on the western continent . . . a city of romance and the gateway to adventure."

No matter, all of this gaudy grandiloquence. Despite the variety and gaiety and hyperbole, San Francisco in 1906 was also in fact a big, dirty, brawling, vulgar, smoggy, sooty, and corrupt town of rather less charm than myth and latter-day boosterism would have us believe. It was a factory town in the south, below Market and Mission Streets, "south of the Slot,"* as the locals charmlessly had it; manufacturing plants and foundries would belch smoke into the air, the boilers fueled by low-grade high-sulfur steam coal. Many of the ships in the Bay burned coal; the houses were heated by furnaces and stoves that burned coal. The railway station at Third and Townsend sent coal-fired steam trains south to Los Angeles in a shower of soot and fire. A yellow-gray miasma thus enveloped the whole city, especially on the warmer days of summer, or when the cold winds were not sweeping in from the ocean.

The streets were filthy, too, covered with the leavings of the thousands of horses that pulled freight and passengers around the city. There were essentially no sewage treatment plants in the city, and foul-smelling fluids poured continually into the Bay. China Basin, an especially unpleasant lagoon south of the city, was described by a ship's captain as "a cesspool, emitting foul odors, especially at low water."

And though in the city center and up on Nob Hill and out at Land's End there were fine buildings, built to impress and to last, farther afield the structures were gimcrack and ugly—shacks and lean-tos and hastily cobbled together cuboids of brick and lath, smoky and insanitary and ill planned and likely to burn or fall down at the slightest excuse.

The houses on Telegraph Hill and Russian Hill and where the Italians gathered in North Beach were almost all made entirely of wood—the classier houses fashioned from California redwood, the humbler homes made wholly of rich and resinous soft pine. This simple fact set some of the less corrupt officials to worrying, not least the city's fire chief, Dennis Sullivan. He had been arguing for years that the city was a tinderbox waiting to be struck. He wanted a saltwater firefighting system—after all, the seven square miles of the city had been built on a narrow peninsula that was surrounded on three sides by water. And he wanted the freshwater cisterns, which a long time before had been built beneath the city streets but had been forgotten and neglected

*For more on the Slot, see note on page 245.

and allowed to deteriorate, to be cleaned, renovated, and refilled with water. But he was ignored.

He must have felt vindicated when, in October 1905, the National Board of Fire Underwriters declared that San Francisco's water-supply system, despite being able to deliver 36 million gallons a day, was structurally in such poor shape that the hydrants would not be able to halt anything approaching a major fire. Chief Sullivan informed the city of the board's analysis, but was studiously and comprehensively ignored once again. Seven months later the contention would be tested, and in all its essentials would be proved fully right, and tragically so.

SEVEN MONTHS AFTER the underwriters' report, and half a world away, a dozen Neapolitan villages were busy being devastated by the eruption of Vesuvius. The great old volcano had started to explode and exhale clouds of gas and lava and dust back on April 6, and was still doing so ten days later. The residents of Los Angeles seem to have been particularly affected by the news—more so, one gathers, were the Italians in the North Beach of the generally earthquake-insouciant, fire-unaware, devil-may-care San Francisco. Down in Southern California they had collected $10,000 to transmit by telegraph wire to the victims in Italy, and had sent it on its way on the morning of Tuesday, April 17.

On that same Tuesday there were two other oddly coincidental events. One was a meeting held in the offices of a U.S. circuit court judge, W. W. Morrow: It had been called quite specifically to consider establishing committees that would be formed in the event of a major disaster or emergency in San Francisco. The other was the delivery that very day of a formal report, completed six months before, by the great Chicago architect Daniel Burnham, on his plan for comprehensively rebuilding the city. The idea was to make it as elegant and planned a city as Washington, D.C., where Burnham had also had a hand.

It was an ambitious, extravagant scheme, to which I will return. It was a plan that called for grand monuments to be sited on the summits of peaks, broad boulevards to be bulldozed through neighborhoods, and scores of parks, fountains, marble piazzas, and wrought-iron elaborations to be built. All this, the critics said, would take far too long to achieve and would present an image of San Francisco utterly at variance with what it truly was. For the city was not a Paris or a Washington or a Buenos Aires; it was a place that made its fortune from making things, importing things, shipping things, and having endless decadent fun with all the wealth that these most basic activities brought. The Burnham plan was too pretty for it, too chic, too frothily pompous; it dressed San Francisco up as though it were Savannah or Charleston, when what it really wanted to be was a West Coast version of New York.

On that Tuesday this grubby, corrupt, decadent young city of 400,000 people was, in other words, considering how to manage itself should it ever suffer the calamity of being grievously damaged; and it was about to consider how it might rebuild itself, the proposed style thought by some to possess an elegance appropriate to its status, and by some to be merely overwrought. And 350 miles to the south a handsome sum of money was on its way to a group of people in Italy who were suffering their way through the aftermath of a mighty volcanic eruption.

But just before dawn of the following day the events in San Francisco brought a sudden and urgent need for all three of the commodities that had been on offer: an emergency needed to be dealt with, a city needed suddenly to be rebuilt, and money was needed, in abundance—but for thousands of victims at home rather than for uncounted numbers thousands of miles away.

As it happens, the cleverest of banking telegraphers managed to see that this money was actually stopped in its tracks, and instead made its way up from Los Angeles to San Francisco. Although, as it turned out, the inhabitants of the city on the Bay were going to need a very great deal more than $10,000 worth of Los Angeles largesse.

NINE

Overture: The Night Before Dark

You could not tell, and yet it looked as if
The shore was lucky in being backed by cliff,
The cliff in being backed by continent:
It looked as if a night of dark intent
Was coming, and not only a night, an age.

ROBERT FROST, *"Once by the Pacific,"* 1928

THE EVENING OF TUESDAY, APRIL 17, 1906, WAS A TIME almost devoid of portents, except for a change in the weather that hinted at the long-awaited end of winter. At four o'clock or so a stiff sea breeze sprang up, driving away the clouds that had dulled a hitherto dreary day. A westering sun could at last be seen, inching its way down behind the Golden Gate, dimming the rocks at Land's End and Point Bonita, and making the Farallon Islands a misty silhouette on the Pacific horizon. This dusk had turned out to be a pretty one, a fitting finale to yet one more Californian day rolling to its contented close.

San Francisco was by now indisputably the greatest city in the American West. And even if there were a fretful few who did think about earthquakes from time to time, only a small scattering of these could ever have had their fears compounded by real experience. The last time an earthquake had hit San Francisco hard was four decades before, and so youthful had the city's population now become that only a very small number who had lived through that earthquake of October 1868 could possibly have been around for the second.

That earlier earthquake was a bad one, right enough. It had struck early in the morning, a few minutes before eight on October 21, and killed thirty-five people. Only five of these victims were in San Francisco itself, however—the remainder died in the collapse of buildings on the far side of the Bay, in what were then small settlements like Hayward and San Leandro. And the reason for this—though it was not readily ascertained at the time—is that the rupture in the earth that caused it occurred not on the San Andreas Fault but on one of its neighbors known as the Hayward Fault, which runs exclusively up the eastern side of the Bay.

It was a warm, humid mid-October morning, the sun just up, when the minute-long shaking began. People rushed out into the streets, many of them "in a state of semi-nudity." They remained there, terrified, while buildings crumbled, streets rose and fell in waves, horses panicked; there was a general air of turmoil and confusion. Their only experience of anything similar had been a relatively small event three years before, also in October, which ruined a handful of buildings; and though several engineers had forecast that more earthquakes might well occur and so had begun to think, at last, about making buildings strong enough to withstand them, the ordinary person—the man out in the street, as it were—found the happenings of 1868 unexpected and terrifying.

The origin of this earthquake was supposed, according to most of the newspapers of the day, to be somewhere on the far side of the Bay, at the base of a range of hills—though the only reports from this unsettled corner of California came from "cow-boys riding the range." Over on the San Francisco side there were enough residents for plenty of eyewitness accounts. The spire of a synagogue on Vallejo Street was toppled, pills and potions were thrown into "a perfect jumble" in W. Pickering's pharmacy on Stockton Street, and the premises of Messrs. Stone and Hayden, saddlers and harness makers, had all their chimneys shattered. The windows of the Empire Restaurant on Sansome

Street were smashed while the patrons were taking their breakfast. A Chinese man named King Young was fatally crushed when the *Scientific American* magazine offices near Leidesdorff Street collapsed on top of him. Brokaw's Mills, the Donohue Foundry, and the City Gas Works were so badly damaged that they had to be demolished. And the front of City Hall looked like "a dilapidated ruin," with the rooms of the Twelfth District Court and the Probate Court wrecked.

AS EVENING FELL on that April Tuesday in 1906, a few may have been abroad who had memories of that last quake. But most people, as usual, were more concerned with the mundane. Everyone seemed to be grumbling that the weather had been poor for so long: It was the tail end of an unusually damp chilly winter, and for most of this April day clouds hung over the hills, replaced later by a cold Pacific fog. But at teatime the wind had sprung up, the mists were blown away, and the skies cleared. The sun briefly warmed things up, then duly set, as a waning moon rose to take its place. People spoke happily of this pleasing conjunction—the clear skies, growing warmth, the setting sun, the pretty moon—as signs of the approaching change of season. Spring was coming; and for this and for a multitude of other reasons, it looked likely to be a pleasing night for many in the city.

There was no doubt about which was the most celebrated event in store for the city's glittering classes: the opera. The Metropolitan Opera of New York was in town for a second visit, and the tenor who had opened every season in New York since 1903, and who would do so until 1917, was in San Francisco to sing the role of Don José in *Carmen.* He was Enrico Caruso, at thirty-three the most admired tenor of his time, and a man amply endowed with all the manners of the *primo.*

He was put up at the seven-story Palace Hotel on Market Street, a short walk from the Grand Opera House on Mission Street—both enormous buildings built in the 1870s. The 800-room hotel, the largest in the country at the time, advertised four "rising rooms," or hydraulic elevators, which at the time were distinctly newfangled. It was also widely regarded as fireproof, with 700,000-gallon iron water tanks built under the roof. All the visiting grandees stayed there*: Caruso was apparently delighted upon being shown to a room once used by President Grant, and it was reported with great deference that he found the marble fireplace, the gold brocade upholstery, the furniture carved from local laurel, and the floors made of pine and redwood entirely to his florid Neapolitan taste.

Yet he was not in the best of moods. He had come out from New York by train, and had found the weeklong journey irksome. Naples, his hometown, was being hammered by the eruption of Vesuvius, and he was fretful about his family's safety. And even though it was said he vowed never to sing in Naples again after the poor review he was once given for *L'Elisir d'Amore,* he displayed a very public sympathy for the city's plight, and had briefly toyed with the notion of scrubbing his West Coast tour and going to see if he could help.

He had been in town a few days now, and was said to be none too impressed with the San Franciscans. The audiences seemed to him largely composed of oafs, crowds very different from the sophisticates back on the East Coast—and the poor reviews of the company's previous night's performance of *The Queen of Sheba* confirmed his suspicion that he was in the midst of a largely artless rabble, no more than provincial arrivistes, well able to pay $10 for a ticket but hardly worldly enough to appreciate what they had come to hear.

And, if all that were not enough to irritate this chubby, generous, driven little man—the third of five children, and not the eighteenth of twenty-one, as legend has it—he had a terrible time at rehearsals with the 200-pound Swedish-American mezzo-soprano Olive Fremstad, who was Carmen to his Don José. She had not sung well—a stagehand had dropped a vase, and the breakage made her lose a note—and Caruso worried that the night would go badly as a consequence.

---

*And not a few died there. President Warren G. Harding did, under circumstances that some still find mysterious; and King Kalakaua of Hawaii—who of course deemed it entirely appropriate to stay in the Palace—can probably lay claim to being the man with the longest name ever to have passed away there: The register clerk called him "Your Majesty," but officially he was David Laamea Kamanakapuu Mahinulani Naloiaehuokalani Lumialani Kalakaua.

Loved, feared, ridiculed, and admired—and always noticed, wherever in the world he might settle: Enrico Caruso was the *capo di tutti capi* of the turn-of-the-century operatic world.

But there was no need for anxiety. In terms of sheer spectacle, the night was a tour de force. Journalist Marcelle Assan published an account of it in Paris two years later:

> The spectacle of the room was wonderful. One would have to recall a similar evening at the Paris opera during the Empire to equal such beauty and majesty: everywhere were diamonds, white shoulders, magnificent eyes, sylph-waisted women wrapped in lace worthy of a Queen, with lustrous Oriental pearls wrapped around lovely

> throats. It was all breathtakingly beautiful. In front of the audience Caruso, the Italian tenor, inspired by the glinting eyes of his rich American admirers, sang as he never had before, his acting imbued with Italian ardor, and elevated the enthusiasm of his public to the point of delirium. People applauded wildly, threw flowers . . .

Charles Aiken of *Sunset* magazine was to write later:

> All society—with a big S—was out in force. Beautiful women gorgeously gowned, with opera cloaks trimmed with ermine, and diamonds on hands and hair; men with pop hats and the conventional cast-iron sort of clothes that mean joyous discomfort; here were wondrous bunches of orchids and roses; the singing and acting that charmed and the deafening applause. Then came the hoarse shouts of the carriage numbers, the strange melody of automobiles, the clang of electric cars; then tuneful orchestras at the Palace palm garden, or at Tait's, or the Fiesta, or Techau's, and oysters *poulette* and Liebfraumilch, Welsh rarebit and steins of Munchenbrau or terrapin Maryland and Asti *tipo* Chianti. And after all came the home-going in the early hours, with down-town streets still crowded, and the dazzling electric signs swinging wide over welcoming portals, making the garish city shame the modest moonlight.

Carmen went as well as it ever had. The audience, stolid and moneyed though they might have been, exhibited raptures of enthusiasm, and Caruso returned their adulation with polite charm. But there was no public afterparty for him: He slipped away, driven off in what passed in 1906 for a limousine, and spent the rest of the evening in a local restaurant drinking, eating pasta, and listening to the twenty-three-year-old Elsa Maxwell—a fat girl from Iowa who would later become America's best-known society hostess—play the piano.

He waited for the early editions of the papers, which, whether out of respect or sagacity, wrote rave reviews—though William Randolph Hearst had issues with the Met and was said to have tried, in vain, to have the exuberant notice in his *Examiner* watered down. After reading the papers Caruso—Erri, as he was known—then called for one of the rising rooms to beam him up to his fifth-floor suite, where he took himself to bed. It was 3:00 A.M. He would sleep for no more than 120 minutes.

John Barrymore was in town as well. This debonair young man—"The Great Profile," as he was known to his fans—was on the verge of thespian greatness, even then. But only on the verge. In San Francisco he had performed few of the Shakespearean roles that later would make him so famous, and had come to town merely to appear in a small-time play called *The Dictator* by the noted war correspondent Richard Harding Davis. The company that had performed it perfectly ably on the Bay was now about to sail off across the Pacific to try it again 7,000 sea miles away, in Australia.

John Barrymore's main interest—both then and for the rest of his life—appears to have been winning the favors of the local chorines, and that Tuesday evening he had found one, a woman who, like him, had come to listen to Caruso. Impeccable in his white tie and tails, he would have been a dashing companion—but it happened that he was low on funds and had to ask a newspaperman for a loan, which he did not get. One might suppose he would have taken the young woman back to his room at the St. Francis, for reasons both of amusement and economics; but he did not, and since he then wrote what he later admitted was an entirely fictitious account of his experiences of the following day, it can never be certain what he did, to whom, and where. All that is really known is that Caruso was at the opera and John Barrymore was in the audience.

After the last arias had died away and the carriages had collected the good and the great, Marcelle Assan headed for the most fashionable restaurant of the day, Delmonico's.

> In its great gilded and mirrored rooms, in its thickly carpeted stairwell, by the light of its electric candelabras, through the crack of the heavy doors left ajar, only joyous parties of men and women could be glimpsed and overheard enjoying themselves.
>
> Champagne corks popped amid bursts of laughter and both sibilant and animated voices. The restaurant was filled on every floor, and everyone was playing all the most exciting games while

tipsy from champagne or cocktails. Husbands and wives no longer knew each other, nor did they want to: it was merriment totally free and American.

There were more modest amusements for the less well-off. The Columbia Theater on Powell Street—convenient for the cable car, which passed right by—was showing the three-year-old (but since then quite indestructible) musical *Babes in Toyland* by Victor Herbert. For a mere dime you could have yourself an evening of vaudeville at the Orpheum. The entirely beloved Australian singer and actress Nellie Stewart was having great success with a production based on the life of Nell Gwynne, *Sweet Nell of Old Drury.**

In the Mechanics Institute pavilion—a huge, ornate, and prettily fretworked wooden structure close to the hospital in the City Hall basement—a local sports promoter was staging a masked roller-skating contest, which continued well past midnight. The doctors in the hospital across the road did a roaring trade from the scores of skaters who managed to land unhappily on the vast redwood floor and found themselves in sudden need of splinter removers.

And for the less physically inclined, there were a dozen smaller theaters ranged along Market Street and Broadway, and hosts of restaurants and clubs and whorehouses, all doing trade that evening that could be described as healthy or not, depending on your point of view. It has been said that Tait's, Delmonico's, Sanguinetti's, the Pup, and Tortoni's were doing good business that night; and that theater patrons crowded particularly into the Alhambra, Fischer's, and the Alcazar—others attending in lesser numbers the little plays being staged at the Majestic and the Valencia.

But Wednesday was a normal working day, and so the saturnalia that invariably gripped the city on Friday and Saturday nights was not

*Later performances became more inventive. When Miss Stewart performed it back in Sydney—to help a charity to buy radium for a local hospital—she took to selling oranges to audience members during the intermission.

in its fullest flower on a Tuesday. Aside from the gaiety of the opera—with well-dressed swells in their phaetons and hansoms and diligences swinging home via the better supper places, thus keeping some establishments going well beyond midnight—a pleasant quiet had settled on the city by one or two in the morning, as it usually did.

There were a couple of fires, which irritated the fire chief, Dennis Sullivan, who had in consequence to attend to duties until at least 3:00 A.M. One was at a warehouse on Market Street, and Sullivan ordered three of his horse-drawn wagons out to extinguish the blaze. The city had eighty fire stations (though none at all among the densely crowded tenements of Chinatown), and 700 men, most of whom were paid and so obliged by more than mere civic duty to fight any fires that might break out.

The insurers had long thought such kinds of preparations insufficient, however. As we have seen, the National Board of Fire Underwriters had remarked only the year before, after an extensive survey, that the city remained a tinderbox, waiting to be consumed once again as it had been six times already in the half century of its existence. Chief Sullivan concurred, often vehemently. Such was the flammability of its structures, the lack of water, the vulnerability of the supply, and the eccentric siting of some of the fire stations that insurers found the risks barely tolerable. There were only thirty-eight steam-powered fire engines in service, and tests had shown they could deliver water at only 70 percent of their rated capacity—much too low for comfort. The men who manned the engines were poorly trained. There were too few hydrants, and the old cisterns that long before had been built to store water below intersections in the city center were rusty and empty. So poorly equipped was the city, the board declared, that it had violated all underwriting traditions and precedents by *not* burning up.

And then there was the wind. In all six of those earlier devastating fires the prevailing wind had been westerly, blowing in from the sea, setting to the east. On this Tuesday night the wind backed slowly during the night from northwesterly to westerly. As Chief Sullivan took himself to his small box bed on the third floor of the Bush Street

Fire Station, he must have noticed it—noticed that if any night was the least ideal for the tackling of a major fire, it would be a night with a wind setting like this.

So the bars closed down, the streets emptied of their stragglers, the lights in the hotel rooms snapped off, one by one. The gas lamps in the city streets hissed and sputtered. The churches pealed the quarters and the hours, announcing the times in a soft clangor of amiable disagreement.

And it is said that some of those who walked home late, or rose unusually early, noticed that a number of animals in the city behaved a little oddly that night. The horses in a livery stable on Powell Street, for example, seemed skittish; and in fire stations, men could be heard behind the stable doors soothing animals that appeared unduly restless. One of the better-known first-person accounts of the earthquake was in the June 1906 issue of *Everybody's Magazine,* by a young Paris-born American writer named James Marie Hopper. He wrote of passing a livery stable on Post Street, and of hearing the obviously unhappy horses inside. He asked the stable boy idling at the entrance about it. "Restless tonight," the youngster is said to have replied. "Don't know why."*

At the Chutes Amusement Park that had been built at the turn of the century in the west of the city, in that area now called Richmond, there were a large number of caged animals. It was later reported by their keepers that they exhibited no peculiar behavior before the event, and that they remained quiet, cowed, and fearful during the mainshock of the earthquake. Once it was over, however, they all roared lustily with relief and puzzled exuberance. And the park's superintendent later related that the animal that led the roaring chorus, and that can thus be said to have exhibited the greatest sensitivity to what was happening

---

*Animals seem peculiarly sensitive to impending seismic doom. A baby elephant penned into a Batavian hotel room went spectacularly mad a short while before the eruption of Krakatoa in 1883; and just before the Sumatran earthquake and tsunamis of December 2004, reports of animals behaving oddly—monkeys chattering with terror, snakes going rigid, cattle bolting—were legion.

beneath his feet, was the animal possessed of the biggest feet of all: the elephant.

But before the earthquake began this elephant was entirely unaware of what was stealing up on him. No one person—and no animal, bird, or insect—had the faintest idea of what lay in store. No credible premonition has ever been reported.

THE SUN WAS DUE to come up that Wednesday at 5:31 A.M., Pacific Standard Time. The sky had begun to lighten about fifteen minutes before five; and by the time the bells of Old Saint Mary's Church in Chinatown had pealed the hour, all of the sky beyond the hills of Oakland and Livermore was lightening fast, limned with the palest, clearest eggshell blue.

The gaslights that had illuminated the deserted streets dimmed and were snuffed out with faint popping sounds at eight minutes past five. At about the same time, an unseen hand in a faraway engine house turned a crank and threw a giant lever, and huge drums began to roll; and so began the clanking grind of steel, steel rope, and ever-turning steel wheels that was then, and is now, one of San Francisco's most haunting and evocative sounds. The cable-car lines were running, and one by one their carriages rumbled out of their barns, ready to haul passengers up and over the city's innumerable hills.

And also one by one, people—men, by and large—began to appear in the still half-dark streets. These were either early starters, idling their sleepy way to their offices or shops, or night-shift workers heading wearily back home. The smell of baking bread, mingled with coffee, was in the air, as was the smoke of early cooking fires. The blue-uniformed policemen, slow and imperturbable, patrolled their allotted beats. The breeze was westerly but light. Dawn was unfolding quietly, serenely. All was perfect peace.

TEN

The Savage Interruption

I have the honour to report that at 5:18 on the morning of Wednesday, the 18th instant, a violent earthquake shock occurred in San Francisco—

From a diplomatic telegram sent on April 25, 1906,
from SIR COURTNEY BENNETT,
His Britannic Majesty's Consul-General,
San Francisco, to Sir Edward Grey, Bart., &c., &c.,
Secretary of State for Foreign and
Colonial Affairs, London

THE RUSTLING OF THE LEAVES

WASHINGTON STREET, LIKE MOST OF THE THOROUGH-fares that cross San Francisco from east to west, does not quite make it all the way from the Bay to the ocean. No one street does. In the specific case of Washington Street all manner of parks, diversions, and doglegs interrupt it, and in the end the huge military reserve of the Presidio blocks it from any western access to the Pacific.

But it is nonetheless a very long road indeed, one of the city's longest, and, on its arrow-straight way from the billets of the soldiery in the west to the docks of the Embarcadero and the storage sheds of the produce market in the east, it scythes past warehouses and office buildings, passes right through the crowded dilapidations of Chinatown, and rises and falls with the hills—Nob, Russian, and Pacific Heights—where the city swells like to live and play—and, in so doing, happens to present an almost ideal cross section of the city. It is in this regard a rather quintessentially San Francisco street, and one that by virtue of all of its manifestations—its houses, its mercantile offices, its slums, its clubs, its hotels, and these days its skyscrapers—offers up the very essence of San Francisco.

So it is perhaps appropriate that at 5:12 A.M. on that Wednesday morning the earthquake that was born out under the ocean beyond the Golden Gate seemed to come roaring into the city, as eyewitnesses like to remember, along the four switchback miles of Washington Street.

It made its entrance in a spectacular, horrifying, unforgettable way. It came thundering in on what looked like huge undulating waves, with the entire surface of the earth and everything that stood upon it seeming to lift up and then roll in forward from the direction of the ocean. The whole street and all its great buildings rose and fell, rose and fell, in what looked like an enormous tidal bore, an unstoppable tsunami of rock and brick and cement and stone. A policeman named Jesse B. Cook was standing at the eastern end of Washington Street, talking to a vegetable seller from the market—and he recalled suddenly stopping, horror-struck, as he saw what was happening along the street ahead of him.

At first there was what Officer Cook called "a deep and terrible rumbling"—presumably the reason that he looked up. Everything by now was fairly well illuminated by the gathering dawn, and so as he looked uphill he could see quite clearly. The ocean itself was spilling what looked like immense waves of water down the street; it seemed as if a huge tide of buildings and pavement had been lifted up and was relentlessly bearing down on him and on all the great commercial palaces of downtown. He spoke of the entire roadway undulating, billowing, and enormous breakers speeding toward him. For the officer, and for those few others who were abroad at that hour, or who were outside or at a window and thus able to see what was happening, it was a scene that had all the makings of an unutterably dreadful nightmare.

Policemen, trained as observers and equipped with notebooks, made excellent witnesses of the events of those first few moments. Officer Cook was one of a handful whose reports have an awful redolence about them:

> The earth seemed to rise under me, and at the same time both Davis and Washington streets opened up in several places and water came up out of these cracks. The street seemed to settle under me, and did settle in some places from about one to three feet. The buildings around and about me began to tumble and fall and kept me pretty busy for a while dodging bricks. I saw the top-story of the building at the southwest corner of Washington and Davis streets fall and kill Frank Bodwell . . .

A few blocks away on Market Street was Officer Michael Grady, who was on his way back to the station after spending the night patrolling in Chinatown. He, too, heard a rumbling sound, like thunder. The ground began to shake, and he was quick-witted enough to run immediately into the middle of the street,* which, in the same way as Washington Street half a mile away, was rising and falling like the sea:

> I thought I was gone when I saw the Phelan Building suddenly lurch over Market Street. But it lurched back again, and as it set back in its place its foundations ripped and cracked and seemed to screech. The tall *Call* Building rocked to and fro from north to south, while the Mutual Bank Building on the opposite side of Market Street, near Geary, similarly lurched and dipped over the thoroughfare.
>
> The shake ended with a violent twist or rotary motion that caused the stone cornices of the building on the south side of Market Street west of the *Call* Building, and therefore much nearer to me, to come crashing down on the sidewalk. At the same time the front of the Oberon Building, on O'Farrell Street near Stockton, and parts of other buildings on the same side of that block, came tumbling to the ground. These were all in plain sight of where I was standing.

---

*Because of its double tram tracks Market Street was the widest boulevard in the city. Horse-drawn wagons occupied the outer parts of the street, and the cable cars were confined to a metal channel, which came to be called the Slot. Before long this became an important social demarcation line in the city: Living "south of the Slot" suggested blue-, rather than white-collar.

> On Grant Avenue the effects were not so startling; but nearly all the plate glass windows sprung and seemed to explode as they were bent and twisted by the force....

There were police aplenty on dawn duty at City Hall; their perspective was thus from rather farther to the west than that of Grady and Cook (they were at the corner of Larkin and McAllister Streets—a mile west of Grady, two miles west of Cook). Edmond Parquette, for example, was visiting what was known locally as the CEH, the Central Emergency Hospital, which was in the City Hall basement.

> I was just stepping through the entrance of the office when the whole place began to shake, and in a few seconds the shaking became so severe that I had to hold on to the door to save myself from falling.... The building was shaking and rolling like a mad thing. The furniture was rolling and hopping about, the plaster and everything else on top was falling. Then there was the roar of the earthquake itself, and the crashes and shocks and rumblings as we felt the walls and pillars of the City Hall bursting and breaking over our heads.

One floor above him was Edward Plume, another patrolman working the end of his shift in City Hall's small police office. He experienced much the same sensation, but noted the time, interestingly, as "5:10 A.M. by the station clock," fully two minutes earlier than the time in those precise reports from central San Francisco and across the Bay (most notably the five mentioned earlier) by which history has since fixed itself.

Quite probably Officer Plume's record of 5:10 A.M. signifies very little—other than that his police station clock was slightly slow. Except—his station was in fact a little closer to the fault and its rupture than the other observers: One could therefore expect the plowshare wave from it to rip through City Hall just a few moments before it reached the waterfront, and significantly before it reached the sleeping Professor Gilbert in his cot nine miles away across the Bay in Berkeley a few moments before. Not two minutes before: With the wave traveling at two miles a second the time difference would be barely noticeable.

The laggardly police-station clock gives us the hint that those in the west of San Francisco experienced the quake first; and it makes it more than likely that Clarence Judson, the workman who was swimming in the Pacific off Ocean Beach before dawn that day, experienced the event a few moments before everyone else did, simply because he was so very much closer to where it all began.

Officer Plume's notes reveal an experience similar to those of his three colleagues, except that they offer rather more of the sense of terror everyone must have felt:

> The noise from the outside became deafening. I could hear the massive pillars that upheld the cornices and cupola of the City Hall go cracking with reports like cannon, then falling with crashes like thunder. Huge stones and lumps of masonry came crashing down outside our doors; the large chandelier swung to and fro, then fell from the ceiling with a bang. In an instant the room was full of dust as well as soot and smoke from the fire-place. It seemed to be reeling like the cabin of a ship in a gale. Feeling sure that the building could never survive such shocks, and expecting every moment to be buried under a mass of ruins, I shouted to Officer Dwyer to get out. The lights were then out, and though the dawn had come outside, the station, owing to the dust and smoke inside, and the ruins and dust outside, was all in darkness.

Officer Plume's view of the likely collapse of City Hall was entirely accurate. Twenty-six years it may have been in the making, and millions may have been spent on rendering its external glories, but the building was like so much of the San Francisco of the day: gaudily finished but shoddily made. Most of it fell that day, and all of it had to be torn down in 1909. The building does still have its defenders, who say it was not that poorly built: They note that a number of city offices remained working in the shored-up ruins until the entire structure was finally demolished. Most, however, including federal government inspectors, condemned its construction, and in no uncertain terms.

Not far away a policeman named Harry Walsh was having much the same experience as Plume: and had already witnessed death and destruction at a machine shop, where he had stopped off to have a cup of coffee with the night watchman. He was appalled to see huge cracks opening up on the pavement of Fremont Street, and then closing and reopening as the shock waves shook everything to pieces. He tried to crank a message to police headquarters on a corner message box, but the lines had already gone down.

What he saw next was memorably awful: a stampede of wild long-horned cattle tearing toward him along Mission Street, from the direction of the docks. It turned out later that a group of Mexican *vaqueros* had unloaded these beef cattle from an inbound ship and were driving them to the city stockyards in the south of town. The moment the shocks began, the drivers took off, leaving their herd to fend for itself—and, as Officer Walsh put it, the cattle promptly "went daft with terror and started running anywhere," continuing:

> While a lot of them were running along the sidewalks of Mission Street, between Fremont and First streets, a big warehouse toppled onto the thoroughfare and crushed most of them clean through the pavement into the basement, killing them and burying them outright. The first that I saw of the bunch were caught and crippled by falling cornices, or the like . . . and were in great misery. So I took out my gun and shot them. Then I had only six shots left, and I saw that more cattle were coming along, and that there was going to be big trouble.
>
> At that moment I ran into John Moller, who owned the saloon. . . . I asked him if he had any ammunition in his place and if so, to let me have some quick. He was very scared and excited over the earthquake and everything; and when he saw the cattle coming along, charging and bellowing, he seemed to lose more nerve.
>
> Anyway, there was no time to think. Two of the steers were charging right at us while I was asking him to help, and he started to run for his saloon. I had to be quick about my part of the job because, with only a revolver as a weapon, I had to wait until the animal was quite close before I dared fire. Otherwise I would not have killed or even stopped him.

As I shot down one of them I saw the other charging after John Moller, who was then at the door of his saloon and apparently quite safe. But as I was looking at him and the steer, Moller turned, and seemed to become paralyzed with fear. He held out both hands as if beseeching the beast to go back. But it charged on and ripped him before I could get near enough to fire.

When I killed the animal it was too late to save the man. . . .*

Then a young fellow came running up carrying a rifle and a lot of cartridges. It was an old Springfield and he knew how to use it. He was a cool shot, and he understood cattle, too. He told me he came from Texas . . . we probably killed fifty or sixty . . . we used the rifle alternately, the Texan and myself.

The celebrity-filled Palace Hotel—where Caruso was staying and had gone to his room at 3:00 A.M.—was to be a victim, too, though, like so many of the city's buildings, it fell not to the earthquake itself but to the subsequent fire. It stood secure at the first shock, though a passerby named Frank Ames, in the hotel's grand courtyard, later reported that the palm trees swayed, the ground dropped beneath his feet, and the horses under the hotel's porte cochere took off in panic. As they did so, he wrote, "I could not help noticing that the beasts' eyes were big with terror and foam was coming from their nostrils." The earthquake, Ames noted, seemed to be running there from east to west—the opposite way from that noted by most of the other witnesses—and "the cobblestones of Market Street seemed alive. Every one of them was moving, and the streetcar rails were twisted from their places."

Other guests spilled out into the hallways, most of them half dressed. One guest said the sound was that of a monstrous train roaring by, and that the hotel seemed to twist on its axis, moaning audibly

*This is a near-perfect example of what is known by the very old and now rarely used English word *chance-medley,* in which a casualty occurs as a result of a confusing set of circumstances in which only a small element is accidental. There was once a crime known as "manslaughter-by-chance-medley," though what took place here was clearly not a crime.

as it moved. Ernest Goerlitz, the manager of the Metropolitan Opera touring company, said the hotel "seemed to dance a jig." One Nevadan staying at the hotel's annex across the road stood up once he thought the worst was over, only to find the floor jerked from beneath his feet, pitching him through the bathroom door and upending him in the bathtub. Another guest reported looking out of the window at the throngs gathering in the streets below; he noted that they included a woman in a nightgown who was carrying a naked baby by its legs, as though it were a trussed turkey.

Caruso was unutterably terrified. His conductor, Alfred Hertz—who of course had to endure the shaking in his own bed, holding on grimly to the mattress supports while the crashing and banging went on and on and on—rushed to the maestro's room. Caruso was sitting bolt upright in bed in his nightshirt, weeping loudly. He wrote later in a London newspaper that he first thought it was all a dream, "that I was on a ship on the ocean, and for a moment I think I am dreaming that I am crossing the water on my way to my beautiful country." But, according to others, he was found in a near-hysterical state, claiming that the quake was a punishment sent to him directly, and announcing that his voice had been irreparably damaged. Hertz insisted that he try to test it: He did, croaked a little nervously, and then all worked perfectly well.

After which Caruso's story branches in so many diverting ways that there is precious little reliability from this point. Did he put a fur coat over his pajamas? Or over his nightshirt? Did he change into formal dress so ornate that, as one writer had it, he looked as though he were about to make a courtesy call at some embassy? He was to write later that he did indeed summon his valet, who dressed him, repacked his innumerable trunks of clothing, and had them manhandled down the stairs (because the rising rooms would now no longer rise or fall). And he sketched, hastily, small pencil pictures of himself watching the fires: They may look a little crude, but they have a ineffable charm about them, capturing as they do this most unusual moment in the singer's life.

When Caruso eventually arrived back in New York, he learned that readers of the penny dreadfuls understood him to have wandered the streets of San Francisco displaying his magnificent embonpoint in his nightshirt, or less. Stuff and nonsense, he retorted. He was fully dressed, indeed really very well so. But then did he breakfast at the Palace? Did he go to the St. Francis? Did he lie down for a short snooze in Union Square while the disaster raged about him? Or, as many say, did he step over the rubble, cigar in hand, and mutter endlessly and to no one in particular, "'ell of a town! 'ell of a town!"?

THERE WAS MUCH MORE certainty out at sea. A scattering of ships experienced the full effects of the rupture, the evidence preserved with dispassion and accuracy in a number of surviving masters' logs.

The great tenor's pencil sketches display a more homespun talent. In the twenties he published a book of caricatures of leading musical personalities and celebrities—many royal—before whom he performed.

Many of the captains of these vessels thought they had run up on the rocks. In the far north, the steam collier *Wellington,* for example, was under way in sixty fathoms off Point Diablo; and the steamer *Alliance* was in similarly deep waters close to the seismically complicated Mendocino Triple Junction. Both vessels felt a sudden huge shock, "as if we had run into rocks at high speed," according to one of the captains. The captain of the *Argo,* ninety miles off Point Arena, thought he had struck a raft of submerged logs. Closer to San Francisco was the schooner *John A. Campbell,* doing seven knots in 2,400 fathoms about 145 miles west of Point Reyes. It was 5:15 A.M., with clear weather and a fresh north-northwesterly wind, when suddenly there came a strange sensation, "as if the vessel struck, slightly forward, and then appeared to drag over soft ground, and when aft a slight tremor was felt; and the whole lasting only a few seconds."

The *Wellington,* entering the Bay at exactly the moment the fault ruptured, was said by her captain to have "shivered and shook like a springless wagon on a corduroy road," even though the sailors at the taffrail looking back at the sea that morning had reported it as smooth as mirror glass.

The pilot boat *Gracie S.* was waiting by the lightship at the bar off the Golden Gate to come alongside the German steamship *Nyada* and guide it to its berth; it felt a shock "as if the chain were running out of her hawser pipe," and the German, when boarded, was white faced and shaking, convinced that he had somehow inadvertently managed to ram into a reef. The pilot boat *Pathfinder* made a similar report: at 5:15 A.M. a heavy bang as if running aground, even though the charts insisted there were 120 feet beneath the keel.

One of the maritime reports, the most colorful, is also a little puzzling: It is that of a man named James Denny, the chief engineer of the steamship *National City,* which early that Wednesday morning was heading south along the coast some twenty-nine miles offshore, well to the north of San Francisco, close to the mouth of the Gualala River. The fault is close to shore there, lying behind a low range of hills. So when the shock hit, it hit hard.

"The ship seemed to jump out of the water; the engines raced fearfully, as though the shaft or the wheel had gone; and then came a violent trembling fore and aft and sideways, like running at full speed against a wall of ice." To add credibility to his report Denny added a courteous and possibly helpful postscript: "The expression 'a wall of ice' is derived from my experience in the Arctic."

The single puzzling aspect of this one report is its timing: The ship's log records the event as having occurred at 5:03 A.M.—nine minutes before the mainshock is known to have originated off Mussel Rock. It has to be assumed that the log is wrong, or that the *National City*'s chronometer was ill set: There seems to be no corroborating evidence of any earlier shock. But, as in all other respects the vessel's encounter seems to have been scrupulously recorded, it does leave one to wonder if perhaps the San Andreas was misbehaving itself locally at earlier moments.

Out on the southernmost of the Farallon Islands, which stand thirty fog-shrouded miles due west of the epicenter—and so just to the west of the fault trace—there was further evidence of chronological complications. The rupture was felt there quite distinctly (and in the usual fashion, with two intense shocks, the second greater than the first), but it was felt at 5:06 A.M. This time was, moreover, confirmed by a secondary source, because at the same moment James Boyle, the U.S. Weather Bureau observer on the Farallons, was talking by telephone to his colleague, a Mr. Legler, in the bureau office in the lighthouse at Point Reyes, twenty nautical miles to the north. Legler experienced the shock wave first, about three seconds before Boyle; both men reported that objects were thrown in an easterly direction—though at the Farallon Observatory a stone weighing a hundred pounds was moved six inches just to the west of south, and it was rotated, very slightly, in a counterclockwise direction. The time for both observers was a puzzling six minutes after five o'clock—still dark out at sea, with only the vaguest hint of dawn coming up from over the land.

THREE OTHER ISLANDS, well known locally, reported the events. On Angel Island—which stands foursquare in San Francisco Bay, and which in the aftermath of the quake was to be the main receiving station for Chinese immigrants and the consequent center of much amusing mischief, as we shall see—there was a Mrs. Nichols, who reported:

> The shock resembled the jolting of a railway train which running at full speed had left the tracks and was bumping over the ties. It was accompanied from the beginning by a loud noise which gradually decreased as the jolting motion ceased. Water standing in a pail was thrown out six feet from northeast to southwest. The clock was stopt. The bay was calm. A cement pavement was cracked. The station was on solid rock.

At the Naval Training Station on Yerba Buena Island—now best known as the midway point on the Bay Bridge—sailors felt a "heavy vibratory shock"; and on Alcatraz—which by then had its Citadel, its lighthouse (the first on the West Coast), and its military prison—the commandant distinctly felt three shocks, the second of the three being the strongest.

Then, from fifty miles south of the epicenter, at Santa Cruz, came an oceanside report from an early riser, an observation that serves as a reminder that the shock was also traveling *southeastward* along the fault track, and causing damage and destruction at places other than on the northwesterly track toward Shelter Cove and Point Diablo confirmed by most of the ships. This report came not from a ship but from the wharfinger on a wooden quay that extended into Monterey Bay from the Santa Cruz docks.

This man was standing on the quay, which extended out to sea in a southeasterly direction, when he heard a loud rumble coming from directly ahead. (The fault is fairly close to where he was standing, about twelve miles due east, close to the garlic-growing town of Gilroy.) He next saw, with perfect clarity, the shock wave coming fast toward him across the water: He felt two distinct sets of vibrations as the waves traveled through, the second being much the greater. The wharf seemed to pitch lengthwise—doing the same bronco-bucking dance that the Washington Street pavement was performing up in San Francisco.

And then the shock waves hit land: The stunned wharfinger could only report that they caused "great rattling and crashing" as they tore through town. The sea, meanwhile, looked strangely unruffled: There was no surf where the shocks passed, and the water's surface appeared only "like that in a tub, when jarred."

The earthquake was indeed felt in Santa Cruz, and in thousands of other communities besides, as it roared its unstoppable way northward, southward, and eastward. In Santa Cruz giant fissures opened up beside the San Lorenzo River, which itself first churned into foam, then sank many inches below its normal level, and finally slowed down to a crawl. A man nearby reported being first thrown to the east, then to the west; and the grove of eucalyptus trees he was idly contemplating in the dawn light started to sway violently in the same direction.

A little farther north, in Los Gatos, people reported smelling sulfur once the two shocks had passed, and a well owned by a local blacksmith produced sulfur-tainted water for several years. Pianos seem to have been thrown around with frenzied violence—an observation that says as much about the popularity of pianos as it does about the strength of the shocks—and clocks by the tens of thousands stopped, something that, since the pendulum has since largely gone the way of the holystone and the whiffletree, would not have happened to a similar degree today.

The reports from the south, from the entire length of what was later to become Silicon Valley, and from around San Jose and back up north among the factories and docklands of the East Bay, display an admixture of gentility and rural charm, with the quake having its effects on a world of manners that is still entirely recognizable today, but that is nonetheless dated, and different.

So the reports are littered with descriptions of how the shock affected people who owned, or did business with, or lived among such things as anvils, wells, grand pianos, pianolas, chandeliers, milk pans, windmills, pilasters, cornices, rocking chairs (which went wild), prune orchards, trestle bridges, billiard tables, pump houses, ice chests, window weights, limekilns, molasses tanks (most notably those belonging to the Alameda Sugar Company in the East Bay), razor strops, terra-cotta chimneys, livery stables, vials of drugs, carboys, gasometers, wire-mounted butchers' scales, gas lanterns, telegraph offices, and hanging lamps with pendant glass prisms (which swung in directions all carefully noted by Andrew Lawson—the head of the state Earthquake Investigation Commission—with the results of his notations plotted neatly onto maps that show how great or weak and in what direction the shocks were, when they came).

And then there was Agnew's State Hospital for the Insane at Santa Clara (its name, in common with others like the Institute for the Feeble-Minded and the Board of Commissioners in Lunacy, suggests suspicion and intolerance with which mental illness was regarded at the time). The hospital dissolved into chaos when the earthquake hit, and it was thought that in the pandemonium hundreds of inmates broke loose and roamed the streets of town, like zombies. No such thing occurred, although a falling clock tower (with the clock's hands stopped forever at 5:12:45) did kill as many as 119 people, rendering the Agnew's tragedy one of the greatest to befall a single building.* The scenes were utterly distressing: horribly mutilated victims, bewilderment, wailing and fear on all sides, violent patients tied to trees with sheets, screeching bloody murder.

Though the Agnew's disaster was especially dreadful, and though there were moderate death tolls in some other southern towns—in San Jose 19 people died and 8,000 were made homeless—to the north of San Francisco things were a good deal worse. Both the force of the shock and the displacement caused by the rupture were considerably greater than in the south, and the experiences were generally much more distressing. The vineyard country of the Napa and Sonoma Valleys—though at the time reeling from the phylloxera infestation—was hit particularly hard, and workers tell of the acres of vines taking on

*One supposed rescue did point up the troubles of some asylum patients. Three days after the event rescuers heard a man crying, "Never mind me, get the others first!" and naturally raced to free this heroic victim. But it was no miracle: The man turned out to be a patient who had escaped from his guards and had *buried himself*, the better to win attention.

the appearance of the ocean, with the rows of grape arbors rising and falling in great waves as the shocks tumbled down the hillsides.

Sonoma County's main city, Santa Rosa, was savaged, to a degree that has been largely eclipsed by the attention given to San Francisco. It was said at the time to be the prettiest community in California, a bustling and contented market town of about 6,000. But it abounds with horror stories, not least because most of the town's brick buildings, held together by a cheap lime mortar made with inferior sand, came tumbling heavily down—while many of the wooden frame houses merely slid neatly off their foundations and stood, sagging inelegantly, out in the streets.

Some of the wooden houses collapsed, too. A man named Duffy was sleeping in one of them, and later said that when the shock came he was thrown out of bed and rushed to go down the stairs, but found the building swaying and shaking so ferociously he could make no headway, and so had to turn back. He threw himself in front of the dresser in his room, trusting it might protect him from the falling timbers. It saved his life. The dresser held up the beams that tumbled over him, and these in turn protected him from the falling mass of debris.

The cities and the villages and all the farms, wineries, asylums, schools, military bases, lighthouses, railway stations, docks, observatories (the Lick Observatory, east of San Jose on top of Mount Hamilton, had its six duty astronomers that morning scrupulously recording every last cat's whisker of a vibration) did indeed all suffer, and, as the foregoing suggests, in numbers far larger and in places more widely distributed than the often-used phrase "San Francisco Earthquake" implies.

At the very outer edges of the roughly circular area where the event was just discernible, and where it was remembered or recorded by humankind,* its shock waves caused no damage at all. The most northerly place where humans experienced the event appears to have

---

*I am not including the places where the event was recorded by machines: Plenty of seismographs existed in 1906, and the San Francisco event was recorded by many, some of them—as we shall see—thousands of miles away.

been the small town of Coquille, Oregon, which lies 390 miles to the north of the Daly City epicenter. One of the town's most prominent men, Judge Harlocker, was woken up just after 5:00 A.M. Mr. and Mrs. Wilson living nearby were awakened and noticed the cord of their electric lamp swinging back and forth. And when Mr. Wilson went to work in his jewelry store, he found that his regulator clock had stopped.

In the far east of the affected region, only one resident of the Nevada desert town of Winnemucca, which lies 340 miles east of San Francisco, became aware that something seismic was up. She was a nurse at the county hospital, just home from working the night shift, and, as she told Lawson's investigators, she was lying quietly in her bed when suddenly she saw her hanging lamp begin to swing violently back and forth. She felt nothing, however—no vibrations, no rocking motion. This single report, from a town of more than a thousand, was about to be discounted simply because it was so singular—until a pair of others, one recording a disturbance in the surface of a water tank, the other a swinging lantern, came in, unprompted, from Nevada towns close by. So, thanks to these three observations, the earthquake now had a definite eastern limit.

And to the south, the town of Anaheim, close to Los Angeles, appears to have been about as far as credible reports of people having felt the earthquake can be found—370 miles from San Francisco. Observers there, in the community south of Los Angeles—and where Walt Disney would later build his first amusement park, in the fifties—speak of twin shocks, the first moderate, the second, as almost everywhere else, very much larger. Someone noticed that a curtain swung in and out of the window frame. A cord with a heavy brass ring swung back and forth, its motion trending from the northwest to the southeast. Wall-mounted barometers rattled against their confining metal hoops. People in the tiny inner-city community of Azusa also felt a very distinct sudden shock—a fact that is possessed of an interesting synchronicity, considering Azusa's later importance, which will become apparent in the further telling of this story.

Averaging these three outer limits—people feeling the quake 390 miles to the north, 340 miles to the east, and 370 miles to the south*—suggests that the effects spread over a roughly circular area of some 400,000 square miles. The western half of this circle lies mainly beneath the Pacific; of the more easterly, land-based half, the 200,000 square miles, a little more than seven-tenths lie in California. And within those 150,000 square miles of the Golden State affected by the events, it is almost too obvious to remark that the consequences become steadily more and more dramatic the closer one gets to San Francisco and the Mussel Rock epicenter.

Although the area where the event was felt was roughly circular, the area where there was real damage was lozenge shaped, with the long axis of the lozenge following, though somewhat unsurprisingly, the San Andreas Fault. So from the outer eastern and undamaged edge in Nevada one has to travel about 300 miles westward, to reach places where the reports of real damage came in. Basically the damage spread through almost every single community between a village called Paicenes in the south and the much more substantial city of Eureka in the north. Not much damage, perhaps—but recordable (and so insurance-claimable) damage nonetheless. Eureka provides a very good example of the kind of edge-of-zone impact that the earthquake had: There were about $5,000 worth of cracked walls and broken windows in town, and the statue of Minerva on the dome of the Humboldt County Courthouse, which had been seen swaying back and forth during the event, finally settled tipped in a southerly position, sloping by 43 degrees.

Closer to the epicenter, more and more of the damage amounted to total ruin; and so, the nearer to San Francisco, there was concomitant mayhem, social dislocation, panic, terror, heroism, and a terrible amount of death. Voltaire's much-derided remark about how knowing all allows one to forgive all springs to mind—though with earthquakes,

---

*These figures are computed from the earthquake's newly established true epicenter beside Mussel Rock in the suburb of Daly City. But, generally speaking, because most of the damage occurred in San Francisco, it is simpler to think of this much-better-known city as the practical center point of the event's effects.

Where the 1906 Earthquake Was Felt

Perceived Shaking	Not Felt	Weak	Light	Moderate	Strong	Very Strong	Severe
Potential Damage	None	None	None	Very Light	Light	Moderate	Moderate/ Heavy
Instrumental Intensity	I	II-III	IV	V	VI	VII	VIII

Subduction Zone
Convergent Boundary
Transform Fault

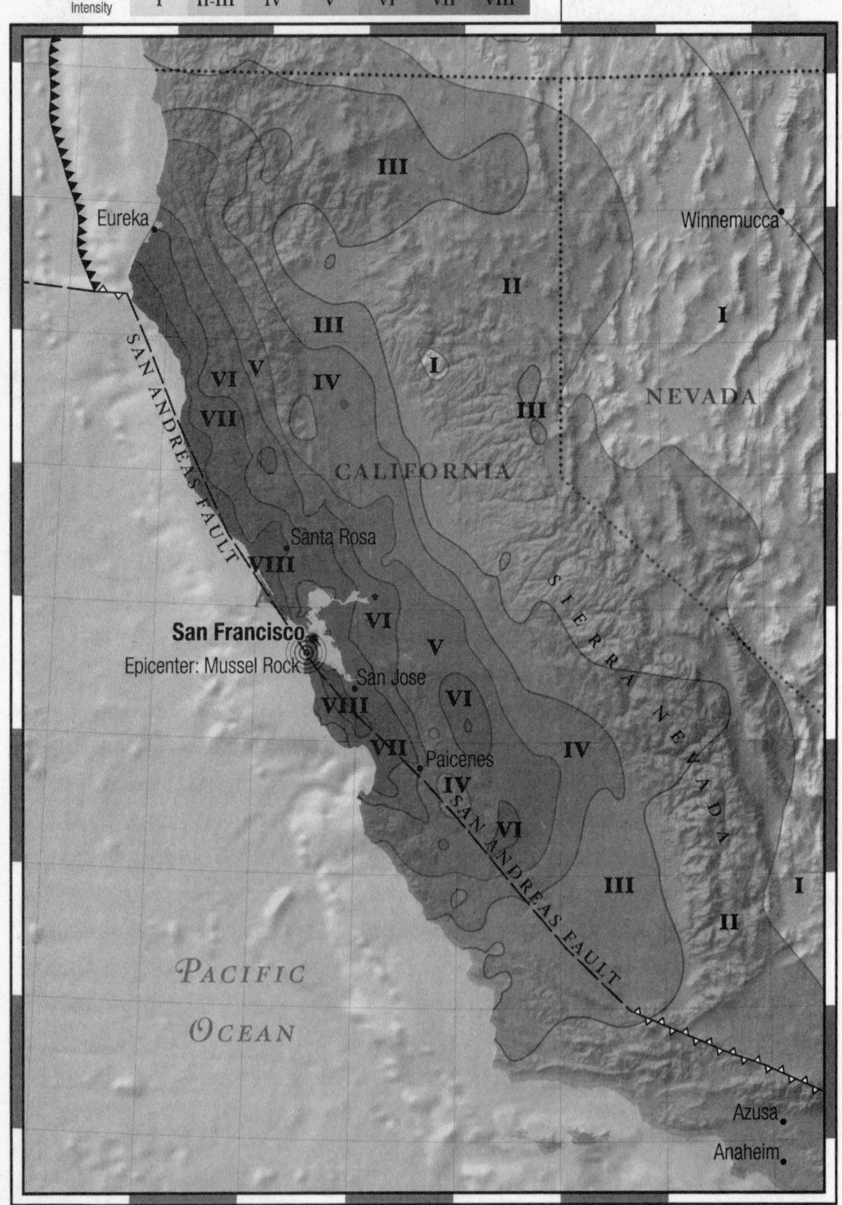

as we shall also see later, knowing a great deal about their mechanics does not seem to mitigate their horrors or help us to avoid them.

The Measurings and the Mechanicals

In the second century of the modern era, when in China it was somewhere near the middle of the Han Dynasty, the astronomer royal to the emperor, a man generally named as Zhang Heng, created an elegantly complicated device known as a *hou fêng di dong yi,* which translates approximately as "earthquake weathercock."

The original has never been seen, only described. It was a brass vessel, rather like a very large wine jar, that sported on the lip of its upper surface the brass heads of eight dragons, positioned equidistantly around the vessel at the major points of the compass. Each of these dragons held in its mouth—and did so very gingerly, by the points of its teeth—a small ball made of jade. Immediately below each dragon was a toad, its mouth wide open, ready and waiting to catch any ball that might drop from the dragons' dental grasps.

No one knows exactly what was inside the jar, but it is assumed that there was some kind of pendulum attached to eight jointed arms. Everything was arranged with mousetrap delicacy in such a way that if the pendulum swung for any reason, the arm that impeded its swing was moved; this knocked the ball from inside the mouth of the closest dragon and into that of the waiting toad. The instrument, designed to be presented to the emperor and set up by his throne, was for the specific purpose of detecting earthquakes.

The device, which is often displayed today to reinforce the idea that the Chinese were the first to create all of the various wonders that mark the track of human progress, does indeed illustrate the ancient concerns about the dangerous caprices of the planet. But seemingly it didn't do very much. If it worked at all, it might have let the emperor know that there had been a distant vibration in his realms; and since the dragons were arranged around the points of the compass, it might also have allowed him to learn more or less in which quarter of his do-

Joseph Needham, the great Cambridge scholar of Chinese scientific history, was one of the first to present pictures of how Zhang Heng's second-century "earthquake weathercock" might have looked. He contends that the instrument was improved and used by subsequent court astronomers in China until at least the eighth century, with the basic idea finding its way to the West by way of Persia.

---

mains this vibration had occurred. It is said that this happened once: The ball dropped and the emperor was able to say that there had been a quake in Hunan Province—and lo, a few hours later a messenger arrived with the news that there had indeed been an earthquake in Hunan.

The device was beautiful, clever, and prescient—but probably so inaccurate as never to have enjoyed more than the most symbolic use. And, like so much that was created in early China, it was neither advanced nor improved upon, at least not by the Chinese.

It took Western science to do that—although it happened that in this particular case it was Western science that was by chance operating in the East. Crude vibration detectors were built in the eighteenth century, but the first modern device worthy of being called a seismograph—an instrument that is able to detect the earth's movements with precision and, unlike the frog-and-dragon-balls affair, to make a record of it at the same time—was invented in 1896 by a Yorkshire wool-dealer's son named John Milne. He invented it while he was studying the earthquakes that occurred with troubling frequency not in Yorkshire but in Japan.

His basic idea has remained the guiding principle behind the instrument ever since, and in essence it is the one that guided Zhang Heng's instrument of sixteen centuries before: A pendulum with a pen is allowed to swing against a paper-covered and clockwork-run recording drum fixed solidly into a case bolted securely to the ground. Any shaking generated by an earthquake will cause the drum and the pendulum to move relative to each other, and the movement will be written by the pen onto the unrolling sheet of paper. (The terminology is more or less self-explanatory; the recording instrument is, strictly speaking, a *seismometer;* the entire package of instruments is a *seismograph;* the record to be studied later is a *seismogram.*)

John Milne constructed his brass, steel, and wire machine while serving as professor of geology at the Imperial College of Engineering in Tokyo, after he found himself frustrated by the inaccurate and shoddily made instruments that were put at his disposal to study the Great Yokohama Earthquake of 1880. And, while his device (sometimes called a Milne Seismograph but equally often called a Ewing, after his colleague-inventor James Ewing) was very accurate and as handsome as most Victorian scientific devices (it was manufactured by the famous, and still existing, instrument-making firm of R. W. Munro), it was soon to be supplanted by scores of instruments with much more exotic European and Asian names.* These would in due course become

*The names were indeed splendid: There were the Repsold-Zöllner Horizontal Pendulums, for example, used at the observatories in Tashkent and Irkutsk; the Stiattesi Vertical Pendulum used in Florence; the Vincenti-Konkoly Vertical

electronic and digital and accurate to unimaginable degrees.

In 1906 there were ninety-six seismographs in existence in the world, many of them Milnes or Ewings. So great was the shock emanating from San Francisco that practically every single one recorded it. Some were nearby—those in San Jose, Berkeley, and Yountville, California, and Carson City, Nevada, were so close to the event that their needles and drums recorded the destruction of buildings just yards away. At Stanford there was a Bristol's Recording Voltmeter: an instrument designed to measure an altogether different kind of phenomenon, but that, since it had a needle, a pen, and a sheet of clockwork-driven paper, also noted the vibration with a spectacular set of indelible black-ink flailings.

Most of the true seismographs, though, were a good deal farther away: There were seismographs of all kinds in an almost alphabetically complete list of places that includes Apia, Bombay, Cairo, Dorpat, Edinburgh, Florence, Honolulu, Irkutsk, Jena, Kew, Leipzig, Mauritius, Osaka, Perth, Quarto-Castello, Rio, Sitka, Tiflis, Uppsala, Vienna, Wellington, the aforesaid Yountville and Zagreb, and, making up for the alphabetical absences of towns beginning with *g*, *n*, and *x*, an additional *z*, the small Chinese city that was possessed of a borrowed Japanese Omori machine, a half-forgotten place then known as Zikawei.

Only half forgotten, however. Zikawei used to be a small town southwest of Shanghai: It is now thoroughly subsumed into its neighbor, bears the name of Xujiahui, and is as filled with skyscrapers, nightclubs, and subway lines as any other part of the world's fastest-growing city. A century ago this village was a place of colleges, orphanages, and hospitals—and, most interesting for this story, it also had a

---

Pendulums of Zagreb (where the observatory director was Dr. Mohorovicic, later to give his name to one of the world's major geological features); the Ehlert Triples employed at Kremsmünster in Austria and at the famous Uccle Observatory in Belgium; and Dr. Hecker's magnificently titled Von Rebeur–Paschwitz Horizontal Pendulum, which apparently performed yeoman service in the main physics laboratory in late-nineteenth-century Potsdam.

large Jesuit mission. The Omori seismograph that recorded the San Francisco Earthquake had been loaned specifically to the Jesuit Observatory of Zikawei—a happenstance that serves to underline the generally forgotten fact that the Jesuits and seismology have enjoyed a long and intimate connection, and that no small number of the machines scribbling down the record of the Californian events of that April morning were in fact maintained and operated by monks and priests ordained in the Society of Jesus.

The society, right from its sixteenth-century beginnings, was predicated on the basis of promoting scholarship and learning, particularly in the fields of mathematics, astronomy, and what were once called the natural sciences. By the seventeenth and eighteenth centuries most Jesuit colleges had departments that studied the earth and the stars; and in China, where the society had long had a foothold, the Manchu court appointed Jesuits to direct the Peking Astronomical Observatory.

The patient, monastic observers who maintained these schools soon noticed the terrible disruptions caused by earthquakes—and once the precursor to the machine that was eventually to be called the seismograph* had been invented in the 1750s, the society's leadership decided, as a matter of canonical policy, that in those areas where earthquakes were common it would be the Jesuits' holy duty to investigate them and, by doing so, perhaps help to mitigate the misery that such unpredictable events brought to God's people.

So the first Jesuit seismograph—a crude two-pendulum arrangement of a type that would these days be somewhat dismissively called a seismoscope, since it never left a written record—was installed at the Manila Observatory in 1868. The next was installed the following year in a monastery in Frascati, Italy. Then one was built in Granada, Spain,

---

*An Irish-born engineer named Robert Mallet came up with the term—from the Greek for "earthquake," *seismos*—in 1854. Mallet was a man of many talents: His legacy includes a steam-powered barrel-washing machine made for Guinness, a pair of powerful siege cannons, critical parts of the Fastnet Lighthouse, and a number of swivel bridges over the river Shannon.

another at Mungret College, County Limerick, another at Stonyhurst College in Lancashire, another on the island of Jersey, the first American instrument in 1900 in a Jesuit college in Cleveland, Ohio. And finally, in 1904, the Omori seismograph from Japan was carefully assembled in a special observatory building at the mission college in Zikawei.

This was the device that registered the San Francisco shock waves—one of four maintained by Jesuits around the world at the time. It was responsible for detecting the impressive set of shocks from which the earthquake's approximate epicenter was determined. (The Shanghai instrument was replaced in 1932 by a newfangled contraption winningly known as a Galitzin-Wilip instrument, and it operated outside the great city until 1949, when the Chinese Communists turfed the Jesuits out.)

The installation of Jesuit seismographs accelerated rapidly after the disaster of San Francisco (with one instrument being installed in Addis Ababa as late as 1957), and for much of the twentieth century in America the Jesuits wielded a considerable influence over the entire seismological community, helping to direct the research that has shaped much modern thinking about the planet's mysterious convulsions. And to this day four large Jesuit institutions—in St. Louis, Bogotá, La Paz, and Manila—continue to observe and conduct research into plate tectonics, keeping the scientific tradition of this unique form of Catholic inquiry very much alive. Most of the small observatories have closed down (though the one at Stonyhurst College still functions, courtesy of the schoolchildren themselves), and the Zikawei machine is, of course, thanks to Maoist suspicions, no more. But their records—the etched traces on smoked-paper drums, the wavy lines of ink on now-yellow graph paper, and, the least scientifically satisfying but the most spectacular of all, ink traces made by a pendulum waving this way and that over a fixed circle of paper—remain.

What was recorded in all of these observatories, Jesuit owned or not, depended on the sophistication of each seismograph. The simple pendulums—like that in the Chabot Observatory in Oakland, just across San Francisco Bay—produced not much more than ragged messes of ink, most of them "too confused to give details." Lawson's report for Oakland presents raw data typical of these primitive instruments: The machine allowed investigators to say little more than that "the movement of the pendulum . . . was in nearly all directions, and more or less irregular, tho' this irregularity was undoubtedly in part due to the pendulum's striking against the side of the case."

But better instruments followed—some in America, some in the Jesuit monasteries, many in Europe. Some of the best were the Bosch-Omori machines in Washington, D.C., at the New York State capital of Albany, and at the U.S. Navy base in Vieques in Puerto Rico. Each of these left a long trace of ink on a time-marked paper roll, and, by poring over the traces later, investigators were able to determine when the first tiny tremors started to be detected; when the secondary set of preliminary tremors began to rattle the needle; when the regular shear waves and pressure waves started to arrive—with the time difference between the two types, as we have seen, enabling the instrument readers to work out how far away the earthquake's originating point was; then when the principal part of the tremor was under way; and finally when the maximum disturbance occurred.

Close modern analyses of the data from all the seismograms show that the earthquake unfolded in a number of stages. All of them are now known to have begun at a point rather less than ten miles south-southwest of where Officer Cook was standing at the eastern end of San Francisco's Washington Street, several miles down in the earth's crust at a site that lies well beneath the Pacific Ocean.

Modern analyses of the information have produced a different set of results from those computed at the time. The very distant seismogram information was not fully appreciated in the months and years that immediately followed the earthquake. It was seen more as a curiosity, as a series of unanticipated confirmations of the event's sheer power. The thinking at the time held that the quake's origins should be deduced from a small number of local recording stations. The more distant records might be impressive, but they were perhaps not truly to be trusted.

Since the recorded time of the shock waves' arrival is crucial to

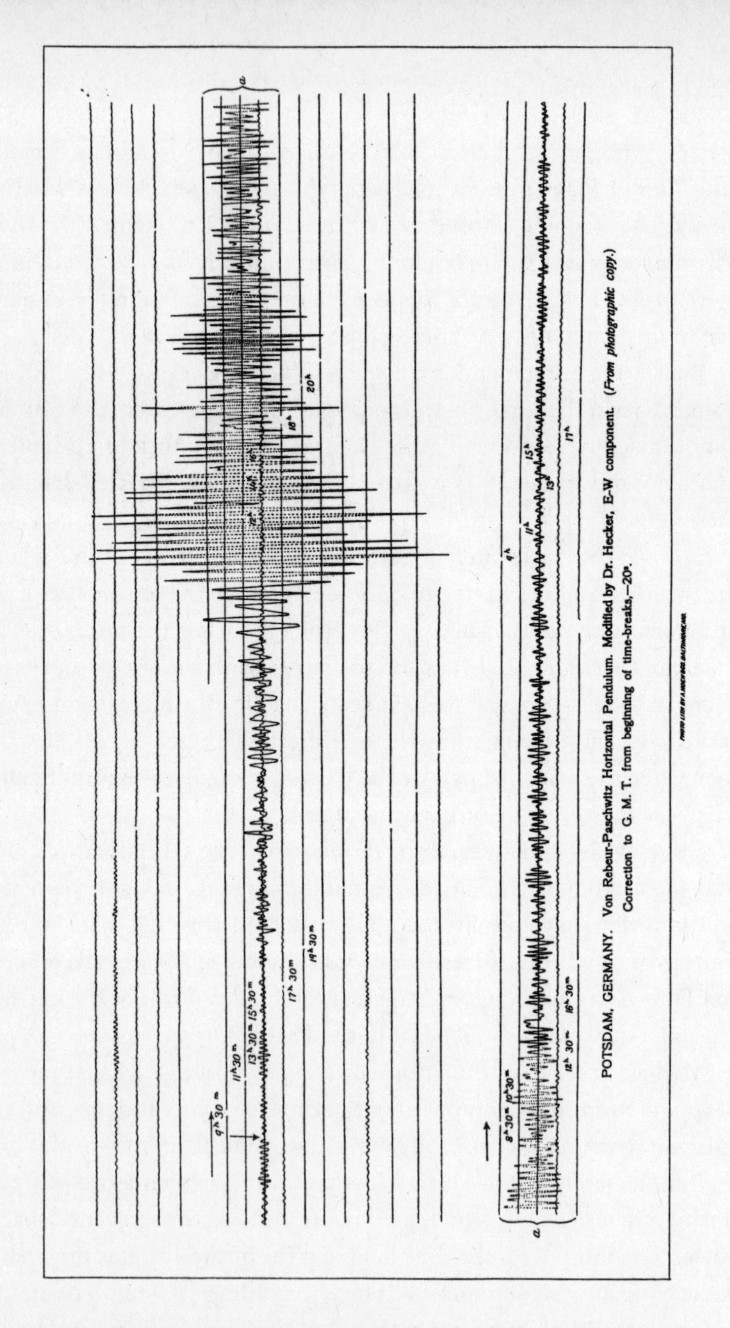

Seismograph traces from the 1906 rupture of the San Andreas Fault near San Francisco display both the enormous nature of the shock and the relatively primitive state of contemporary seismograms.

finding out where the earthquake happened, the Lawson team became obsessively concerned with determining exactly when the event was felt, detected, and recorded in places very close indeed to its supposed point of origin. Buried deep within the commission's great report are interviews galore with those who felt and recorded the two most distinct phenomena of the earthquake: the onset of the first perceptible vibrations, and the arrival of the truly great and destructive shock. Typical of the reports that Lawson collected is one from a Professor T. J. J. See,* the appropriately named director of the (equally appropriately named) Mare Island Naval Observatory on San Pablo Bay:

> I had been sleeping downstairs, lying with my head to an open window, which faced the south, and as the house was not seriously endangered at any time I was favorably situated for making careful observations of the entire disturbance. I had been awake some time before the earthquake began and, as everything was very quiet, easily felt and immediately recognized the beginning of the preliminary tremors.
>
> *The exact time of the phenomenon.* This was found by the stopping of two of the four astronomical clocks at the Observatory. The violent shocks were so extreme that the pendulums were thrown over the ledges which carry the register for measuring the amplitude of the swing. The standard mean time thus automatically recorded was: by the mean time transmitter 5h 12m 36s; by the sidereal clock 5h 12m 35s. The yard clock at the gate, which is simply an office clock, though electrically corrected from the Observatory daily, and therefore approximately correct, gave the time as 5h 12m 33s. The agreement of all these clocks is very good; but I think the best time is the mean of the two astronomical clocks, viz 5h 12m 36s. I estimate that the error of this time will not exceed about 1 second.

---

*Thomas Jefferson Jackson See was a brilliant and notoriously ill-tempered astronomer, a maverick who is generally reckoned to have squandered his boundless potential by opposing Einstein's theory of relativity and espousing the so-called wave theory of gravitation.

Thus were the rigorous minds of science applied to this one small problem of when the quake occurred—because it was realized that by knowing *when,* they could work out *where.* So an obsessive, super-pedantic dedication to precision, similar to that of Professor See's, seems to have informed all of Andrew Lawson's witnesses—such that in the end, and without fanfare, he was able to announce his own considered view of when the shocks first began, and when the biggest of all the shocks occurred, and where these two events had their origins.

He crunched all the data from four close-in observing stations—Mare Island, of course; George Davidson's observations at his home in Lafayette Park, San Francisco (an account of which is in the prologue); the Students Astronomical Observatory at Berkeley; and the Lick Observatory at the top of Mount Hamilton near San Jose—and from one rather distant site, the International Latitude Station* in the town of Ukiah, about eighty miles north of the city.

From all of this data he was able to declare, definitively, that the shock began at 11 minutes and 58 seconds after 13 hours, Greenwich Mean Time (GMT)—in other words, two seconds before 5:12 A.M., Pacific Time, plus or minus three seconds. The place where this opening salvo was triggered was computed to be under the sea, just a short distance off the Golden Gate. And the depth of the vibrations' beginnings—all computed from the simple geometries of the observations from these five sites—was about twelve miles, give or take about six.

And then there was the biggest shock—the terrifying "second shock" that so many of the observers could not fail to notice, whether they were rocked softly in their beds far away up in Oregon, or down in Anaheim, or over in Winnemucca, Nevada, or whether they were knocked flat by it in San Francisco, Santa Rosa, or San Jose. According

*This was one of six stations sited precisely on latitude 39°08' N, designed to measure how much the North Pole wobbled around it axis (which would cause latitude to shift as it did so). The other stations were in Misuzawa, Japan; Tschardjui, Russia; Carloforte, Italy; and near the American cities of Cincinnati and Gaithersburg.

to Lawson, it came exactly thirty seconds later, at 13h 12m 28s GMT, or at twenty-eight seconds after 5:12 A.M., Pacific Time. This shock, he went on to say, took place somewhere between the surface and twelve miles down, and at a spot close to the village of Olema, which has long considered itself—wrongly—to have been the earthquake's epicenter (and where the locals, somewhat shiftily, still peddle hats and T-shirts to try to underline the point).

Olema was the place where, from the evidence of broken fences and torn-up roads and ruined barns—it appears that the surface of the earth was ruptured to a greater extent than anywhere else along the entire fault trace—as much as twenty-one feet. This was a pleasing happenstance for the scientists, because they were able to see that their calculations proved—or so they thought—that the most violent shock took place just below the point where the earth itself was most dramatically torn apart.

As it happens, Lawson and his fellow researchers were quite wrong—understandably so, given the paucity of the information they chose to examine. Had they looked closely at the entire seismic record they had available to them, and been inclined to focus on the data they had collected from all the other ninety-six seismographs, then they might have been able to make a more accurate fix. But there was a problem, in that they were not necessarily fully aware of what it was exactly that they were *seeing* in all of these records. In those seismically unsophisticated years some of the records must have seemed little more than runes or hieroglyphs or impenetrable codes. Only recently has the science advanced to the point where the records, combined with the anecdotes and the records of earth movement, can be annealed into a comprehensible whole. And from all this has now emerged a rather better understanding of just what took place beneath the ground on that fateful morning.

Specifically, and as already noted, the place where it all began is now known to have been just a few cables northwest of the tiny guano-crusted island of Mussel Rock, off what were in the early years of the century thousands of acres of empty and open grasslands, with clifftop meadows dotted with dairy cows. The channel between the island and the mainland marks where the San Andreas Fault passes by: The island stands on the Pacific Tectonic Plate, and the mainland on its North American opposite number. The fault that divides them is the line along which the Pacific Plate moves, northward, by as much as thirty-five millimeters, one and a half inches, every year.

And at 5:12 A.M. that day, along that very line, just a few hundred yards off the island and some miles down deep within the zone of crushed rock where the rocks had been grinding together for decades, something suddenly gave.

With a lurch, the pent-up energy of all the years when the Pacific Plate (on which Mussel Rock stood) had been moving northward past the North American Plate (on which the grasslands, meadows, and cows all stood), was instantaneously released, and a foreshock, like a sudden ignition of a pile of explosives, vibrated out and through the rocks, causing the entire area's underneath to pulsate in a huge, sonorous, basso profundo roar. That was what the policeman heard: The announcement of the foreshock, the enormous snapping sound of the rocks at the hypocenter breaking, lurching forward, and suddenly setting themselves free.

Once free, they moved. The pent-up forces that had held the Pacific Plate static for the better part of the previous two centuries were unleashed. The elastic rebound was now free to occur, and it did so in a jolt that some calculate as moving northward and southward at two miles a second. The rocks lying above the Pacific Plate then shifted, some of them by scores of feet. The vibration caused by this motion spread outward like a ship's bow wave, and it set the earth above the north-shifting plate rearing up and outward, rippling and bucking just as Officer Cook had reported. It spread outward in different ways, depending on the nature of the rocks through which it spread: Sometimes it moved slowly and trembled fairly minimally; at other places it shot out with terrifying speed and caused immense rocking and crashing, which triggered immense destruction. Sites like Olema, where the quake seemed to go especially wild, are known as asperities—patches of geological roughness that corrupt the otherwise smooth progress of the outbound shock waves.*

All of a sudden the inches-a-year motion of the plate that had been locked for so long was suddenly translated into something much, much, faster: A ripping, tearing movement that was unimaginably fast, as though a glacier had suddenly superheated and been turned in an instant from solid ice into a raging flume of water. The huge bow wave of shock and motion spread up and down the city, up and down the state, roaring along at 7,000 miles an hour, and affecting ever more northerly and southerly towns and villages as they appeared on each side of the spreading rupture in the surface.

One by one the streets of San Francisco rose and fell under the influence of the spreading wave; and one by one the nearby towns and villages were affected and afflicted too. It was as if a plowshare were being driven through their countryside, with the soil on each side of the blade turned up and over, carrying all before it and tossing it contemptuously to each side.

On the western side of the fault rupture, all of this motion occurred beneath the waters of the Pacific—at least, it did where the fault ran close to the ocean, or ran below it, as it did near San Francisco. On the eastern side, however, the bow wave roared beneath the land, passing through houses and streets, across roads and bridges and railway lines, through fields, farms, and fences, and tearing through hamlets, villages, towns, and cities, without care or interruption, and leaving the most terrible and unforgettable damage and destruction in its wake.

---

*Because the local damage is so severe the aberrant behavior of a place like this can also make it appear, temptingly, as though it *deserves* to be the epicenter. This is the kind of red herring that modern geologists take care to avoid.

And the Walls Came Tumbling Down

> The disturbance lasted 48 seconds. On looking from the window of our hotel, which was badly shattered—two men in fact to right and left of our rooms were killed—I saw the whole City enveloped in a pile of dust caused by falling buildings.
>
> *From a diplomatic telegram sent on April 25, 1906, from* Sir Courtney Bennett, *His Britannic Majesty's Consul-General, San Francisco, to Sir Edward Grey, Bart., &c., &c., Secretary of State for Foreign and Colonial Affairs, London*

Official pronouncements from the governments of the world's great cities are documents seldom marked by excess of any kind, and, being written by bureaucrats, are rarely colored by sentiment, hyperbole, or dash. But the Municipal Report of the City of San Francisco for 1907, published by the city's Board of Supervisors—as the law required it to be, each year—has a power and pathos all its own. Paragraph by paragraph, the book-length report recounts the ruin of the great metropolis in tones of unrelieved and detached misery, culminating in a single haunting sentence: "The greatest destruction of wealth created by human hands was that which resulted from the fire which occurred in San Francisco on April 18, 1906, and the three days succeeding."

Whenever one considers the earthquake that resulted from the rupture of the San Andreas Fault that April morning, one must take into account what followed almost immediately after the trembling of the ground had stopped—and that was a quite astonishingly destructive and long-lived fire. The two calamities have since been almost inextricably linked as the "San Francisco Earthquake and Fire," and the conflation of the two events has, so far as narrative and investigation is concerned, made for both convenience and confusion.

So when fatalities are spoken of, we have to ask: Did they die as a result of the quake, or in the subsequent blaze? Or when we refer to structural damage: Were the buildings destroyed because of the shaking or because of the fire? Many made much of what could have been a rather fine distinction—not least the insurers, some more eager than others to weasel out of their contractual responsibilities, their obligations to pay up. On all too many occasions homeowners would be icily informed that the damage their particular house had sustained had been due to the earthquake—against which, *with profound regret, but do please read the fine print*—they had not been insured. On the other hand the fire against which they *were* financially protected had arrived some hours later, laying waste to what was an already injured and uninsured—and hence suddenly quite valueless—building. And thus, from the insurance point of view, it was all bad news, and such *hard luck*. There would, thanks to this fine distinction, be no payment.

The simplest way to examine the physical damage caused by the ground shaking is to chronicle the destruction suffered by communities where fire was not so great a factor—at Stanford University, for example, where spectacular harm was done to the newly built university,* all of it very evidently wrought by the shaking of the ground, and yet without a single flame to add confusion or misery to the scene. But up in San Francisco it is well nigh impossible—except later, and forensically—to separate the two events. So it is perhaps best to rationalize this by accepting that everything—the earthquake, all the consequent destruction, all the death, and the outbreak of all the all-consuming fires—had as its ultimate cause the movement and the ground rupture from the fault. The earthquake shook the ground, it broke the buildings, it killed and maimed, and it spawned the fire.

IN THOSE FIRST few bewildering, sleep-fuddled, terrifying moments, scores upon scores of buildings, some grand and famed, most ordinary and unsung, crashed to the ground. Buildings that had long supplied familiarity and comfort and scale to the urban environment

*Among the most poignant images of Stanford's ruin was a marble statue of the great biologist Louis Agassiz, which tumbled from a great height, speared its way, headfirst, into the cement courtyard and stuck there, helplessly pinned. There was much greater damage than this; but somehow a figure of great learning thrust so ignominiously to earth struck a chord. It also prompted the university's then-president to remark that he had always thought more easily of Agassiz in the concrete rather than the abstract.

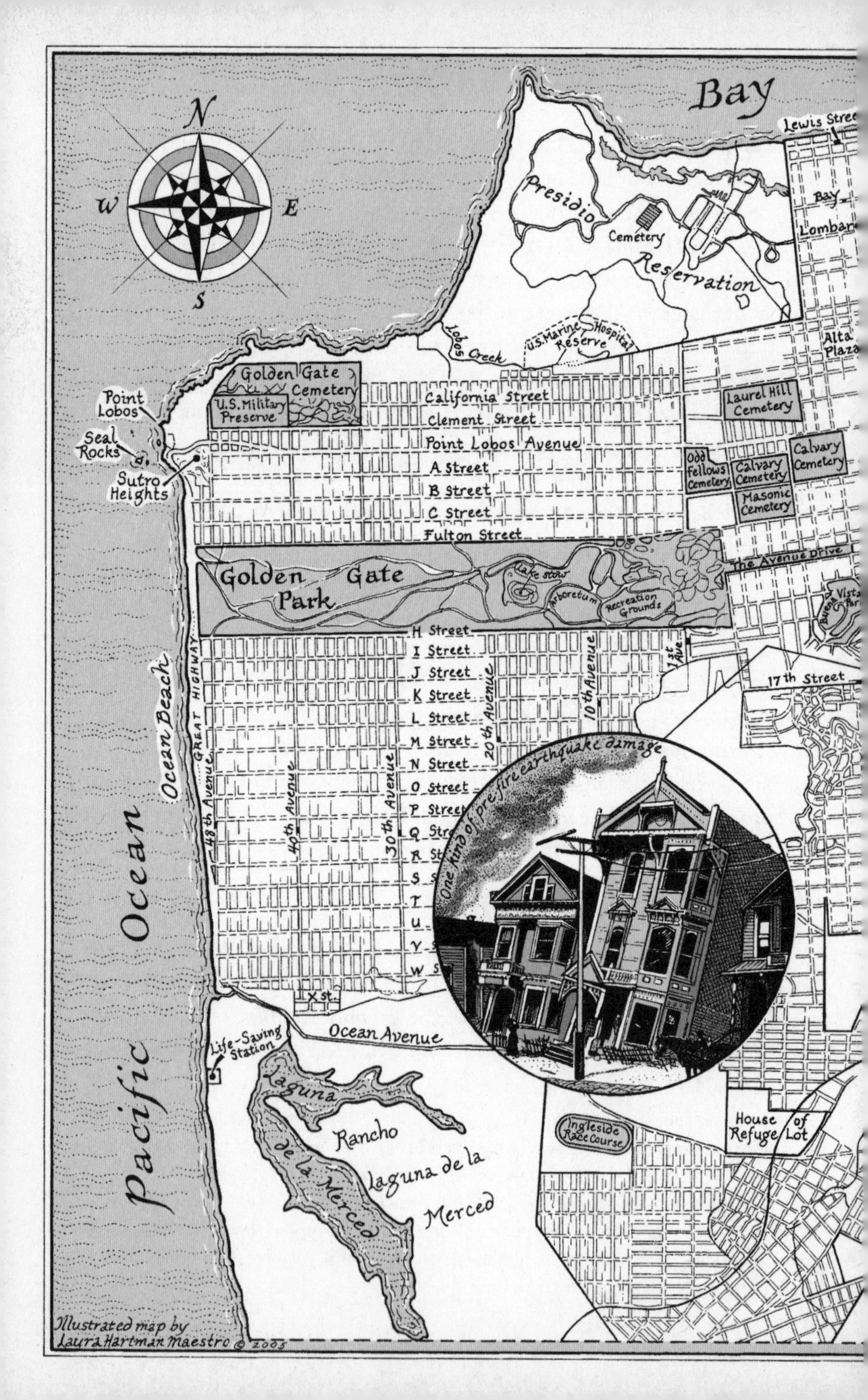

Bay
N
S
E
W
Lewis Street
Bay
Lombard
Presidio
Cemetery
Reservation
Alta Plaza
U.S. Marine Hospital Reserve
Lobos Creek
Golden Gate Cemetery
U.S. Military Preserve
Point Lobos
Seal Rocks
Sutro Heights
California Street
Clement Street
Point Lobos Avenue
A Street
B Street
C Street
Fulton Street
Laurel Hill Cemetery
Odd Fellows Cemetery
Calvary Cemetery
Calvary Cemetery
Masonic Cemetery
The Avenue Drive
Golden Gate Park
Lake Stow
Arboretum
Recreation Grounds
Buena Vista Park
H Street
I Street
J Street
K Street
L Street
M Street
N Street
O Street
P Street
Q
R
S
T
U
V
W
X St.
1st Ave.
10th Avenue
20th Avenue
30th Avenue
40th Avenue
48th Avenue
Great Highway
Ocean Beach
17th Street
One kind of pre-fire earthquake damage
Pacific Ocean
Ocean Avenue
Life-Saving Station
Laguna de la Merced
Rancho Laguna de la Merced
Ingleside Race Course
House of Refuge Lot
Illustrated map by Laura Hartman Maestro © 2005

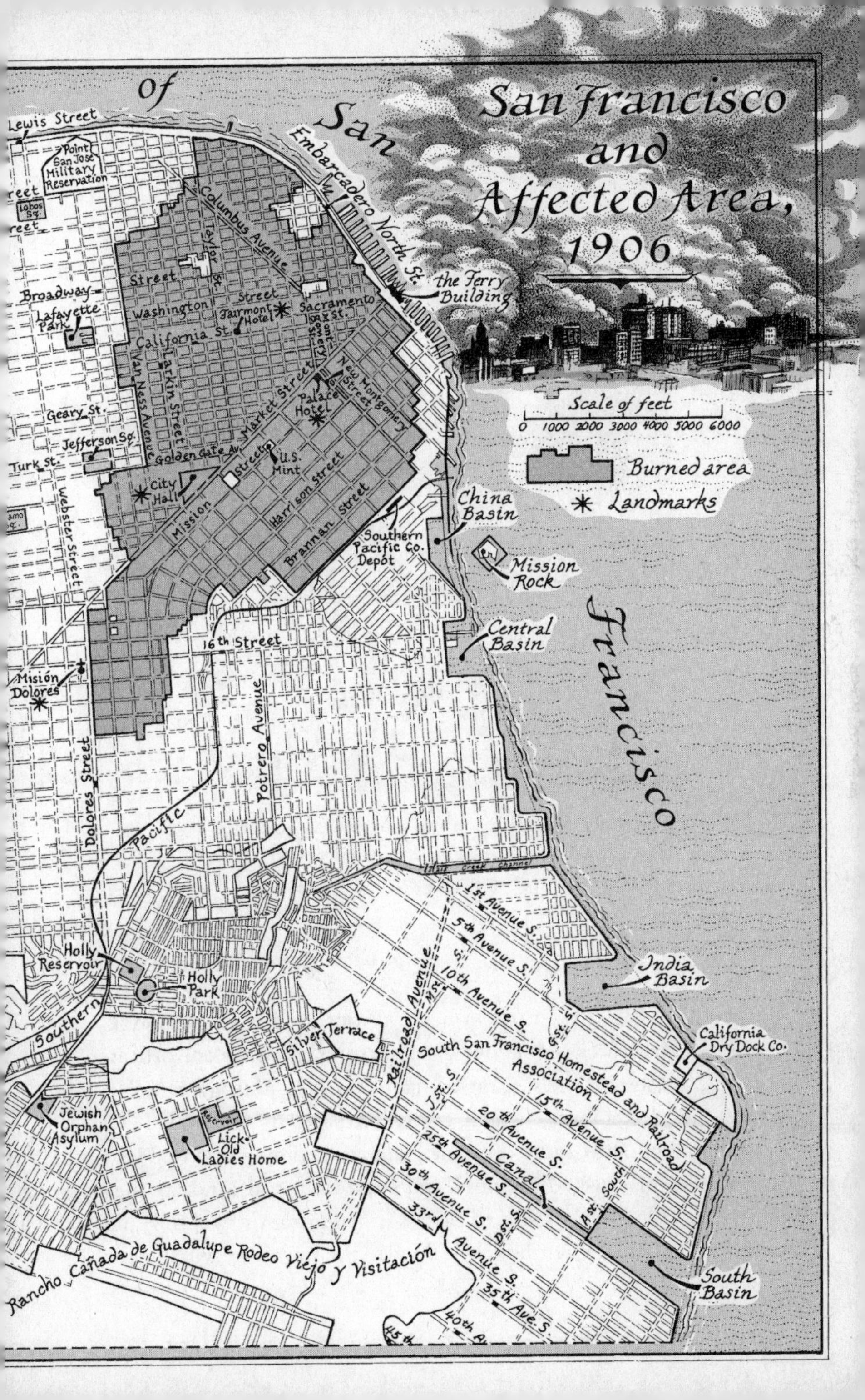
San Francisco and Affected Area, 1906
Scale of feet
0 1000 2000 3000 4000 5000 6000
Burned area
Landmarks
of
San
Francisco
Lewis Street
Point San Jose Military Reservation
Lobos Sq.
Embarcadero North St.
Columbus Avenue
Taylor St.
Street
Washington Street
Broadway
Lafayette Park
Fairmont Hotel
Sacramento St.
Montgomery St.
Battery St.
the Ferry Building
California St.
Van Ness Avenue
Larkin Street
Market Street
New Montgomery Street
Palace Hotel
Geary St.
Jefferson Sq.
Golden Gate Av.
U.S. Mint
Turk St.
City Hall
Mission Street
Harrison Street
Brannan Street
Webster Street
China Basin
Southern Pacific Co. Depot
Mission Rock
Central Basin
16th Street
Misión Dolores
Potrero Avenue
Dolores Street
Pacific
Islais Creek Channel
1st Avenue S.
5th Avenue S.
10th Avenue S.
Railroad Avenue
M St. S.
G St. S.
India Basin
Holly Reservoir
Holly Park
Southern
Silver Terrace
South San Francisco Homestead and Railroad Association
California Dry Dock Co.
J St. S.
15th Avenue S.
20th Avenue S.
Canal
25th Avenue S.
Jewish Orphan Asylum
Reservoir
Lick Old Ladies Home
30th Avenue S.
A St. South
D St. S.
33rd Avenue S.
South Basin
Rancho Cañada de Guadalupe Rodeo Viejo y Visitación
35th Ave. S.
40th A
45th

The shock tumbled Stanford's marble statue of the famous zoologist Louis Agassiz from its position on a parapet down into the cement of the plaza.

cracked, listed, and then fell in heaps of dust and shattered brick. Countless structures—churches, hotels, stores, government offices, libraries, statues, brothels, theatres, mansions, jails, Masonic temples, art galleries, restaurants—which, even if unvisited by most, had been passed by thousands and so became landmarks, simply vanished into jagged piles of crushed masonry and contorted iron. And, as they passed, this most remarkable of cities seemed instantly to shrink and to shrivel. When the morning light flooded the town, one witness was moved to remark that all of a sudden it seemed "no distance, between points formerly too far to walk. Squares thought commodious ... dwindled to insignificant enclosures."

Streets were choked with clouds of impenetrable dust from the falling brickwork, and for a few minutes after the initial shock no one could be certain of anything. As the shaking continued, people generally stayed put—those who tried to move would be shoved back and forth across their rooms, unable to withstand the enormous forces from below. But the moment that there seemed a lull, or perhaps the blessed peace at the end of the shock sequence, they poured out into the streets, stunned into what all who remember that moment insist was an eerie silence. When people did talk to one another, it is said they did so in whispers.

The miasma of brick dust tended at first to muffle the screams of the trapped and injured; and buildings, weakened and tottering, continued to fall, crushing people, horses, cattle, and any other living creatures that happened to be in the open, no matter that the quake itself, in a proper sense, was now over. Later on people expressed their relief that the quake had happened when so many people were still indoors; had it occurred at noon, when crowds thronged the streets, they would have been hit by the falling brickwork, making the casualty figures unbearable. But at the time there were no thoughts of relief—instead the whole world seemed to consist only of the stunned silence of the mob, the dust-laden darkness, the half-heard cries of pain, the crashing of masonry as cornices and chimneys and walls kept on falling, falling, falling.

Police Officer Edmond Parquette,* as we have seen, had experienced the first shocks at the Central Emergency Hospital in the basement of City Hall. Now it was busily collapsing all around him: And twenty years later, the image still vivid, he wrote in a publication called the *Argonaut* about the destruction that was evident in those immedi-

*Or Edmund Parquet—the spellings of victims' and witnesses' names have become hopelessly confused in the years since.

ate postquake moments. The hospital patients, particularly those in the locked insane wards, were not nearly so quiet as the mobs out on the streets:

> Even when the quaking and twisting ceased, the lumps of masonry still kept falling; and above all these noises of crashing and breaking, and the bellowing and thundering of the quake itself and the thuds of the pillars and the cornices as they hit the ground, there were the shrieks and yells of the lunatics, and the moans and cries of the other patients. Everybody seemed to be yelling and shrieking at the top of his voice.
>
> Very quickly after the shocks ceased, the dust began to clear away or settle down, and stopped choking me. The cries died down too, though many of the poor creatures kept on shrieking from terror or moaning from hurts and apprehensions. As soon as these cries and howls had quieted down a little, and I could see a bit through the gloom—the dawn light was able to filter the chinks in the fallen masonry, and so through some of the upper parts of the windows—I made my way as best I could to the room of the matron, Mrs. Rose, whom I found safe and unhurt.

The U.S. Geological Survey, which later sent its best sleuths into the area to take a dispassionate look at the disaster, succumbed, just as the Board of Supervisors had, to uncharacteristic hyperbole. "The whole civilized world stood aghast," their team of scientists wrote, "at the appalling destruction." But they quickly pulled themselves together, and in short order produced a sensible analysis of how the various buildings had fared. Such was the violence of the shock, the report's authors wrote, that

> only structures of first-class design and materials and *honest workmanship** could survive. Flimsy and loosely built structures collapsed like houses of cards under the terrific wrenching and shaking, and many of the structures which withstood the earth-

*My italics. The dismal reputation of the city and its practices had spread even into the dusty distant halls of the Geological Survey.

quake were subjected to a second test in a fire which surpassed all the great conflagrations of recent years. Some of these structures that successfully withstood the first test failed signally under the second, by reason of inadequate fireproofing. A very few withstood both tests successfully.

The shoddy construction of City Hall—it had taken twenty-six years to build, had cost $6 million, and was by far the biggest and grandest structure west of the Mississippi—was dramatically revealed by its spectacular collapse during the quake. Graft, corruption, swagger, and show had all been combined in the city's wrecked official offices: The slow and trying replacement of the building came to symbolize the new attitudes of the leadership of postearthquake San Francisco.

In this regard concrete proved to be king. There were precious few buildings in San Francisco built of this newfangled material, in part precisely because it *was* so newfangled. (The local Bricklayers' Union took a dim view of it, imagining it part of a plot to take work away from its members.) The first high-rise concrete building had been constructed in Cincinnati only three years earlier, and, though the century would later be dominated by the demand for reinforced and prestressed concrete, in 1906 it was all very new, and even if employed in the East, little known in the West. Nonetheless, those buildings that were made of concrete tended to survive well. If there were tie bars and metallic reinforcements in a structure, then it survived also. If the building was on firm ground and its foundations were solidly anchored, then—even though the shaking inside might have been terrifying, with most of the belongings inside the house broken—the structure as a whole quite probably survived, too.

But otherwise San Francisco was a devil's playground. Its countless unreinforced brick buildings, its scores of city blocks crammed solid with gimcrack wooden buildings (which, as a local insurance company put it, had been "hastily built, to sell"), its thousands of structures that had been hurriedly hauled up on landfill (which had been carted into the dozens of streams, rills, and nullahs flowing down from the hard-rock hills of the town, in order to increase the acreage on which developers might build)—all leveled on every side, within microseconds of the shock wave hitting.

Chimneys were the most frequent villain. Most buildings had one—open fires and furnaces and boilers were invariably fueled by wood, or soft and sooty coal—and they were a menace. Tall, unsupported, weathered, ugly, and fragile, these great hollow columns of brick shook, wobbled, and fell in the thousands. The enormous chimney, for example, that loomed above the cable-car powerhouse at the corner of Washington and Mason Streets, where just a few moments before an unseen man had turned a crank and set the cables running, broke and crashed through the building's roof, wrecking it. Time and again this happened—so often that fully 95 percent of all San Francisco's chimneys collapsed, the War Department later said in its own review of the event. The experience of the newly appointed diocesan of Grace Church,* Bishop William Nichols, at 2515 Webster Street, was typical:

> I arose and made a hasty inspection . . . the house had stood solidly, though as developed later one chimney had tossed out so that the top went through our roof, and the center through the Beaver's roof next door; the two other chimney tops were moved from the base, and several courses of bricks were thrown out of the front gable, caroming on the stone parapet outside Clare's room and going through the open window into her room, fortunately all choosing that open entrance rather than the two closed windows beside it. . . .
>
> A look out of the windows brought evidence of the disasters abroad, in streets littered with fallen bricks, tall chimney stacks toppled over, the streets ominously astir with refugees from houses, and a general sort of anxiety in the air. And when I went to take my bath and no water ran, another phase of what had happened dawned on me.

His Eminence got off fairly lightly. Falling chimneys killed innumerable sleeping men and women that morning, most infamously (and, in a practical sense, most inconveniently) the city's fire chief, Dennis Sullivan. It will be remembered that he had presided over the fighting of two smallish fires in the wee hours of the morning, and at about 3:00 A.M. had gone back to his wife, Margaret, and his bed in his small third-floor apartment on Bush Street between Grant and Kearny Streets, which housed on the first two floors the city's Chemical Company No. 3.

When the shock wave hit, the fifty-four-year-old Sullivan was out of bed in a flash, and he ran to his wife's room, presumably to help her to safety. But then the enormous mass of towering chimneys that dominated the turreted roof of the California Hotel next door toppled and hurtled down sixty feet on top of the little fire station—covering the

---

*Among the early rectors of this Gold Rush relic was the Reverend James Bush, whose great- and then great-great-grandsons went on to be American presidents.

Sullivans with bricks and smashing them and everything else down through to the ground floor. Margaret Sullivan survived, little injured; but the fire chief, his skull fractured, his lungs punctured by broken ribs, and, worst of all, his entire body scalded by steam from a broken radiator on which he landed, was fatally hurt, and died four days later in the Southern Pacific Railroad Hospital. A memorial plaque still stands on Bush Street; and, at a ceremony held at dawn every April 18, the Fire Department dips its ladders respectfully to the memory of this thoroughly agreeable man. "Dead on the Field of Honor" says the plaque, subscribed by grateful Californians. "Her Hero and Her Son."

In the rough-and-ready residential quarters south of the Market Street Slot, the flimsy clapboard row houses slumped and lurched drunkenly as if they were toys tossed carelessly around by a child in a tantrum. Outwardly, many seemed little harmed—just dramatically moved, or tilted to one side, hopelessly out of plumb, or shifted wholesale from one side of a street to the other. But in fact inside they were wrecked, their floors torn and broken, their backs fractured, their residents trapped under beams and spars. And besides, it would not be long before the fire would get them, and turn recognizable ruin into charred and blackened wreckage.

North of Market Street, where the city's commercial buildings were concentrated, brickwork by the thousands of tons was hurled into the street, as building after unreinforced building was broken and brought crashing down.* City Hall, the grandest structure in the city, dominated the ruin, the sides of its tower torn away, its cupola precariously balanced on top of a spiderweb of bent iron, its elaborate arrays of columns, lintels, cornices and caryatids cracked and smashed and flung into the street. The Geological Survey report on its state was almost contemptuous, noting a basis of construction somewhat less

---

*The task of clearing away these bricks brought out a curious assortment of bedfellows. John Barrymore was found beside one ruined building, stacking some in piles under a soldier's orders. A friend promptly remarked of the notoriously idle actor that "it took a convulsion of Nature to get Jack out of bed, and the United States Army to get him to work."

noble than that of the Sistine Chapel. The grandest-looking of San Francisco's monuments was revealed in fact to be no more than "a brick building . . . consisting of steel floor beams and corrugated-iron arches with cinder-concrete filling" that was (*deservedly,* one can hear the report's authors implying) first wrecked, then gutted:

> A prominent feature was a central tower, surmounted by a dome built over a structural steel skeleton. Grouped around the dome were a number of cast-iron columns of half-inch metal filled with brick concrete supported on brackets. Some of these columns in falling broke into small pieces. The brickwork was completely shaken from the central tower. The cement-plastered brick walls were laid in lime mortar of generally poor quality and without adequate tie to the steel work. In some places there was an absence of any mortar, but in others it was very good, the brick walls falling in large masses and the broken bricks showing the mortar to have been the stronger. The massive architectural ornaments were top-heavy and lacked adequate bracing. The ceiling was formed of corrugated metal against which the mortar plaster was pressed, with intermediate brick partitions where the span of the beams was too great. The expansion . . . caused . . . the arches to fail. The building was a monument of bad design and poor materials and workmanship.

The rest of the report offers a grim catalog of structural failures across the city. The Academy of Sciences building was "completely destroyed." The Aetna Building, occupied by a hardware wholesaling firm, stored tons of tin on an upper floor, and it all came crashing down to the basement. Parts of the Bullock and Jones Building were "haunched" and "badly spalled"; the ironwork had "buckled." The frames of the Jackson Brewing Company's plant, which was being built at the time, were held together by "an insufficient number of bolts," fell apart and left the plant completely wrecked. The Hall of Justice was entirely ruined, the skeleton of its iron dome dangling horizontally over the wreckage.

In several buildings huge iron safes—the most popular being those of the locally based Hermann Safe Company, which operated until the

1980s—fell through floors, gathered momentum, and went rocketing downward, smashing huge holes in the structures as they did so.* The Majestic Theater, "particularly bad in design," was practically demolished; the Post Office, "badly racked"; the Rialto Building, "badly racked . . . cracked . . . brick facings thrown off." The Wells Fargo Building, despite being grotesquely twisted out of true, its terra-cotta spalled, its marble wainscoting thrown down and its cast-iron stairways considerably damaged, somehow stood, six storys tall and defiant.

Some buildings were very lucky indeed. The infamous Monkey Block of low-rise stores and bars—the hangout of many of the city's best-known writers and more amusing scoundrels—got away unscathed. There was much amused perplexity at the survival of A. P. Hotaling's whiskey warehouse on Jackson Street, down by the waterside—the amusement derived from the evident caprices of a deity who permitted so plainly immoral a business to keep on running, while the city itself was punished for its self-evident sinfulness.† And the U.S. Mint, which had $300 million in specie in its vaults, survived—being both well built (its nickname was the Granite Lady) and provided with a well, which provided water when it was most needed. (The building, its massive blocks fire singed, remains standing today, on the 500 block of Mission Street. But it is empty and unused, because the

*The Geological Survey report noted that fully three-quarters of the city's iron safes failed to protect their contents from the subsequent fires. Most tragically of all, merchants often opened safes that were still red hot, with costly consequences: The moment the superheated insides were exposed to fresh air and oxygen, the paper contents—Treasury bills and negotiable securities—burst into flames, bringing sudden financial ruin to owners who had been optimistic but who now looked helplessly on at the blaze. After this, prudent merchants waited for as much as two weeks before allowing safecrackers to free what they hoped would be their now properly cooled treasures.

†It led a local poetaster to compose what has become a famous stanza: "If, as some say, God spanked the town / For being much too frisky. / Why did he burn the churches down / And save Hotaling's whiskey?" But A. P. Hotaling himself did not get off entirely: His house in Pacific Heights, near the junction of California and Franklin Streets, was dynamited as a firebreak.

city of San Francisco cannot afford to bring it up to today's earthquake-proof standards—an irony that historians of the 1906 events will find more than a little droll.)

THE THOUSANDS of black-and-white photographs of the event's aftermath, both from San Francisco and from communities well beyond, have a sad eloquence all of their own. This, after all, was the world's first major natural disaster to have been extensively photographed: It was the seismic equivalent of the Civil War, a tragedy made more poignant and immediate because the whole world was able to share its images within moments—more easily once the city's broken telegraph links with the outside had been repaired*—of their being taken.

The photographic community did not get off scot-free. Many studios were smashed, cameras ruined and photographers hurt—not the least of them being one who, for entirely understandable reasons, was actually *not* involved in capturing images of this particular event. Ansel Adams broke his nose that morning. He was four years old.

The Adams family lived in the far west of the city, in a house set down quietly among the sand dunes. His experiences of the quake remained with him all his life—the waking, the shaking, a vast noise, chimneys falling through the greenhouse roof, an eerie silence except for the sea, which was suddenly louder because all the windows had broken. Kong, the Chinese cook, tried to start a fire in the kitchen stove, but was ordered out into the garden with the rest of the family, where he prepared the food. Ansel was in the charge of his nanny, Nelly. He remembered enjoying the garden, waiting for aftershocks, and learning how to interpret the signs of their imminent arrival.

And then, just as Kong had finished cooking and there came the

*All internal communications links, and most external ones too, had been broken by the earthquake. Local telephone exchanges were hastily abandoned. Only one secure military telephone circuit, two commercial telegraph lines that went across the Bay to Oakland, and the trans-Pacific cable to Manila were working after the shock waves hit.

call to breakfast, Ansel was running toward the table when there came a violent shock: "I tumbled against a low brick garden wall, my nose making violent contact, with quite a bloody effect." The doctors advised against fixing it, and it never was. His famously chiseled profile was created in that instant.

In considering the damage small and great, from the nasal septums of artists to the pilasters of City Hall, one sobering reality is often overlooked: The direct and immediate effects of the earthquake itself, while great in absolute terms, were, in relation to the disaster as a whole, relatively small. It was estimated by the federal government at the time that only between 3 and 10 percent of the damage done to San Francisco was directly attributable to the earthquake. It was the subsequent, indirect effects—a distinction of which the insurance companies made much, as we shall see—that brought about the greatest loss by far.

For it was these subsequent and indirect effects that caused the bulk of the destruction of half a billion dollars' worth of property,* killed at the very least 600 people (some estimates today put the figure as high as 3,000, and there are accusations that the casualty figure was massaged downward to ensure San Francisco enjoyed a good reputation for investors) and rendered more than 200,000 homeless, many for months and years to come.

And the first of these subsequent and indirect effects came within

*Many will ask: What is this worth in today's money? It is a complex question, since countless factors are at play that alter the real and perceived value of money over time. Economists have, however, developed a number of algorithms that can assign value: and so, for example, the $500 million in 1906 dollars is said in terms of the Consumer Price Index to be worth $10 billion today; in terms of the nominal Gross Domestic Product, $8 billion; when measured by comparing the price of unskilled labor now and then, $45 billion; when rated by GDP per capita, $57 billion; and when measured by comparing the total values of GDP, $195 billion. So the modern value of the destruction can be said to range across two orders of magnitude—almost as much a variation as the suggested magnitude of the earthquake itself.

The proboscis that made Ansel Adams so distinguished in profile was born in the San Francisco Earthquake: He broke his nose when he was thrown to the ground during one of the tremors.

seconds of the first vibrations, when, all over the afflicted area, the cast-iron water pipes broke.

The three-foot main that brought water in from the reservoir lakes, which, somewhat ironically, lay in the very valleys that had been created by the fault line to the south of town, all fractured and split; and those few that were carried on trestle bridges across streambeds and other small and inconvenient valleys ruptured when the bridge supports gave way. Then the smaller-diameter water pipes that routed this same water through the maze of streets inside the city limits broke also, turning thousands of water taps—and all of the hundreds of cast-iron street-corner hydrants—suddenly and creakingly dry.

At the very moment this was happening, and as the breakages were suddenly denying water to those who had no thought that they might need it, the second wave of subsequent and indirect effects of the earthquake hit. All across the city utility poles broke and toppled, and the high-tension electrical wires they were carrying fell to the ground—where they writhed and hissed "like reptiles"—then crossed and arced. Gas pipes broke, chimneys fell and spilled hot coals onto hundreds of wooden floors, cooking ranges tipped over, fuel tanks ruptured and spilled thousands of gallons of oil, gasoline, and kerosene down stairs and into basements and toward the arcing wires—and in a matter of moments, in locations from one end of the city to the other, the region erupted in fire.

By this point the earthquake's shaking had ceased, and the main-shocks had wrought their immediate damage. There would be after-shocks, which would collapse still more weakened structures and further induce fear in the terrified and demoralized citizenry. Now the longer-term side effects of the earthquake, of which the fire was by far the most infamous and the most lethal, would begin to compound the misery of the moment—and over the next three days they would compound it by what would seem a thousandfold.

The Stricken City

> And within five seconds of the end of the shake twenty or thirty fires broke out all over the lower parts of the city. On leaving the hotel I found that the earthquake had smashed the water mains in the upper part of the city and that therefore the splendidly equipped and drilled fire brigade of San Francisco was utterly powerless. There was however no fire near the Office, and it appeared to be in no immediate danger of burning, but it was badly cracked. A little later, however, a fresh fire started in the rear of the Office and in a few moments the Consulate-General

> was a mass of flames and is now a wilderness of brick and crumpled iron girders. Just as the building caught Mr. Chambers, the Chief Clerk, arrived on the scene and rescued a few papers, ships' articles, Registers, Fee Book, &c. These he turned over to me and they were placed in the safe of my Hotel which subsequently burned down. What is their fate cannot yet be known. With this possible exception nothing was saved from the wreck. The loss of furniture &c. was mainly covered by insurance. I have made arrangements for guarding the safes, &c., which contain many confidential documents. I am without fee stamps, official seals &c . . . I now have temporary offices in Oakland.
>
> *From a diplomatic telegram sent on April 25, 1906, from*
> SIR COURTNEY BENNETT, *His Britannic Majesty's Consul-General, San Francisco, to Sir Edward Grey, Bart., &c., &c., Secretary of State for Foreign and Colonial Affairs, London*

"There is no such thing as a fireproof building," declared the U.S. Geological Survey's *Report on the Effects on Structures and Structural Materials,* which exonerated the firefighters of all blame for what happened as the San Francisco fires raged, at first wholly unchecked, for the next three days. For, though the conditions in the city after the quake were, in the survey's words, "highly unusual"—with, thanks to the broken pipes, no readily available water for the firefighters to use—the situation was made impossible because of the sheer number of fires that erupted all across the city almost simultaneously. No fire department anywhere in America, or probably anywhere in the world, could have possibly dealt properly with this conflagration, had they all the water that they could use. The 1906 fire was essentially uncontrollable, somewhat akin to the firestorms that ruined Dresden and Tokyo, and that raged in the aftermath of the atomic bombing of Hiroshima. The San Francisco fire burned across almost 2,600 acres, utterly destroyed 490 city blocks, and brought America's greatest West Coast city almost to its knees.

It began quickly, within seconds of the arrival of the shock waves. It may seem presumptuous and more than a little risky to turn to fiction for an illustration of what occurred, but there is a scene in the climactic moments of *San Francisco,* W. S. Van Dyke's well-regarded 1936 film, that manages to distill what seems to have happened into one unforgettable and precisely observed moment.

In the film the earthquake has just taken place, surviving after-theater party guests have all run screaming into the streets, and a bewildered, bloodied, and distraught Blackie Norton (Clark Gable) is searching for Mary Blake (Jeanette MacDonald)—he suspects that she has died in the ruins. He is clawing back bricks from where he supposes she is buried. For a moment there is quiet, save for a haunting scene of a nameless man running by, screaming out, "Irene! Irene!"

And then there comes an aftershock. People shout a warning. Blackie steps back from his frantic clawing—just in time to avoid being crushed by yet another fusillade of falling bricks, but not in time to avoid being swallowed by a deep chasm that suddenly opens up in the roadway behind him, and in which one sees, vaguely, pipes being broken and water cascading into the fathomless pit.

Blackie—being Clark Gable—is naturally pulled to safety, to act out the film's climax. But, the moment he reaches the surface, another pipe breaks, this time gushing town gas—and, as it snaps, a pair of falling power lines cross, there is a cannonade of sparks, the gas ignites, and a huge fountain of orange flame hurtles up into the air, scorching the timbers of a still-standing building like a flamethrower. From that moment on, Blackie Norton's search continues against a background of unending inferno, with flames, smoke, ashes, and sparks in every scene, making the broken city resemble nothing more than a hellish vision of war.

Many of the early fires had names when they were born. There was the Chinese Laundry Fire, the San Francisco Gas and Electric Fire, the Girard House Fire. And, most infamous of all, the Hayes Valley Fire—better known locally as the Ham and Eggs Fire, if only because of the seeming need for storytellers everywhere to bring human scale to some almighty tragedies, to catastrophes too large to comprehend. It helped that the Great Fire of London of 1666 was comfortably begun in a place called Pudding Lane. The Chicago Fire of 1871 was supposedly started by Mrs. O'Leary's cow. The Baltimore Fire of 1904 began when someone tossed a lighted cigar down through a grate, and it was stopped most agreeably down beside a waterfall. And here in San Francisco the Hayes Valley Fire was begun by a woman at 95 Hayes Street who was cooking herself breakfast in a building with a badly broken chimney (or so it is said—there is only circumstantial evidence, and not very much of that). The cozily domestic name allows the fire to be remembered today almost with affection, though it was a particularly vicious blaze, among the very worst.

Over the city the wind was westerly, and, as the sun came up, it began to freshen. And the fires, some of them big and named, others small and of seeming insignificance (at least fifty were identified as having broken out in those first few hours), took hold and began to join hands with one another. And as these isolated fires became larger and larger, and joined one another to grow larger still, they began to suck more and more oxygen from the atmosphere and to create winds of their own, eddies of superheated air that sucked ever more of the islands of fire together. By midday there was a wall of flame a mile and half long to the south of Market Street, and the wall of smoke rose at least two miles up into the sky, visible across all the counties of the bay and horrifying thousands.

An early map purporting to show the spread of the blazes looks like a diagram of the Maelstrom, with a turmoil of fire engulfing almost everything in the bustling eastern quarters of the city. In the slums south of Market Street, where whole wooden buildings had slithered out onto the streets, the fire that would soon present itself to that mile-and-a-half-long front is shown roaring its way eastward, which would be expected, as the wind was westerly. Residents, terrified and stunned by the collapse of their homes and hotels, stood helpless out in the streets. Then, as the fires came, they scattered and began to flee, not knowing which way to run to avoid being trapped by a wall of flame. A writer named Charles Caldwell, a man clearly unaccustomed to seeing the poor up close, described them in unkindly fashion as looking "like rats startled out of their holes, this beer-sodden, frowzy crew of dreadful men and even more dreadful women. A breed that has passed out of American life completely—red-faced, bloated, blowzy."

Jack London, as vivid a writer as they come, and a Sonoma Valley neighbor, described matters for *Collier's* a month later:

> By Wednesday afternoon, inside of twelve hours, half the heart of the city was gone. At that time I watched the vast conflagration from out on the Bay. It was dead calm. Not a flicker of wind stirred. Yet from every side wind was pouring in upon the city: East, west, north and south, strong winds were blowing upon the doomed city. The heated air rising made an enormous suck. Thus did the fire of itself build an enormous chimney through the atmosphere. Day and night the dead calm continued, and yet, near to the flames, the wind was often half a gale, so mighty was the suck.

Immediately north of Market Street the blazes coalesced and spread eastward; around City Hall they formed vortices that raged in all directions; on Russian Hill they stormed westward, directly under the influence of the prevailing wind of the day; and in North Beach, Chinatown, Telegraph Hill, and along the Embarcadero they seemed to burn back inshore again. It was all very confusing for both the residents and the firefighters. Time and again, since almost every one of the hydrants proved to be dry, the firemen could only look on impotently and suffer the jeers of the crowds, which at first just could not understand why nothing was being done to contain the inferno.

Most dreadful, though, was the plight of those who were pinned and trapped in the wreckage and thus unable to get out of the path of the inexorably advancing flames. Stories abound of such victims being put out of their misery by passersby with guns, policemen especially. The coroner's reports, however, remark on the almost total lack of corpses with gunshot wounds, casting a certain skeptical shadow over these suggestions. It is generally accepted, though, that bodies consumed by fires that often reached more than 2,000 degrees would barely be recognizable as bodies at all.

The four-year-old Ansel Adams, bruised but unburned near the Golden Gate, knew from the family cook, Kong, that Chinatown had been particularly badly hit. This dignified old man had gone off to look for his family moments after preparing breakfast for his employers, and probably after helping to nurse young Ansel's newly broken nose. Hours later, bone weary, he returned to the relative peace of the great house among the dunes. "He had found no one, and fire was everywhere," Adams later wrote. "He never discovered what happened to his family." Probably they died; what happened to other Chinese families in the aftermath of the fire was to have, as we shall see, considerable consequences for American society.

ONCE AGAIN it is the pictures that offer most poignantly the portrait of a city being rapidly burned to death. There are thousands of them, made as amateurs and professionals alike—for simple cameras were inexpensive and photography a growing hobby—snapped at the blazes, turning tragedy into spectacle, and in doing so perhaps minimizing, at least for themselves, its impact at the time.

And the odd thing that many noticed was that a large number of those watching the fires rage (and perhaps taking pictures) seemed to be not so much perturbed by the unfolding events—unless they had suffered an irreversible personal loss—as they were awed by them. The esteemed Harvard psychologist William James was visiting Stanford University and very much hoping—professionally speaking—for an earthquake; he was delighted to be caught up in one.* He wrote later that people did indeed remark on how "awful" and "dreadful" the event was; but they were nonetheless full of some kind of wonder at being able to be part of so majestic a catastrophe. They eagerly watched it, James noted; they took pictures of it; and they thought themselves lucky to be enfolded in an event of truly historic significance.

Four years later James went on to write the famous paper "On Some Mental Effects of the Earthquake" and claimed, among other things, that, by taking photographs of the realities of the event, the city residents were undertaking what nowadays are called "coping strategies,"

---

*So delighted, in fact, that he caught a train directly into the city that very day—luckily making the only one that ran.

placing their agonies in an easier perspective, by way of the recorded reactions of light on silver nitrate–covered paper. There is one image in particular that captures this notion to perfection: A pair of young women, pretty, behatted, and beaming, pose to have their portrait taken while the sky behind them is a pall of rising smoke. They might as well have been standing in front of Half Dome, or under the Eiffel Tower, or on the Great Wall: Their city was being destroyed, yet the catastrophe—at least for that brief moment—was for them little more than *background.*

Some of the pictures are on an epic scale—most of them panoramas taken from rooftops, some managing to make the tragedy look both gargantuan and human, awe-inspiring and prosaic. One taken from the roof of the St. Francis Hotel on Union Square, looking northeastward along Market Street, toward downtown and the docks, shows immense billows of smoke rising like a huge volcano, enveloping everything in the middle distance. But then, close up, one can see the billboards of merchants—billboards that the day before might have read as jaunty and proud proclamations, but that now, since their owners are about to be burned out, seem no more than bitterly pathetic.

THE BIGGEST FURNITURE HOUSE ON THE PACIFIC COAST announces one huge banner on top of a store shortly to be consumed, and housing, no doubt, chairs and tables and overstuffed sofas that would provide a most agreeable fuel. LENGFELD'S PHARMACY proclaims another, on top of a four-story building right on the square on Stockton Street. And then, most wretched of all, there is the wall of a thus-far-untouched building on Powell Street, off to the right of the panorama, with an immense painted notice announcing that GOLDSTEIN AND CO., THEATRICAL COSTUMERS, NOW AT 733 MARKET STREET, WILL OCCUPY THIS ENTIRE BUILDING MARCH 1ST. Needless to say, they didn't, or for not that long. Neither 733 Market nor this building on Powell would survive the fire that swept in later on the day that the picture was taken.

(But it is worth noting, according to the San Francisco *Crocker-Langley Directory* of the following year, that both businesses did survive, with Goldstein moving to new premises on Van Ness Avenue, and Lengfeld's* to Fillmore Street, both well to the west, and away from the charring and the ruin. The California Pacific States Telephone Company announced that the exchanges for many of the new businesses would be called Temporary: One asked the operator for TEmporary 839, for example, to reach the Pinkerton's Patrol Center, TEmporary 3780 for the Union Lithograph Company.)

Perhaps the most famous of all the panoramic photographs of the event was that taken in the middle of Wednesday morning by Arnold Genthe, and titled simply *San Francisco. April 18th, 1906.*

Genthe was the German-born draftsman-artist (though he had a Ph.D. in philology) who was later to make his name with the 1908 book *Pictures of Old Chinatown.* He had taken these photographs much earlier in the century, out of frustration at being endlessly forbidden from sketching the mysterious Celestials, as they were regarded, who inhabited the nine square blocks west of Portsmouth Square. He had abandoned the all-too-visible pursuit of sketching, and instead bought himself a tiny camera with a decent Zeiss lens; like Cartier-Bresson half a century later, he took his pictures quietly, fading himself into the background.

Unlike Cartier-Bresson, however, Genthe often retouched his pictures, to remove such things as might dilute the Chinese-ness of the images: All English-language billboards and electrical wires had to go, for instance. The pictures were published after the fire's near-total destruction of the ghetto: They have an elegiac quality to them, and the portrait they offer of the vanished world of Tangrenbu—of men with pigtails and old women with bound feet and the rows of trussed ducks and strange market fare and monstrous Manchu guards on sentry duty outside the dozens of opium dens—is intensely romantic, offering a shot of Oriental *galangal* to viewers in today's vanilla America.

The earthquake shook Genthe awake, as it did almost everyone

---

*Famous as "Purveyors of Parisian Skin Lotion, Carson's Cough Cure, Opera Cream and—Free from Anything which can Possibly Injure the Skin—Opera Face Powder."

else in the city. He was something of a swell—a well-born Berliner, the son of a classicist—and had finely decorated bachelor quarters on Sacramento Street, as well as a Japanese manservant. (Genthe never married.) In the first few moments after the earthquake had tumbled him from his bed and broken most of his impeccable collection of Chinese porcelain, and after Hamada, his imperturbable and admirably foresighted Jeeves, had announced he was heading off to stock up with food, he pondered just what might be the most suitable attire for attending so astonishing an event. He settled on khaki riding clothes and, after checking on his studio (it was ruined, though a sixteenth-century Buddha had fallen and was sitting in the midst of the wreckage looking "serene and indifferent of fate"), headed downhill with a pair of friends to see if they all might get breakfast at the St. Francis Hotel. It was here he supposedly met the fur-coat-and-pajama-clad Caruso and heard him make his "'ell of a town!" remark. There was hot coffee, and a simple cold breakfast was being offered free, so long as the food held out.

Then he decided he should be taking pictures—except that he swiftly realized he had no camera. So he went to his dealer, a man named Kahn on Montgomery Street, and asked to borrow one. Kahn was only too well aware of the fires licking hungrily toward him, so told Genthe to take anything he wanted—anyway, it would all be molten scrap in a few hours at best. And so Genthe took a 3A Kodak Special, hurried off up the hills that looked down on the city-center destruction, and began to work. Later he wrote of the one picture taken from the upper end of Sacramento Street, close to where his house would soon be consumed by fire. He was peculiarly fond of it:

> There is particularly the one scene that I recorded the first morning of the first day of the fire (on Sacramento Street, looking toward the Bay) which shows, in a pictorially effective composition, the results of the earthquake, the beginning of the fire and the attitude of the people. On the right is a house, the front of which had collapsed onto the street. The occupants are sitting on chairs calmly watching the approach of the fire. Groups of people are standing in the street, motionless, gazing at the clouds of smoke. It

> is hard to believe that such a scene actually occurred in the way the photograph represents it. Several people upon seeing it have exclaimed, "Oh, is that a still from a Cecil DeMille picture?" To which the answer has been: "No, the director of this scene was the Lord himself."*

Genthe's picture is interesting on many levels—not the least being the haunting, captivating contrast between the calm voyeurs occupying the foreground, and the terrible and spreading calamity in the back. But it also offers evidence of the peculiar behavior of the fire, which is here known to be heading west in spite of the morning's fresh westerly wind. In the picture the *smoke* from the blaze is actually shown being blown away, wafting *eastward* toward the Bay, as it should be under the wind's influence. The fire, on the other hand, is creeping inexorably *westward,* toward the watchers and toward the place where Genthe is standing—in fact by late afternoon it would reach halfway up the hill.

The picture thus illustrates the often unrealized truth: that fire in cities is not necessarily blown by wind but often spreads, because of its ferocious concentrations of heat, anywhere nearby that has combustible material. In the specific case of this photograph two things can be said: First, there is nothing further to burn in the direction that the smoke is blowing, because that is where the city ends and the Bay begins; and second, the hungry fire will find a veritable banquet of delights in the direction of the photographer—domed buildings, wood-framed houses, wall-less rooms bulging with sofas and pianos and pictures and beds, barrels of oils and bottles of liquor and overturned wooden drays, all of them unprotected and ready to lure the flames westward. So this was the way the fire chose to spread, notwithstand-

---

*Since DeMille did not begin to make films until eight years after the earthquake, this may well have been an apocryphal story. But Genthe was very much involved in Hollywood in later years, and the dark and brooding publicity portraits of Greta Garbo that he made in 1925 are widely credited with jump-starting the career of the newly arrived actress.

ing the direction of the winds. In forest fires, winds counts for much in the way a fire spreads; but in cities a host of other factors comes into play, and this is assuredly what happened in San Francisco.

The fires lasted for three days. They were sometimes beaten back and checked at their perimeters, but rarely extinguished by the men trying to fight them. They eventually burned themselves out on Saturday, and only after everything flammable had been consumed. They were prevented from spreading still farther by the efforts of the soldiers who had been called in to create firebreaks with dynamite; by the efforts of firefighters who had made some saltwater hydrants work; and by others who discovered the very few freshwater hydrants whose supply pipes had not fractured. And then, at the end of the week, the weather turned damp and cold, as it so often does in spring in San

No picture distills the approaching terror of the San Francisco fires better than this one, carefully composed and shot with a borrowed Kodak camera by the scrupulous German philologist-photographer Arnold Genthe.

Wagons—carrying refugees, belongings, or goods to trade—appear on Market Street soon after the fires are out, as life starts to return against a panorama of devastation. By now the insurance companies are at work, trying to determine which ruins were caused by fire and which by the much-less-insurable act of God. Arguments over niceties of timing raged for years, with some companies—most notably German-based firms—notoriously eager not to pay.

Francisco, and then it began to rain, and the fires were slowly snuffed out despite the mud of black ash everywhere and the dreary appearance of thousands upon thousands of gutted and ruined buildings, and despite the dripping misery of hundreds of thousands of refugees in their tent cities and out on the grasses of the parks. At last the city could begin to think again of what it might do next.

It is known that 28,188 buildings were destroyed; almost 500 people were officially said to have been killed—a figure that has risen over the years to well over 3,000; and of the 400,000 people whom the last city census had counted as living in San Francisco, 225,000 were homeless. The great majority of these last were men, women, and children seeking refuge—men, women, and children who were, in other words, now of a class that the "promised land" had never imagined it might see created within its own domains. They were Americans seeking refuge from the calamity, and thus they were American *refugees.* Not until the migrations enforced by the midwestern miseries of the Dust Bowl would such wretchedness be seen again.

THE HUMAN RESPONSE

> Considering the catastrophe that descended upon San Francisco without a moment's warning, conditions here are simply marvelous. . . . So far from being prostrated by misfortune, the citizens have banded together in a determination not only to reconstruct, but to establish a San Francisco that will be known as the most beautiful and attractive city in the wonderland of California.
>
> *From a front-page editorial in the* San Francisco Examiner, *Sunday, April 22, 1906*

Seldom does an entire and very large urban community fall victim to utter disaster. Most great catastrophes tend to be relatively local—an explosion will devastate an awesome number of city blocks here, a fire will wreck a neighborhood there, a flood will inundate the lower-lying parts of a town, terrorists will wreak mayhem in a crowded urban quarter. But once in a mercifully rare while there are those events that enfold and ruin in full the complex engine work that is an established, fully developed urban society. The dropping of the atomic bombs on Hiroshima and Nagasaki are among the most obvious. The Great Fire of London in 1666. The Black Death. The wartime destruction of Berlin and Dresden. The volcanic ruin of Santorini, of Pompeii and Herculaneum, and Martinique's St. Pierre. The huge earthquakes in Lisbon and Tangshan—and then, in 1906, in San Francisco.

The biggest of these cities survived. The smaller communities—Pompeii and St. Pierre, for example—lost their raison d'être once their buildings were gone, once their monuments were buried and their byways obliterated. But the world's big cities generally exist for reasons that go far beyond the accumulation of buildings that is their outward manifestation. Their presence in the place they occupy is invariably due to some combination of geography—they lie by a river crossing, in a bay of refuge, at the mouth of a mountain pass—and of climate, together with some vague and indefinable organic reason that persuades humankind to settle there.

Trials of any kind—war, pestilence, natural or human violence, with wholesale death or total physical destruction, or both, being the harshest of all—may slow that growth or cause some other setback; but such things are just setbacks, and before long the original reasons for a city's existence reassert themselves. Life returns, buildings and roads are rebuilt, new monuments spring up or old ones are found and dusted off, and before long the city returns to its old self, ready to see what more fate can hurl at it, to challenge and strengthen and temper its will to survive. It may not always entirely regain its predisaster status—San Francisco had to cede much to Los Angeles, for example. But generally, so far as their respective quiddities are concerned, great cities always recover.

It is a different matter when there are no cities—when disaster afflicts rural communities that lack settled centers of great consequence, as happened when the Sumatran tsunami swept across the Bay of Bengal at the end of 2004. In cases like this the return to normality is more shaky, less certain. A city has an uncanny ability to shrug off catastrophe; a devastated countryside, without a city to help it pull itself up by its collective bootstraps, can remain ruined for generations.

So whether it is Manhattan, Falluja, Warsaw, Coventry, or Hiroshima, it seems true that though cities may on occasion lose their heart, they seldom also lose their soul; and San Francisco was no exception. All that its shattered, wearied, and suddenly impoverished citizens needed was leadership, someone to take charge, someone to lift the demoralizing burdens of wreck and ruin from their shoulders, and show them the possibilities of remaking the place that they had called their home.

The leader who first emerged was a forty-year-old career soldier, a braggart and bully with a controversial record of recklessness and impetuosity—just the right man for the job, some would argue—named Frederick Funston.

It was only by the purest chance that Brigadier General Funston was on duty at the time. The resident commander of the American army's Pacific Division, based in the San Francisco Presidio beside the Golden Gate, was Major General Adolphus Washington Greely—a man who had ample personal reason to hope for the chance of meeting a challenge, such as restoring order after an earthquake. Twenty years before he had led a military expedition to the Arctic that had gone badly wrong—nineteen of its members had starved to death, and there were rumors of cannibalism, kangaroo courts, and drumhead justice. Greely's reputation and career had in consequence gone into a decline, and he must have known that skillfully organizing a military response to a major catastrophe would do much to restore his standing. But it happened that in April 1906 he was away on leave in Chicago, at his daughter's wedding, and he was able to get back to his command only by way of a slow transcontinental train, four days later. By this time General Funston, his deputy, had matters pretty much in hand.

Funston had a lively reputation. He had won his spurs both in the Cuban revolutionary army and in the Philippines, where he led so cunning and courageous a campaign against the nationalist leader Emilio Aguinaldo that he was awarded the Congressional Medal of Honor in his thirties and promoted to one-star general. But there were always dark rumors: that he took no prisoners, that he engaged in summary executions, that he looted Catholic churches for the amusement of his Southern Baptist troopers, that he had an "advertising bureau" to promote his image as a man of courage and dash.

There was a brief suggestion that he might be Teddy Roosevelt's running mate in the 1904 election;* but the bombast and braggadocio he displayed during a whirlwind speaking tour across America—threatening to establish "bayonet rule" in the Philippines, for example—put an end to that, and it brought a warning from the White House for him to pipe down. He was then sent off to be Greely's deputy commander in San Francisco, in the hope that he might fade, if briefly, from public view. To his presumed delight, and to Greely's presumed chagrin, the earthquake saw to it that he did no such thing.

Funston and his family had quarters on Washington Street, halfway up Nob Hill. On being awoken by the shudderings he ran to the summit of the hill, then down the steep slope of California Street toward the flatter reclaimed land beyond Montgomery Street. As he went, so he saw the fires beginning—smudges of smoke, flickers of flame, some of them already starting to coalesce into larger and more dangerous blazes. He quickly surmised that the water mains had broken; he imagined that very shortly there would be overwhelming demands made on the city's civil authorities—the police, ambulance squads, fire department. He knew that the city telephone systems were down, and that there was no immediate hope of making contact with either Greely or any other senior federal official.

So he was the senior soldier in the area, in charge of many hundreds of well-drilled soldiers. He suspected he might have to use explosives to limit the spread of fire, and knew full well how to get his hands on them and the military engineers who could use them; and he understood that he would have to commit his troops, to ensure they were fully armed to deal with who knew what, and to commit them without further ado.

A policeman he encountered told him of the ruin of City Hall, and cleverly predicted that it was most likely that the mayor, Eugene Schmitz (a man who also rose to the occasion, and organized matters

---

*Sensibly, those who feared a superabundance of machismo on the ticket chose instead an anodyne Indiana senator named Charles Fairbanks, remembered only by Alaskans, whose second city is named after him.

The U.S. Army commander in San Francisco, Adolphus Washington Greely, was away in Chicago when the earthquake struck, and so it was down to his deputy, the swaggering and politically ambitious Brigadier General Frederick Funston, to take temporary charge. By all accounts he rose memorably to the occasion, and though Greely—who arrived three days later by train in Oakland—took eventual control of the situation, it is Funston's tireless command that is now most widely remembered.

quickly and very handily), would use the so-far-little-damaged Hall of Justice Building on Portsmouth Square as his operational headquarters. And so, almost without thinking, General Funston acted.

He needed first to send out orders, and from an army base. He tried to flag down cars and wagons that were speeding by, but was ignored. (He was a moderately sized man, and at this moment was in mufti, without much appearance of authority. And besides, most drivers had more urgent business.) He ran back up the ten steep blocks to the army stables at the corner of Pine and Hyde Streets, and from there wrote hasty messages to his two senior commanders—a Colonel Morris at the Presidio main base, and a Captain Walker at the smaller army detachment at Fort Mason. He gave the notes to his coachman, saw him mounted on a carriage horse, and told him to ride like fury. The messages read: "Immediately send all available troops at your disposal to the Hall of Justice, and make them at the disposal of the mayor and of the chief of police."

All went like clockwork. The first troops left their barracks at 7:15 A.M. and arrived half an hour later—initially Captain Walker and 155 soldiers, all in field equipment and each with twenty rounds of ball ammunition, reported to Mayor Schmitz at 7:45 A.M. They found him in a fighting mood—one born perhaps out of his initial bewilderment at what was going on. The city clerks who had turned up at the wrecked City Hall at 6:00 A.M. were astonished that their mayor wasn't already there. They drove with difficulty to his house three miles away on Fillmore Street, then back to City Hall, then on to the Hall of Justice. From this point on the mayor was seized with determination: Clearly, those around him said, he was not going to let his early puzzlement inhibit his ability to issue commands.

And so he did, with staccato efficiency. If Funston took immediate charge, Mayor Schmitz promptly joined him in equal stature. He first declared that soldiers should cordon off all burning areas and keep onlookers away. A strong detachment would go immediately to City Hall, bayonets fixed, to secure the Treasury, with its millions of dollars in cash and specie. He ordered that looting, instances of which he had already seen on his drive to work, was henceforward a capital offense, and that soldiers would be ordered to shoot on the spot anyone they found stealing from ruined stores and houses. Saloons would close. The sale of liquor would be prohibited. Naked candles would be banned.

It was a draconian pronouncement, and for years later it was assumed—because soldiers did indeed execute looters on the spot—that martial law had been declared. But it never had been. A city mayor has very limited authority—only the state governor can call in National Guardsmen, and only the president can summon federal troops. And, in any case, the civil courts were still in theory operational in San Francisco in 1906. So, although soldiers were given a free hand for a short while, it was at the behest of the city's civil authority, which was always in unofficial, though ultimate, control. Later Schmitz explained: "While the orders issued at the time were perhaps without legal authority, and were extraordinary, they were accepted by the people with good nature and good will, and there was a general desire to carry out the suggestions made in my written and verbal messages."

But he demanded that the orders be printed. The most controversial of these was rushed out by the Altvater Printing and Stationery Company, whose works down in the Mission District were still functioning, though without electric power. Soldiers commandeered passersby to operate the treadles. The mayor had 5,000 handbills letterpressed, each sternly (and technically illegally) telling the citizens:

> The Federal Troops, the members of the Regular Police Force and all Special Police Officers have been authorized by me to KILL any and all persons engaged in Looting* or in the Commission of Any Other Crime.

---

*The *Socialist Voice,* eternally primed to pounce, condemned Schmitz's pronouncement, less for its self-evident illegality than for its hypocrisy. The mayor, said the journal, had been conducting his own looting for years, so "why should the poor little looter who is scraping something together to live on for a little longer, be shot to death?"

Eugene Schmitz, who had come to San Francisco as a violinist and rose to prominence in the local musicians' union, was handpicked to serve as city mayor by the corrupt (and later convicted and imprisoned) railroad lawyer Abraham Ruef. Despite such unpromising credentials Schmitz earned praise for the decisiveness and efficiency that he displayed during the crisis—and won a reputation that has generally outlasted the various controversies that swirled about him.

Very few people were actually shot on Schmitz's order. When General Greely's train arrived and he reassumed command on the following Monday, he began an inquiry but came to the conclusion that only two men had been executed by soldiers. And in May the Coroner's Office reported that of the 358 corpses thus far recovered and examined, only one had a gunshot wound. While the figures were probably higher—bodies, it was said, were often thrown into burning ruins in order to be incinerated beyond recognition and forensic inquiry—it does appear that more was made of the intent of Schmitz's pronouncement than of any employment of its explicit threat.

Somehow it is indeed quite hard to imagine anyone—no matter how well armed or endowed with authority he may be—feeling anything but the utmost reluctance to carry out a summary execution in the midst of a tragedy like this. William James observed that people involved in the tragedy were awestruck by the unfolding events—their awe compounded by the fact that there was no one to blame, that everyone had suddenly become a victim of a vast natural circumstance. There was neither anger nor envy in the air, just an all-consuming need for survival. How difficult must it have been for a soldier to shoot someone dead, particularly some shabbily dressed thief who was, in all likelihood, a victim of ruin just trying to survive?

Schmitz's other proclamations were less contentious. He warned people that the local gas and electrical companies had been ordered to suspend their services until the mayor's office said otherwise. "You may therefore expect the city to remain in darkness for an indefinite period." He ordered a nighttime curfew and ended with a warning about broken chimneys and leaking gas pipes. There was no bombastic rhetoric, no call to arms, nothing to inspire or encourage. Mayor Schmitz simply forbade, warned, threatened, and instructed—a strategy that, by all indications, seemed to work well, and quickly.

He used such military telegraph links as were working to signal both Governor Pardee in Sacramento and the mayor of Oakland, demanding help: food, fire engines, hoses, dynamite. And from all over the state, assistance began to pour in remarkably swiftly: A train from Los Angeles with wagonloads of packaged food and medicine arrived by midnight, a scant eighteen hours after the quake.

General Funston sent cables to Washington, demanding tents, rations, medicine. William Howard Taft, the secretary of war—and the next American president—responded with exemplary promptitude. An order went out at four the following morning—Congress would pass an emergency enabling resolution the same day, making everything legal and fiscally proper—and 200,000 rations were dispatched immediately from Vancouver Barracks, on the Canadian border in

Washington State. The next day more trains were ordered to leave army bases in Texas, Pennsylvania, Nebraska, Iowa, and Wyoming, all loaded with equipment. Before long every single tent in the military's possession was in San Francisco, and the largest hospital train ever made had been sent out from Virginia. A few weeks later fully 10 percent of the standing army was there as well—an enormous commitment of men and matériel, and an immense expenditure of federal funds.

Later on the day of the earthquake, as the first troops were fanning out and the general was ordering still more from other bases in California and beyond that were in his command district—he had 1,500 soldiers deployed by noon—the mayor moved to set up a formal committee that would reestablish order and good government in the shattered city. The committee was made up of fifty supposedly upstanding citizens, all of them men. Many were political cronies and yet, by all accounts, not nearly as corrupt as might have been supposed. Printers were swiftly set to work issuing pronouncements about the committee's existence and its responsibilities.

The body's very first meeting was something of a shambles: It had to be moved from the Hall of Justice after soldiers outside dynamited a nearby building and all of the hall's windows blew in. Schmitz, who at the time was in midstream with a powerful piece of oratory further demanding the shooting of "all and any miscreants who may seek to take advantage of the city's awful misfortune," asked that everyone immediately move outside to the grass in the middle of Portsmouth Square—the old plaza, in the time when the city was born—and carry on their meeting in relative peace and quiet there. The Committee of Fifty, Schmitz said, would meet twice a day for as long as the crisis lasted—and so they did, though hastily moving their points of rendezvous whenever fires bore down on them or the soldiers' dynamiting became too deafening.

Fresh soldiers moved in—the newest from small bases on Angel Island and Fort Miley—and took up positions, on Funston's orders, guarding the U.S. Mint, the Post Office, and the Appraisers' Building—all three federally owned, and all survivors of the quake. (The Appraisers' Building was demolished later.) The National Guard came in as well, though leaderless, as their commander was away in the north of the state. Governor Pardee, at whose behest the National Guard would serve, turned up from Sacramento in the small hours of Thursday morning.

The U.S. Navy sent in a destroyer, a fireboat, and a number of heavy tugs, together with doctors and nurses from its base on Mare Island. The Coast Guard, called the Revenue Cutter Service at the time, helped with evacuations, taking people on its boats from the Lombard Street slips to safety across in Oakland, Berkeley, or points still farther east along the Bay's more distant shores.

And many hundreds of pounds of high explosives were used, most of it brought in on army orders from the California Powder Works in Pinole, at the north end of San Francisco Bay. This company, built by Chinese workers at the close of the previous century, would in due course become the biggest manufacturer of TNT and dynamite in the world: At the time of the earthquake there were six plants, three at Pinole producing black blasting powder, dynamite, and guncotton for the U.S. Navy, and three near Santa Cruz making specialist powder for artillery pieces and breech-loading cannon, as well as a cartridge that was guaranteed never to "hang-fire" in a sportsman's shotgun.

Almost everything the company made was pressed into the service of destroying buildings and widening streets to create the desperately needed firebreaks. This sometimes led to more dramatic explosions than had been predicted. Early in the saga some soldiers, using a mysterious granular form of dynamite made for special quarrying tasks, tried to knock down a pair of houses at the corner of Kearny and Clay Streets, but both buildings were promptly set ablaze, then blew up and sent burning wood across the street, setting adjoining buildings on fire, too. Granular dynamite was never used again.

But in time the soldiers learned how to blow up things properly,*

---

*San Franciscans are still angry about the wholesale use of dynamite in 1906, claiming that it seemed to do more harm than good—destroying much more than was needed in order to curb the fires. And even today, playing "what if" with all the various factors of the firefighting in the city can be counted on for spirited dinner-party conversations.

and in due course the fires were damped down and snuffed out. The weekend after the quake was clear but for silvery feathers of steam rising from the ruins. Long trestle tables went up on the cleared streets, and were used for public feeding programs for the displaced and dispossessed. Well-known restaurants, now burned-out shells, opened under canvas awnings and began offering their familiar fare at knockdown prices. And all over the city the homeless were beginning to be settled into their immense wildernesses of tented encampments—which would be home to scores of thousands for many months.

The mayor and his team made sure the camps were well set out: All soon had running water, bathhouses, sewers, and drains* and were protected from any further fire. Before long, small green-painted officially issued earthquake cottages, nearly 6,000 of them, replaced the army's much-needed canvas tents, and so comfortable and cozy were they that many were inhabited for several years to come. The camps—Harbor View Camp and Lobos Square Camp among them—became like small cities, with governments and social stratification and reputations: Harbor View was said to be where the city's hard men were sent, and there were small outbreaks of crime. But generally there was good order in the camps, and, far from there being any eruptions of contagion (always expected when sewers break and rats run wild and the people are fatigued), the health of the city actually markedly improved. People found that being compelled to live and work in the fresh air, barred from drinking hard liquor, and forced to survive on rationed food and tobacco kept them fitter and leaner than they had been for years. Hardship, to a measured degree, can be beneficial to at least some aspects of society: as with Britain during the Second World War, so with San Francisco after its great earthquake.

Sympathy for the city's plight was boundless. Money flowed in from outside, the sums soon amounting to $8 million: The city of

---

*And mule-drawn latrine-emptying wagons belonging to a newborn firm called Odorless Excavators, Inc.

Boston gave half a million, the Dominion of Canada wired in $100,000, the Businessmen's Association of Helena, Montana, sent $6,000, the staff of Barnum and Bailey's Circus gave up a day's takings—$20,000, and one hitherto unsung C. E. Wilson of Clinton, Iowa, gave $500, one of the hundreds of widow's mites, all happily received by the mayor and the governor and everyone in the city who would fain spend it on making things better. "The nation's heart was wrung," said *Harper's Weekly.* And was it any wonder—the San Francisco people seemed "born for rejoicing," and they should be permitted to rejoice again, as quickly as possible.

And slowly and steadily the city began to struggle back onto its feet. Many symbols have been selected over the years to illustrate the city's unflagging spirit. There is Francis, the St. Francis Hotel wine steward's faithful terrier, who hid for five days in the cellars and emerged as perky as ever. And eight days after the quake, immense crowds thronged to the city's one unburned theater to see vaudevillians and burlesques, and to watch a one-act précis of *Carmen** for no better reason than *pour mémoire.* And to further inflate the city's improving morale, there came the probably poetically enhanced story of a cheerful and fat Irishwoman who suckled an infant for an Italian lady too terrified to produce her own milk—thereby, in the overwrought phrasing of *Harper's* magazine, making certain that "Italy drained the milk of human kindness from Erin's fount."

For rising most splendidly of all to the occasion the laurels should go to one federal institution, the U.S. Post Office. For not only did this body manage to recover in double-quick time its ability to function—to collect, transmit, and deliver the mails—but managed to do so without losing one single item that was being handled at the moment the earthquake struck.

The newly built Main Post Office building, a magnificent Beaux Arts palace close to the Mint on Seventh Street (and that still stands

*But *sans* Caruso, who had left for New York as soon as he could find a suitable train.

today, though not as a post office, and until lately with a spectacularly quake-cracked terrazzo floor), was saved by the superhuman efforts of the forty-six men who worked there—all of whom refused military orders to leave in advance of the fires. Just like their federal colleagues at the Mint (though without having the luxury of an en suite artesian well), they battled back the blaze, plugging the windows with mailbags soaked with water from the huge iron tank above the freight elevator. It took all day and most of the night to keep the blaze at bay; but, once the worst of it had passed and showed no sign of returning, the soot-stained staff picked up the toppled furniture and returned to vertical the endless racks of sorting pigeonholes. By daybreak on Friday, two days after the San Andreas Fault had ruptured, the U.S. Post Office was open for business once again.

The city postmaster promptly announced his overarching desire that the citizens of San Francisco who used his service should be able "to tell their friends and loved ones of their condition and their needs." Since all wires were down, there was no telegraph system worth speaking of, and telegrams inside the city were actually being delivered by the post office, on payment of a two-cent stamp. Clearly the human mail carriers were more reliable than the overhead copper cables. And so a major effort was mounted, to see that outgoing letter mail was collected properly and on time, and that it was sent across the Bay to catch the outbound postal trains for points east, north, and south.

Temporary tented post offices were set up in the refugee camps, to ensure that those without either homes or addresses were able to send letters; and William Burke, the postmaster's secretary, recounted what happened when he took a U.S. Mail sign from a streetcar barn and mounted it on the top of a car he had pressed into service to collect the mail.

> The effect was electrical. As people saw the machine bearing the mail coming, they cheered and shouted in a state bordering on hysteria. We told them where the collections would be made in the af-

> ternoon and asked that they spread the news. As we went into the Presidio there was almost a riot, and the people crowded around the machine and almost blocked its progress. It was evidently taken as the first sign of rehabilitation and, as it proceeded, the mail automobile left hope in its wake . . .
>
> . . . when I went back in the afternoon . . . to collect mail from the camps, the wonderful mass of communications that poured into the automobile was a study in the sudden misery that had overtaken the city. Bits of cardboard, cuffs, pieces of wrapping paper, bits of newspapers with an address on the margin, pages of books and sticks of wood all served as a means to let somebody in the outside world know that friends were alive and in need among the ruins.

And there was one additional aspect to the marvel that was the San Francisco Post Office.

The postmaster also decreed that, so far as private letters were concerned, it would not matter one whit if they bore a stamp or not. Maybe it was illegal; maybe it was a violation of a post office decree, or bylaw, or mantra. But, as Burke was to write,

> No man without money and without stamps who trusted to the postal service of San Francisco and dropped his unstamped letter in the mails found he had trusted in vain. Thousands of these unstamped letters, typewritten and bearing well-known return cards, were caught and returned to the firms that sent them. But no unstamped letter that bore evidence of coming from a man who could not afford stamps was delayed in transit.

The banks, too, performed impressively—working in collaboration with the Mint, which had gold coin untouched and in abundance in its vaults. It was agreed that any bank customer needing money, and who was known as a bank customer, could be given scrip certificates that the Mint would honor in gold (which still backed currency in those days, with America firmly on the gold standard). And, thanks to this cunning device, bank business was just as busy as ever in little more than a month. The week of June 23, just eight weeks on from the disaster, saw more money change hands through the banking system than it had in the equivalent early summer week the year before. Matters were indeed going back to normal.

So, in terms of cash and communications, San Francisco was rising fast and furiously. And yet overall the task confronting the city was prodigious. The immense scale of the disaster was fully appreciated within days of the quake, and all was made officially clear in the municipal report for the financial year that ended in June, just ten weeks after the event. Items chosen at random hint at the size of the challenge that the city had to meet:

> City Auditor: All records and data pertaining to the financial condition of the Municipal Government . . . destroyed.
>
> County Clerk: Notwithstanding the loss of blanks, books, filing and records, the County Clerk has been continuously open, and has transacted business since the memorable eighteenth of April, without a moment's inconvenience to the Bench, Bar, litigants or the general public.
>
> Sheriff: Since April 20, 1906, there are but two jails, the one on Broadway Street having been dynamited and destroyed by fire . . .
>
> Department of Elections: All records of this office, save affidavits . . . which happened to be in fireproof vaults . . . destroyed. [But there was a ruling made that accepted the profound changes in where many voters perforce now lived. In part it read: "it is the right of an elector to abide in any habitation he chooses. A tent, a cabin, a cave . . . if it be fixed as the abiding place with intent that it constitutes a home . . . is permitted for use as a residence."]
>
> Public Library: Main library . . . totally destroyed . . . irreparable loss to the files of San Francisco newspapers and periodicals dating from the time when California first began to attract the attention of the world . . .
>
> Board of Education: Number of Schools Destroyed: 31.

WITH THE CITY SUDDENLY facing demands like these, it is scarcely surprising that not a few Cassandras expressed their doubts as to the ability of either the city fathers or the citizenry to bring San Francisco quickly back to life. The British consul general, Sir Courtney Bennett, was gloomy in his assessment, made in an official telegram sent "under flying seal"* to London in September. It is a document that also displays, in its closing paragraphs, the remarkable detachment from reality that was an abiding characteristic of many diplomats of the time.

> In previous reports I have expressed the opinion that the rebuilding of the city of San Francisco will be a task of many years, and that the extent of the disaster of April 18th was either not recognised, or was intentionally underestimated by the inhabitants. I see nothing now that can lead me to a different conclusion; indeed events subsequent to the Earthquake make me think that my first impressions as to the time required for recuperation were even too hopeful.
>
> As the weeks and months pass by it is seen more and more clearly that the process of recovery will take much longer than was at first thought. For this there are several causes.
>
> Five months have now elapsed, and there is very little sign of permanent construction. The mass of debris, millions of tons of it, remain more or less where it was. The amount removed is barely noticeable, though in the aggregate it is large. Some of the principal streets are cleared sufficiently to allow of the passage of Electric cars and wheeled traffic, but except in some of the main streets the pavements are still more or less obstructed. Business is active . . . but yet, even this statement must be qualified.
>
> The nature of goods sold is not what it was before the disaster. Everything is of a cheaper kind. Luxuries are not to be obtained. Men who smoke shilling cigars now smoke three-penny ones. Men who wore six or eight suits of clothes made to order at seventy or eighty dollars a suit now buy ready-made clothing at about twenty five to thirty dollars a suit, and order half the number of suits or less. Economy is practised in every direction. Perhaps no clearer proof of this can be found in the fact that the leading fashionable

---

*This pleasingly archaic term, or its French equivalent, *sous cachet volant,* was once much used by diplomats. In meant that the message carried a seal, but was not closed up by it, so that anyone to whom it was sent could read the contents and then send them on to their final destination. It indicates the very lowest level of official secrecy.

> clubs have reduced their monthly dues and now give a *table d'hôte* lunch for fifty cents, whilst beer is drunk instead of wine. Formerly the members of these Clubs spent five or eight dollars for a lunch, frequently more, where now they part with a humble dollar.

The question of whether the extent of the disaster was, as Bennett asserts, "intentionally underestimated by the inhabitants" has been a subject of much lively discussion in the years since. There is no doubt that city boosters wrote a good deal of hyperbolic nonsense—such as that in this section's epigraph, about the state of the city being so "marvelous" by the following weekend. There were magazine articles and advertisements and morale-boosting performances, all fashioned along the phoenix-rising-from-the-ashes theme. San Franciscans were proud of their city, as they remain today: Any injury to the place is felt by the entire community, and there was much can-do optimism sprinkled about by civic and commercial leaders, the better to keep the citizenry content with the pace and direction of rebuilding.

But there is a good deal of evidence that some commercial companies were almost too anxious that the city—despite having been located with what seismologists would regard as wanton foolishness—should not become widely known as somewhere that was riskily earthquake prone. That it was, there was no doubt; nor is there any doubt that it remains a very dangerous city in which to choose to live today.

In 1906, however, official advocacy of new building codes and revised firefighting contingencies took second place to a public-relations campaign—not named as such but recognizable by today's standards—that sought to prevent people from leaving, to induce would-be immigrants to continue coming, and to persuade investors that San Francisco was still the ideal place to sink uncountable sums of money.

A singular effort was launched to convince people across America and around the world that the city's suffering was not the result of an unpredictable caprice of nature—that, in other words, its misery had not been caused merely by an earthquake. Quite the contrary: The destruction of this great imperial city was caused by fire. It was a direct consequence of the all-too-preventable carelessness of humankind, which, by its greed, insouciance, and neglect, had let the only slightly damaged city burn itself into ruin.

Moreover, the weakness of so many of the hastily erected buildings, the vulnerability of the fast-built system of water mains, and the widespread and self-evidently unsafe habit of throwing up structures on landfill—when so much solid rock was readily available—showed that humankind itself had not thought through the implications of living in a city so interestingly sited as San Francisco. There was no doubt about it: Humankind was to blame; nature was by and large exonerated.

Thus was the earthquake officially demoted, and further implications were made crystal clear: *Since what had caused the crisis was the mistake of humankind, then all could be improved by humankind.* Things could be bettered by a simple act of the common will. The city could be handsomely rebuilt, and anyone going to live there, or wanting to place his money there, could be assured that such a human-made disaster—since humankind had the ability to control how it arranged matters—would never happen again. Suggesting otherwise, that San Francisco had been ruined by a force of nature, was disloyal, unpatriotic, and just plain wrong.

Geologists were furious and bemused by turns. Writing two years later in the *Bulletin of the Seismological Society of America* (a body set up as a direct consequence of the 1906 event), John Branner said that

> a major obstacle to the proper study of earthquakes [was] the attitude of many persons, organizations and commercial interests . . . to . . . the false position . . . that earthquakes are detrimental to the good repute of the West Coast, and that they are likely to keep away business and capital, and therefore the less said about them the better.
>
> This theory has led to the deliberate suppression of news about earthquakes, and even the simple mention of them.
>
> Shortly after the earthquake of April 1906 there was a general disposition that almost amounted to concerted action for the purpose of suppressing all mention of that catastrophe. When efforts were made by a few geologists to interest people and enterprises in the collection of information in regard to it, we were advised and

> even urged over and over again to gather no such information, and above all not to publish it. "Forget it," "the less said, the sooner mended" and "there hasn't been any earthquake" were the sentiments we heard on all sides.
>
> There is no doubt about the charitable feelings and intentions of those who take this view of the matter, and there is a reasonable excuse for it in the popular but erroneous idea prevalent in other parts of the country that earthquakes are all terrible affairs; but to people interested in science and accustomed to the methods of science, it is not necessary to say that such an attitude is not only false, but it is most unfortunate, inexcusable, untenable and can only lead, sooner or later, to confusion and disaster.

The principal purveyor of this spin-doctored view appears to have been the Southern Pacific Company, which, since it owned or leased all the main railroad lines into San Francisco, had a lot to lose if the city was regarded by potential investors as having gone off the boil. The company was heavily indebted after building costly new rail lines and dams and bridges all across California. A sharp decline in its stock price in New York, where it had its headquarters, would seriously inhibit the company's ability to service these debts.

The stock market had already been hit as a result of the earthquake. Naturally, some had made money, as some always do. The notorious speculator Jesse Livermore sold Union Oil stock short on Wednesday, some minutes before anyone else on the floor heard the news, and he made a small fortune.* But there was a more general stock market decline as well, one that went beyond a slump in prices of the locally based businesses that might be expected to find themselves in trouble. The nationally known insurance companies who were soon being pressed to pay claims as a result of the disaster had to raise lots of cash quickly, and many of them sold off holdings of blue-chip stocks, which consequently went into a steady decline for several

---

*Livermore was also said to have made $100 million selling short at the time of the 1929 Wall Street crash. Not the most content of men, he frequently went bankrupt, spent wildly, married often, and killed himself in 1940.

weeks following. The Dow Jones Industrial Average, then a decade old, fell 10 percent in the month after the earthquake, worrying investors mightily.

Though there is no firm evidence that there was ever a fully fledged conspiracy—either to bowdlerize the word *earthquake* from the lexicon or to massage casualty figures down to more comfortable levels—it is abundantly clear that there were informal discussions about the problem, and the setting of a tone of rather forced jollity. Ernest Bicknell, a Chicago charity organizer who would later become a leading figure in the American Red Cross, heard something of the gathering mood when he was traveling out to offer official help on behalf of the state of Illinois. He first heard talk in the train corridors about the harm that might be done by idle chatter about San Francisco's reputation as an earthquake-prone city. It was on the afternoon of the second day out of Chicago, as the train was weaving its way through the Rockies, that the talk began to coalesce:

> As many men as possible were herded into the sleeping car for a mass meeting. After half an hour of vigorous speeches, a resolution was adopted to great cheering and entire unanimity, announced as a fact beyond dispute that the disaster in San Francisco was due solely to fire, to such a calamity, in short, as might occur in any well-ordered city and that *the slight tremor which preceded the fire** had nothing to do with the tragedy beyond, perhaps, breaking gas mains or water mains here or there.

In some instances formal moves were taken. The San Francisco Real Estate Board, for example, fretted that customers might not buy houses if they thought that they would be periodically flattened. So they passed a resolution a week after the earthquake insisting that in all publications there should be pains taken to see that the phrase "San Francisco Earthquake" was replaced by "San Francisco Fire."

In recent years there have also been suggestions that the earth-

---

*My italics.

quake casualty figures were made more palatable—softened, minimized, massaged—under pressure from the Southern Pacific and like-minded commercial firms. There is no evidence, once again; but the official figures, and the new calculations, are very different, and by a full order of magnitude.

William Walsh, the city coroner, issued a report that was a close analysis of the 2,195 questionable deaths that had occurred during the year from July 1905 to June 1906. Of those—which also included 343 accidents, 36 murders, 177 suicides, 11 continuing mysteries, and 1,200 deaths by natural causes—some 428 were officially classified as "deaths from shock from earthquake and fire." For years that figure remained unchallenged. But during the 1990s a former city archivist named Gladys Hansen began to collate additional reports of deaths that occurred within a year of the earthquake and that could, in the opinion of some, be ascribed indirectly to the events of the previous April. Using this yardstick, Ms. Hansen has come up with a figure in excess of 3,000 deaths—which probably, but not definitely, reflects more accurately the toll taken by the disaster. The precise number will never be known; and it may never be proved that the Southern Pacific (which produced a book called *San Francisco Imperishable*) and its colleague companies ever conspired to keep the numbers acceptably low. Certainly no memorandum has ever been found, and, given the mergers within the railroad industry in the years since 1906, it is improbable, even if any exists, that such a document will ever be found.

Interestingly, the number of suicides in San Francisco dropped rather significantly in the year after the quake, from 177 to 131. Other than health problems, the principal reason for taking one's own life in 1906—stated simply as "drink"—became one of the least common in 1907. Although twenty-four drink-afflicted men and women took their own lives in 1906, only seven people killed themselves in 1907 for this reason, just one more than the total of all those troubled that year by "love," "jealousy," and "business reverses." One has to suppose the long period of prohibition had something to do with the change. "Death of sweetheart"—which was a reason for two of the suicides in 1906—quite possibly had a rather more direct connection to the quake. One assumes so, at least, because the statistical tables show that no San Franciscans died for this reason a year later.

ACCORDING TO the coroner's report, "financial troubles" provided a common reason for men and women in San Francisco to do away with themselves in both years. In a few of these cases there is probably some connection—difficult to prove, however—with an aspect of the tragedy that still nags, even today: the number of bitter arguments that broke out between owners of buildings and businesses that were damaged during the disaster, and the attitude of the insurance companies that had written the policies designed to protect them.

The notion that, for a price, a company with great financial resources might secure another firm, or an individual person, "against pecuniary loss," as the *Oxford English Dictionary* has it, "by payment of a sum of money in the event of destruction of or damage to property (as by disaster at sea, fire, or other accident), or of the death or disablement of a person" was born in the seventeenth century, in London. It was an idea that spread to America almost as soon as the colonists were properly settled and, following the establishment of the first fire insurance companies in New York and Philadelphia in the late 1750s, it has been an integral part of American capitalist life ever since.

Some might inveigh against the idea—pointing out the vast present skyscrapered wealth of modern insurance companies, agencies, brokerages, and reinsurance firms and noting with scorn and envy the mysterious web of commercial complexity that surrounds them; but few would argue that insurance companies have not, generally, had a largely beneficial effect on society. All manner of social reform occurred at the urging of those companies who decided to shoulder the risks of society's multifarious spheres of activity. Properties were insured against fire—but only if the property owners could show that "they cleaned their chimneys regularly," thus making fire-plagued cities rather safer. When Lloyd's began to offer smallpox insurance at the beginning of the last century—"but only to those who were vaccinated"—it prompted people to get their shots, and the world became a little more healthy in consequence. When insurance companies inserted clauses into their policies demanding crash helmets for motorcyclists or seat belts for drivers, the traveling public became a little less vulnerable to injury, too.

The shared risk that these insurance companies assume—by way of their stockholders or their reinsurance agencies or their syndicates of underwriting "Names," as they are called at Lloyd's—truly is shared, as it has to be. A once-wealthy friend of mine who lives in California became a Lloyd's Name when he had sufficient cash in his bank to do so, and always scoffed genially at the idea that from that moment on he was, theoretically, at personal risk down to his last shirt button. But the scoffing stopped in September 2001, with the World Trade Center attacks, and the heart-stopping realization that his Lloyd's syndicate had underwritten coverage for both of the buildings and for all four of the crashed aircraft. His personal financial liability, per his understanding when he became a Lloyd's Name, was from then on—and not theoretically but in practical fact—unlimited.

He lives 3,000 miles away from the tragedy, in a small village beside a lake, a place where crime and violence of any sort are almost unknown. But since that September seldom has a month gone by without my friend—who is now long retired, in his eighties, and living with his wife what he once imagined would be a peaceful and prosperous old age—receiving from London an urgent demand for funds. The language of each message is always the same, always spare, exquisite in its blunt delivery of pain. Kindly wire, it always begins, a quarter of a million dollars—the sum is invariably the same—for settlement of such and such a claim, by noon on this coming Friday. No ifs or buts, no excuses, no delay. And so, quarter million by quarter million, his long-accumulated wealth has been steadily drained away. He has sold all of his houses but for one modest cottage. He may have to sell this, too, and move into a rented apartment. He is now on the verge of total ruin. He seems outwardly phlegmatic, but his lip quivers when he tells his story. Yet this is what he had signed up for, he tells me; he hoped it would never come to this; but now that it has, he has no choice but to do as bidden, and it does not help matters to complain. Besides, he adds, so many others suffered so very much more.

And, very generally speaking, such also appears to have been the attitude of most of the insurers in San Francisco in the days and weeks that followed the ruinously costly disaster of April 1906. The tone was best set by the famed British insurer named Cuthbert Eden Heath, the tall, deaf son of an admiral who joined Lloyd's at eighteen and, in a matter of five years, began to turn the Victorian insurance industry on its head, taking on risks that no one, up to that time, had ever thought worth insuring against.

It was Cuthbert Heath who first had the idea, for example, of offering fire victims insurance against the loss of profits they might suffer. It was Heath who took Lloyd's into the American market for risks other than the traditional shipping ventures. It was Heath who came up with the smallpox-if-vaccinated insurance plan in 1901, who in 1887 had offered jewelers policies that covered any of their diamonds that might be lost in transit, and who in 1914 would cover people against damage from air raids (their premiums becoming higher as the German bomb aimers became more skillful). It was Heath who set out the first plans for workmen's compensation, and who in 1907 offered the first Lloyd's policy covering American drivers and their cars—and in 1895 it was Heath who first offered, in America, insurance against damage caused by earthquake.

The message he sent to his American agents and claims representatives in the aftermath of the April 18 event was short and simple, and has passed into insurance legend and lore: "Pay all your policyholders in full, irrespective of the terms of their policies." It was a generous offer, honored perhaps more in the breach than in the observance by some; and it came at a very difficult time for the industry.

For in the years just prior to the San Francisco Earthquake there had been a number of very costly disasters. In 1903 some 600 people had died in Chicago at a fire in the Iroquois Theater. A blaze aboard the steamship *General Slocum* on the East River off Manhattan killed 1,021 people in 1904. And the enormous fire in Baltimore the same

Cuthbert Heath, the legendary London insurance underwriter whose "pay all claims" telegram did much to help rebuild the city as well as enhance the reputation of Lloyd's.

---

year wrecked the city's business center, with losses of up to $90 million. The ruin caused by the earthquake and then the fire in San Francisco was the last thing the insurance industry needed in 1906. The total value of property destroyed was perhaps as much as $500 million—of which the insured liability was estimated at around $235 million. About a hundred insurance companies, some big and national, some tiny and local, had written policies for householders and businesses in San Francisco. Of the total risk, British underwriters had written about a fifth, and German insurance companies stood to be out of pocket by a similar amount.

Claims were made by 90,000 people and companies within a day or two after the last fire was out. But the fires had caused unforeseen complications: Many people had lost their policies, and many insurance companies had lost their offices. Confusion and consternation erupted on all sides.

Then there began as a consequence a great deal of undignified haggling—though, from a moral viewpoint, most now think there should not have been—as company after company tried to force policyholders to accept far less money than they had insured for. Some firms argued in a quite arbitrary fashion that one risk was insured and another was not; and often technical arguments went on for months over whether it was a fire or the earthquake that destroyed a particular building, because one risk was reasonable and covered, the other an act of God and not. In particular there was much heated debate over whether to invoke the so-called fallen building clause that was now standard on most American policies, which held that "if a building, or any part thereof, fall except as a result of fire, all insurance by this policy on such building or its contents shall immediately cease." Only three American companies that used the standard form, and that could therefore by rights apply the clause, decided to ignore it and to pay up in full. Most of the rest dithered, argued, stalled.

Many others cunningly suggested a dastardly solution: They offered what they christened a "horizontal cut," proposing to each policyholder that since it was impossible to determine if it was fire or shock waves that had caused his building to collapse, why not take a 30 percent deduction and say no more about it? There was an almighty outcry over the sheer impertinence of this suggestion; but in the end the horizontal cut was applied by a large number of companies, though the discount was reduced to 10 percent, and those firms that applied it became moderately less unpopular as a result.

The reputation of the insurance industry as a whole suffered grievously from the poor behavior of these companies, particularly after the topic became a national issue when a local congressman stood up before the House of Representatives in Washington and inveighed against the number of blatantly dishonest and evasive companies that had been trying, as many began saying, to weasel out of their obligations.

A scant six of the one hundred firms involved were said to have performed impeccably, paying all of their policyholders in full and on time. Four of these were American—the Aetna of Hartford, the California of San Francisco, and the Queen and the Continental of New York. The other two, the Royal and the Liverpool London, were British. The Hartford essentially paid in full as well, deducting just a small percentage for those who demanded to be paid in cash. Some thirty-seven firms had no choice but to go back to their stockholders and pass on the liability to them—to "assess" them, to use the formal phrase, a total of $32 million in claimed money. Twelve companies went entirely broke.

There was enormous praise for the Fireman's Fund Insurance Company, based in San Francisco, which faced more than $11 million in claims, with only $7 million in assets. Fireman's took the unusual step of forming an entirely new company—paying out all the cash the old firm had on hand, and offering policyholders stock in the new firm in lieu of the cash they were owed. A body called the Credit Men issued a report some while later saying this company above all had "proven itself entitled to the confidence, good-will and patronage of the insuring public."

The same could not be said of fully fifty-nine other firms—a technical majority of the total of one hundred companies that were known to have written policies, though not representing the greatest part of the total value they had insured. Of these pariah companies, all of which argued, prevaricated, demanded huge deductions for paying out cash, and who in some cases defaulted altogether, the very worst appear to have been those based in Germany and Austria. Six of these firms denied all responsibility and simply closed up shop and went home. The Hamburg-Bremen Company seems to have been a special villain. It demanded discounts of between a quarter and a fifth for nearly all of its $4 million in liabilities; and it was excoriated for "insulting and discourteous treatment" of its customers, and for misleading

New Yorkers—potential new customers—by running advertisements claiming, falsely, that funds had been sent over from Hamburg to pay San Franciscans in full.

AND THEN THERE was the question of Chinatown. It had been almost entirely destroyed. As anticipated by the insurance board, fires had roared along the tiny alleyways and consumed almost every single one of the tenements of its nine crowded city blocks; and though Chinese casualties were relatively light, thousands of Chinese men and women, made comprehensively destitute, immediately set to roaming the city in search of shelter, food, and work.

At first the city was, if sotto voce, delighted. There was still a powerfully racist element to San Francisco—at least, so far as the Celestials were concerned—and not a few thought the fires a blessing. Now many residents breathed quietly, the Chinese could push off elsewhere, and the slums where they had practiced their peculiar arts could be replaced by office buildings or houses for more respectable folks. A committee was very quickly formed—within six days—to decide where they should be put. No thought was given at first to where the Chinese themselves might want to live; it was merely assumed by members of the committee (which was led by a Methodist minister named Thomas Filben, and had members named Deneen, Ward, and Phelan—not a Chinese among them) that they knew what was best for this most alien of communities.

By now the terrified Chinese had scattered, many taking boats across to Oakland and settling in what was perceived as relative safety. The committee demanded they come back: first to a temporary camp set up for them at the foot of Van Ness Avenue and then to what had long before been planned for them, an "Oriental City" out at Hunter's Point, a bleak peninsula to the southeast of the city. An industrialist named John Partridge had been pressing for the Chinese to be moved to this distant redoubt, well away from the city center, for some time; now, it seemed, there was an opportunity to achieve his ambition.

But the Chinese were having none of it. They wanted their old community rebuilt, and they wanted to live there. They swiftly won the backing of their government at home in Peking—and by mid-April the Chinese legation in Washington had made it abundantly clear that no less a figure than the Dowager Empress herself, speaking from deep within the Forbidden City, had demanded that her people be housed where they had long wished to be housed. To underline the issue, the legation (which sent an official party to San Francisco) pointed out that the Chinese government owned title to land on Stockton Street, in the heart of Chinatown, and fully intended to rebuild its consulate there. To suggest they do otherwise would offend the Forbidden City, would damage relations between China and the United States, and would damage, no doubt, the lucrative trade across the Pacific.

Faced with such dire consequences, the committee backed down; and Chinatown—though later to be subjected to trials of an altogether different kind, as we shall see—began to rebuild itself in its selfsame sixteen blocks, where it remains today, aromatic and mysteriously and defiantly different from all the rest of San Francisco.

FIVE MONTHS AFTER the quake the British consul general, Sir Courtney Bennett, was in what at this remove seems a gloomy mood when he wrote a lengthy assessment for his superiors in London. He had a prescience about him that comes across today in one passage, close to the end of a lengthy telegram. He had written copiously about the insurance debacles, about the strikes and riots that he felt were now gripping the city, about the fractious and disputatious mood of the place, and of how even the local press was abandoning its eternal optimism and beginning to ask questions about the city's long-term future. And then he began:

> The moral effect of the earthquake has been great, and would-be investors are wondering whether there are not places other than

> San Francisco where money might not be more profitably invested. A man naturally hesitates to put up a million dollar building when it may be shaken down at any moment.
>
> ... a building could be put up in Los Angeles ... for thirty percent less than in San Francisco.

And this, in essence, is what then happened. Though the city was rebuilt—hastily, without regard for the kind of planning that might have given it the appearance of the imperial city that so many longed for it to be—and though its commerce did return and its courage was recognized and has been celebrated ever since, one reality obtained. And Sir Courtney, perhaps unwittingly, had forecast it: This was the end, slow in coming and slow to be noticed, of San Francisco's supremacy among the cities of the American West.

The torch would in due time be passed southward, to Los Angeles. And though many will argue that such a transfer of power and standing would have occurred anyway, for other, organic reasons, it remains an unarguable fact that San Francisco's crown began to slip immediately after the disaster of 1906. And the city has never regained its status, nor will it ever.

Large cities survive great natural disasters, true; but earthquakes, of all such trials, tend to have a very different kind of effect upon the future of the cities that survive them. The very fact that a city falls victim to a huge shaking of the earth on which it is built tends to suggest an unanticipated flaw in the process that brought the city into existence in the first place—a sudden revelation that the land on which the city was built held some dark secret, which lay cunningly unrevealed to those who first arrived.

When the first settlers put down their tent poles beside the meadow of sweet-smelling yerba buena, and when they saw how safely and prettily their schooners rode out in the calms of the harbor, there seemed no end of logic and good sense to building a homestead; no one could possibly have imagined the turmoil that was lurking deep within the earth. There was no clue, no hint, of any trouble to come. And then came 1906, and all those comfortable assumptions were shattered, and deep-seated confidence was replaced overnight with an equally deep-felt anxiety.

Earthquakes can as a consequence have effects on cities and city populations that may not become clear for years, or decades, but that in time are realized—as is recognized here—to be profound indeed. A deep anxiety for the future conspired with a host of other circumstances to ensure that, over the decades following 1906, the lure of San Francisco ebbed away, to be replaced by the undisputed magnetic appeal of the gigantic and seismically rather safer city that was then starting to grow 400 miles to the south.

Los Angeles has earthquakes, true. Everywhere in California that lies close to where the two tectonic plates are scraping past each other is, when compared to stable places like Kansas, Nebraska, Siberia, Queensland, or the Canadian Shield, relatively seismically unsafe. It is all a matter of degree. San Francisco was in 1906 almost obliterated by a quake; Los Angeles has, by contrast, never been more than scarred.

And this has all to do with proximity. The change of relative status of the two cities is a result of one immense and often unstated factor: the relative closeness of each to the track of the San Andreas Fault. San Francisco is no longer America's principal western city because, quite simply, the fault runs directly underneath it. Los Angeles, on the other hand, has taken over as the now unassailable capital of California and of the American West because, equally simply, it does not.

ELEVEN

Ripples on the Surface of the Pond

And there were voices, and thunders, and lightning;
and there was a great earthquake, such as was not
since men were upon the earth, so mighty an
earthquake, and so great.
And the great city was divided. . . .

Revelation 16:18

The Vengeance of the Lord

Something so big, majestic, powerful, and inexplicable as this great shaking of the earth could only have been brought about by the hand of God. For years afterward hundreds of thousands of sensible people across America believed this simple explanation, and a considerable number in all likelihood still do today. In churches across the country, from Oregon to Florida, from Maine to Arizona, American citizens who went to worship the following Sunday prayed energetically for the souls of the San Francisco dead, for the relief of the wounded, and for the revival of the city—while at the same time stoically accepting, by and large, the cruel mysteries of the ways of the Creator.

The depth to which this belief was held varied from place to place. A clever and a skeptical few looked for a natural cause, while the faithful and the pious accepted without question that this was divine fate, nothing more or less. A pamphlet distributed from the then-village of

Mountain View told its readers that without doubt the event had been retribution for "the sin of man." This view was firmly held around the planet, too: The great British expository preacher George Campbell Morgan, who could draw crowds of almost 2,000 at Westminster Chapel in London, had no hesitation in ascribing blame and cause. The earthquake and the fire that followed it, he told his enraptured congregation soon after the dust had settled, was the judgment of God, visited on a wicked city.

But few believed more dutifully that all was the mysterious workings of the Divine than a small gathering of faithful who had met in Los Angeles for the first time just a few days before the earthquake. They were in the main black men and women, and they had gathered to profess their faith—in a very new and somewhat surprising way—in a disreputable-looking two-story mud-colored clapboard building, a former church that had fallen on hard times and was being used as a boarding stable.

The building was huddled into a short and otherwise forgettable alley in an industrial quarter of the city, an alley that is as famous to many ardent believers around the world as Golgotha, Medina, or the Hagia Sophia are to those of other creeds. It was called Azusa Street—and the beliefs that sustained the small congregation that met in what had once been the Azusa Street Methodist Church were to have consequences that would bring about a profound alteration of American society, one that has continued right up to the present day.

They called themselves Pentecostalists, in remembrance of an ancient Jewish feast day and in honor of the Holy Spirit, whom they worshiped with unalloyed enthusiasm. They clapped and cheered and waved their hands and they spoke in tongues; they interpreted the words of the Bible strictly, as fundamental laws; and they claimed they were guided by unmistakable signs from heaven to perform the Lord's work on earth. The San Francisco Earthquake was the first sign this congregation had ever seen, and coming so early in the movement's history it was universally interpreted as a message of divine approval, a spur that would henceforward send out the Pentecostalists fearlessly on their way.

The intellectual roots of Pentecostalism can probably be traced to John Wesley and to the Methodism that he founded in the middle of the eighteenth century—a church for ordinary people, a profession of Christian faith that had been stripped of its deadening burdens of propriety and pomp. Although Wesley's thinking had been well regarded in America for some time, as a movement it took off rather later, flourishing essentially after the Civil War; it has been viewed as a reaction among the less well off to the cold formalism of most Christian church services of the times. Moreover, the poor and the dispossessed and the neglected of America were eager to take matters rather further than their British counterparts: They wanted to participate in what was called "heart religion," to profess their faith with uninhibited enthusiasm and drama, and to do so in "Holiness" churches that recognized miracles and otherwise inexplicable happenings that could only be the work of the Almighty.

They finally got what was needed when, in 1900, a preacher named Charles Parham, teaching at a seminary in Topeka, Kansas, triggered an extraordinary reaction in a young female student—soon to be as famous among charismatic Christians as Bernadette was to Catholics—named Agnes Ozman. During the school's New Year's Eve service, Miss Ozman asked Parham to baptize her. As he did so, he later wrote, so "a glory fell on her, a halo seemed to surround her head and face, and she began speaking the Chinese language and was unable to speak English for three days. When she tried to write in English to tell us of her experience she wrote the Chinese."

Agnes Ozman had demonstrated the miracle of xenolalia—a sudden ability to speak a foreign tongue of which she had no prior knowledge.* Parham and his followers promptly saw this as an unambiguous revelation, a sign from the Holy Spirit that he had visited and baptized the young woman. The Holy Spirit had done so, moreover, in precisely the same manner as he had done—according to the Bible, in

*Glossolalia, another form of speaking in tongues that Pentecostalists take as a Spirit-induced sign of their rapture, differs from xenolalia in that it is comprehensively unintelligible, dismissed by skeptics as gibberish.

Acts 2:1-4—for the twelve disciples on the Day of Pentecost, fifty days after the second day of Passover. That being so, the followers of the new religion that the Reverend Parham now saw revealed to him would call themselves Pentecostalists.

Within days other students, too, began speaking in languages unfamiliar to them—Swedish, Russian, Bulgarian, Hungarian, Italian—and in ways that independent experts confirmed as accurate and authentic. Parham now had no doubt: The Holy Spirit was being revealed in these miracles, an age-old prophecy was being fulfilled, the new church must be founded, spirited missionary work must begin. And so, religious fervor being what it is, within months thousands of adherents were swarming to Parham and his charismatic church—so much so that he and his acolytes quickly established themselves in Texas and Missouri, two of the states that bordered Kansas, as well as in Florida and Alabama.

And there matters might have rested—with these new Pentecostalists no more than another small group of flamboyant Christian charismatics and evangelists tucked safely away among the cornfields of America's southern plains and the bayous of the Gulf Coast. But in fact matters unrolled quite otherwise.

A one-eyed black preacher named William Seymour, who had listened, enthralled, as Charles Parham spoke at one of his outreach missions in Houston, decided late in 1905 to take the Pentecostal Word to California—and, specifically, to bring it to the poor quarters of the fast-growing city of Los Angeles. Seymour, a short and stocky son of slaves, and who was by no means an arm-waving orator—some described him as meek and having "no more emotion than a post" when he gave his sermons—at first had a difficult time: The pastors of one church padlocked the doors to prevent him from spreading his exotic message there, and he had to resort to asking believers to join him for worship in the living room of his lodging house.

But slowly, steadily, a congregation formed—made up at first mostly of black men and women, and then with a small number of white adherents, too. Everything changed when, on April 9, 1906, Seymour baptized a man named Owen Lee, who promptly began to speak in tongues, prompting the pastor's supporters to declare a miracle. Seven others in the audience, duly amazed by Lee's linguistic performance and gripped by religious ecstasy, started to speak and howl in tongues as well. A woman named Jennie Moore promptly began to play the piano and sing sweetly in Hebrew—even though she had never played a piano before, and knew not a single word of the language.* And then so many curious people thronged from the slums to the lodging house, and danced and shouted and sang on the porch, that the foundations gave way, the porch collapsed, and everyone was tipped out into the streets. No one was hurt—which was taken as a sign as well—and William Seymour (or Elder Seymour, as he now was known) had to go and find himself another church.

Which is how he came to Azusa Street. In April 1906, number 312 was no more than a partially burned-out clapboard stable, marked for sale and half filled with debris (it had once been a church and a remaining window in the Perpendicular Style reflected this use). Seymour made a deal with the owner, moved in nail kegs for chairs and redwood planks for tables, and spread straw and sawdust on the floor. He then announced he would hold services there every following day from 10:00 A.M. until midnight. As word got around, the place became gripped by fervor, as the *Los Angeles Times* remarked in an article on the front page of Wednesday, April 18, 1906:

> The devotees of the weird doctrine practice the most fanatical rites, preach the wildest theories and work themselves into a state of mad excitement in their peculiar zeal. Colored people and a sprinkling of whites compose the congregation, and night is made hideous in the neighborhood by the howling of the worshipers, who spend hours swaying back and forth in a nerve-racking attitude of prayer and supplication. They claim to have the "gift of tongues," and to be able to understand the babel.

The city officially frowned on the group, which grew steadily and frighteningly larger and larger. The police had to be called to break up

---

*Jennie Moore went on to become Seymour's wife.

This forlorn clapboard building—once a stable, long disused—on Azusa Street in the center of the young city of Los Angeles was chosen in 1906 as the local headquarters for the new Pentecostalist Revival. The earthquake 350 miles to the north galvanized activity among the congregation, and led to an extraordinary burgeoning of what has since become one of the most powerful Christian movements in America.

---

the crowds that gathered outside the church. The Child Welfare Agency, such as it was, tried to close it down because unsupervised youngsters were attending meetings at all hours. The Fire Department had to be called because of unexplained red glows seen inside the building and the sounds of explosions disturbing the neighborhood. The City Health Department decided that too many were crammed into the building, that it was insanitary, and that it ought to be shut down.

And then, on the very same day that the readers of the *Times* first read of these strange goings-on at 312 Azusa Street, San Francisco was hit by the earthquake. Frank Bartleman, a wandering preacher who had come to at Azusa Street and was helping Seymour deal with the throngs, knew immediately what the onset of seismic mayhem truly meant. The Book of Isaiah, chapter 26, verse 9, he declared, offered the reason: "When thy judgments are in the earth, the inhabitants of the world will learn righteousness." The earthquake convinced him of the wisdom and truth of the Pentecostal approach. "I seemed to feel the wrath of God against the people and to withstand it in prayer," he wrote.

> He showed me he was terribly grieved at their obstinacy in the face of His judgment on sin. San Francisco was a terribly wicked city. He showed me all hell was being moved to drown out His voice in the earthquake. The message he had given me was to counteract this influence. Men had been denying His presence in the earthquake. Now He would speak. It was a terrific message He had given me. I was to argue the question with no man, but simply give them the message. They would answer to Him. I felt all hell against me in this, and so its proved. I went to bed at 4 o'clock, arose at 7, and hurried with the message to the printer.

Thousands of copies of Frank Bartleman's hastily written tract, in which he claimed excitedly but with impressive sincerity, that the earthquake was "the voice of God," gave Seymour's group an aura of respectability and of divine imprimatur. He was able to cite no fewer than eight earthquake-related passages in the Bible,* together with a seismically relevant jeremiad from John Wesley. All of a sudden this strange gathering of the gibbering and gesticulating faithful could not be so easily dismissed as merely a mob of hysterical fanatics. The new adherents who flocked to Azusa Street and to the dozens of sister churches set up to accommodate them became part of a true revival of sincere charismatic worship—and the Pentecostalists of America and the World Pentecostal Movement have never really looked back from that moment on.

---

*Job 9:6; Psalms 18:7; Isaiah 2:19; Isaiah 13:13; Isaiah 24:1; Isaiah 26:5; Nahum 1:5; Revelation 16:18.

With Azusa Street as its first spiritual headquarters, the church then spread like a forest fire in summer. Branches were opened in Tennessee, North Carolina, Norway, Oregon, South Africa, Brazil, Chicago, the Ukraine, Korea, Colorado . . . such that by the middle of the century there were branches and followers by the tens of millions almost everywhere. The names of the leading practitioners of this peculiar kind of Christian worship become ever more familiar—Aimee Semple McPherson in the early days, and more recently colorful figures like Oral Roberts, Jimmy Swaggart, Dennis Bennett, Jim and Tammy Faye Bakker, Pat Robertson, Steve Hill. This exuberant evangelism became so powerful that politicians, particularly in America, and particularly conservatives, looked to this faith above all for support. The Pentecostal Movement is these days a thing of formidable power and wealth and influence and spiritual importance—a galvanizing force for America's underprivileged like few other movements in the country's history.

It is risky to attempt to forge a direct link between any natural occurrence and the growth of any subsequent religious or political movement—the fundamentalist Islamic movement that some will argue was spawned by the eruption of the volcano Krakatoa in 1883, for example, probably came about for a host of other reasons, too. Much the same can be said about the Pentecostal Movement in America, and the possible triggering effect that the San Francisco Earthquake had upon it, in its early days. The roots of the movement had, after all, already long been sown—by John Wesley in Britain, by Charles Parham in America, and by hundreds of other adherents who wished to break loose from the rigid practices of settled Episcopalianism. All these people needed was a sign, a catalyst that bore the signature of the Divine: What they got, in 1906, was the San Francisco Earthquake.

The Coming of the Paper People

From all along the curving northern coastline of San Francisco—from most of the Embarcadero, from North Beach, from Fort Mason or the Marina or Crissy Field, from the Presidio or up by the tollgates at the Golden Gate Bridge—the hills of Marin County lie green and billowing beyond the choppy waters of the Bay. There are pale scatterings of houses and dark patches of forest, and the peninsulas and embayments and hills on which they stand distinguish themselves from one another only by a careful scrutiny of their geography. The view has all the appearance of a diorama, with pastels done on drop cloths, the fly curtain pulled aside. Mount Tamalpais looms tallest, to the left and in the far background; somewhere within its folds is Muir Woods, with its ancient trees made memorial to the long-dead Scotsman who was the archdruid, so-called, of the Sierra Club; in front are Sausalito and Tiburon and the low hills of the Headlands, and beyond them all there is San Quentin and Bolinas and Corte Madera, and then the low hills and valleys of Napa and Sonoma, warm and sheltered places where wine is made.

In front of this scene stand two islands—of which only one appears properly insular, since it is possible to see from all along the city shore the sea that entirely surrounds it. This is Alcatraz, the incorrigibles' prison island, with its ever-flashing lighthouse and the sand-yellow fortress that stands on top, its giant water tower attendant, and all now for the tourists. The other island, however, is very much larger, with a grassy flattish-topped peak called Mount Ida* that looks quite similar to the Marin Hills that range beyond it. Since from the south—from San Francisco, in other words—it is impossible to glimpse the sea behind, it is easy to assume it is a part of the mainland only a little closer than the rest. But is in fact quite separate, has for the last two centuries been called Angel Island, and it is a place that enjoyed a short

*The top was shaved away by bulldozers in the middle of the last century to allow a nest of Nike missiles to be placed there during the cold war, one of several missile sites designed to protect the city from Moscow's potential mischiefs.

and poignant spell of importance as a direct result of the events of 1906.

It was all to do with one peculiar coincidence of two of the episodes of ruin: The first, the utter destruction of almost all of Chinatown; and the second, the burning of all the records relating to the thousands of immigrant Chinese who lived there. From April 1906 onward no one in official San Francisco had any significant idea of who had lived in the city's Chinese quarter—and this degree of ignorance had important implications for those back in China who wanted to emigrate, to join their relations already living in what was still known from its Forty-Niners-era reputation as Jin-shan—the "Gold Mountain City."

For them, everything depended on the precise wording and interpretation of the 1882 Chinese Exclusion Act—its title and its intent so nakedly racist as to be barely credible today. It began:

> Whereas, in the opinion of the Government of the United States the coming of Chinese laborers to this country endangers the good order of certain localities within the territory thereof:
>
> Therefore,
>
> Be it enacted by the Senate and House of Representatives of the United States of America in Congress assembled, That from and after the expiration of ninety days next after the passage of this act, and until the expiration of ten years next after the passage of this act, the coming of Chinese laborers to the United States be, and the same is, hereby suspended; and during such suspension it shall not be lawful for any Chinese laborer to come, or, having so come after the expiration of said ninety days, to remain within the United States.

Over the next two decades a number of amendments were added to the legislation: The most important, so far as it affected matters in 1906, was a liberalizing clause that allowed relatives of Chinese people who were already legally settled in America to go there to settle, too.

Until the spring of 1906 all would-be Chinese immigrants had been processed and interviewed in a small two-story shed that belonged to the Pacific Mail Steamship Company on the San Francisco waterfront. It was there that immigration officials, armed with index cards giving the details of every Chinese-born American citizen legally in the country, conducted Exclusion Act interviews with the arrivals, and did so with clinical efficiency: If the arrival turned out to have a father or brother already rightfully in America, then the official stamped them in; if not, they were first put in the lockup, then marched down the docks and put in the hold of a China-bound ship without further ado.

But after the earthquake and fire there was no more two-story hut—and there was a sudden rise in immigration from Canton and Shanghai and the so-called coolie ports on the coast of Fukien, all of it brought about specifically by the rumor that it was suddenly going to be very much easier to get into America. Word went around that all would-be immigrants had to do was invent their own family trees—to become what were known as "paper sons" and "paper brothers," or, for those young women claiming that fiancés were waiting on the California harbor fronts, "paper brides." What began after 1906 was, then, the invasion of the "paper people." Official America's job was now to identify any fictions and to prevent their authors from coming, settling, and establishing a beachhead in San Francisco—hence the elaborate game, serious and bizarre by turns, that was to go on for the next three and a half decades.

Maxine Hong Kingston summed it up impeccably in her celebrated book *China Men,* published in 1980. "Every paper a China Man wanted for citizenship and legality burned in that fire," she wrote. "An authentic citizen, then, had no more papers than an alien. Any paper a China Man could not produce had been 'burned up in the Fire of 1906.' Every China Man was reborn out of that fire a citizen." It was swiftly realized that everyone wanting to come into America from China could quite easily invent his own history. He could give himself a brand-new genealogy, in which he could claim a relationship with someone who was already settled in San Francisco. Since there was now in the City Hall files no evidence to the contrary, how could any immigration official properly deny the immigrant's stated story as the unvarnished truth?

It was on Angel Island that the confrontations caused by this dilemma took place—with the stakes being both simple and life-changing. Those who convinced the skeptical authorities won the right to remain in America, and those who did not were told to leave. The end results of the confrontation games were sometimes tragic, sometimes inexplicable, sometimes prosaic and unexceptional: In all cases they left an indelible imprint on the face of Chinese-American society.

The Bureau of Immigration had always planned to create an Ellis Island West, as it were, on the hitherto almost deserted thousand acres of Angel Island.* The initial decision had nothing to do with China. Rather, it was assumed that once the Panama Canal was opened, Europeans who had an eye on living in the western states would buy their sea passages to California rather than to New York, and would undergo the immigration process in San Francisco, conveniently close to their expected front doorstep, as it were. The idea was thus to create on the Pacific Coast as welcoming an environment for Europe's "huddled masses yearning to breathe free" as already existed on the Atlantic seaboard.

The earthquake, however, changed all that. The buildings that had been begun in 1905 would now be pressed into service not to welcome Europeans but to prevent tale-telling Chinese from getting in. Precisely because of the Exclusion Act, and because of the new suspicion that untold numbers of Chinese immigrants might now try to outwit the system, Angel Island fast became notorious as the place where the American government tried hard to identify them, exclude them, and not welcome them at all. (Besides, as it turned out, barely any European immigrants-to-be ever used the Panama Canal. But for the Chinese and a smattering of other arriving Asians, the Angel Island

---

*Previously little happened on the island but for the occasional duel (until dueling was banned in 1854). The best-known contest, following a bitter argument over slavery, pitted a circuit court judge against a state senator: The senator lost, and died.

Processing Center might never have been used at all. As it happened, from 1910 onward they had the island almost all to themselves.)

Central to the technique employed by American officials to trip up each would-be immigrant was a lengthy and detailed interrogation. Back on Ellis Island the typical immigrant was asked around 30 questions, and all of them fairly perfunctory. But most Chinese arriving at Angel Island were often asked as many as a thousand questions—certainly 200 at the very least, and all of the answers to be corroborated by those on the mainland to whom they were supposedly related. Each experience was miserable in the extreme—with Angel Island now mythologized among most of today's Chinese as a place of "great sadness and pain."

Once they had landed and passed through the undignified rigors of deratting and delousing—Angel Island was also a quarantine station, with body searches and lice dustings—the Chinese were separated from any other Asians, such as any Punjabis or Siamese, who had been on the voyage,* and then divided by sex: Generally speaking there were, at any one time, as many as 300 Chinese men in one barrack block and 50 Chinese women held separately in another, behind doors four inches thick. They were told that it would be folly to try to escape: The currents around the island were fierce and the waters unusually cold. Just as with Alcatraz, few tried to leave.

Each day they were called in, alone, for questioning—usually by a pair of white officials helped by an interpreter and a shorthand reporter. The questions to each applicant invariably related to the supposed relative who was already living in America. Exactly where in China did that relative come from? How big was the family house? How many steps do you think it took to go from the front door to the rice bin? What was the eldest brother's favorite breakfast? Who sat where in the mah-jongg games? The answers, laboriously written down, were taken back on that evening's ferryboat and given to offi-

---

*Any white passengers had their passports inspected, with courtesy, onboard ship.

cials, who took them on to the target relative—who may well have lived in New York or Dallas or Kansas City. They then tried to see if the answers were right.

It often took months for inquiries to be completed. Several detainees were held for years, always under lock and key, in their barracks. Coaching was forbidden; visits were banned; all kinds of efforts were made by officials to ensure that the applicants were surprised by the questions, and by the applicants and their friends to ensure that they were not. One woman who was prepared for the interrogations while still in China made copious notes, turned them all into a lengthy song and—during the weeks that her ship was sailing to America by way of Manila, Honolulu, Yokohama, Shanghai, and Hong Kong—learned the song by heart. And the kitchen staff who prepared the (reportedly virtually inedible) food that was served to detainees could be bribed to pass messages back and forth between the Chinese in the city and those waiting on the island.

The detention center functioned until 1940, when a fire—ironically—destroyed it. The law that permitted its existence was repealed in 1943, and then only because China decided to become America's ally in the Pacific theater of war. Even then, white American antipathy to the Chinese was not entirely extinguished, and only 105 Chinese were permitted full immigrant status each year. Today matters are more equable, relations more benign. The story of Angel Island and the role it played following the earthquake has a continuing resonance—not least in helping to ensure, through some kind of social Darwinism, that those Chinese who did manage to get in over the Exclusion Act's innumerable hurdles were either very clever and capable, or very cunning indeed.

But there was one other legacy, discovered by chance by a National Park Service warden only in 1970. The wooden walls of the detention blocks, he noticed, bore dozens of carved inscriptions, Chinese characters written in vertical lines, seven characters to a line, the lines usually grouped in fours. Interpreters called to inspect them discovered them to be poems—all brief and elegant expressions of misery that

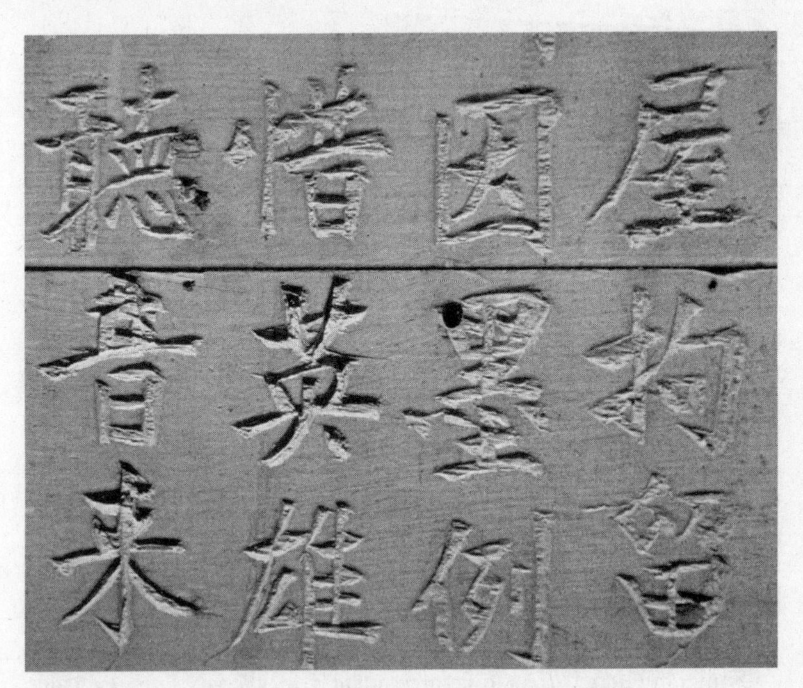

A fragment of one of the many mural poems found carved into the wooden walls of the Angel Island immigrant detention center, which, after the earthquake, became the main West Coast port of entry for would-be Chinese immigrants. This was written by one of the "paper people," migrants who tried to prove their qualifying identities, despite the loss of their relatives' documents in the fires.

---

had been carved into the wood, in sets of near-perfect ideographs, by many of the lonely, angry, and frustrated Chinese men who had been held in the barrack blocks. None of the poems comes up to the standard of Li Bai, maybe; but all have a certain elegance that lifts them well above doggerel. All are unsigned, and many make reference time and again to the sadness of the "wooden buildings" in which they are held, and to the "Land of the Flowery Flag" in which they still hoped to be allowed to settle:

Imprisoned in the wooden building day after day,
My freedom is withheld; how can I bear to talk about it?
I look to see who is happy but they only sit quietly.
I am anxious and depressed and cannot fall asleep.
The days are long and bottle constantly empty;
My sad mood even so is not dispelled.
Nights are long and the pillow cold; who can pity my loneliness?
After experiencing such loneliness and sorrow,
Why not just return home and learn to plow the fields?

And in some of the poems there were thoughts expressed of, one day perhaps, demanding recompense for all their suffering:

If the Land of the Flowery Flag is occupied by us in turn
The wooden building will be left for the angel's revenge.

Some might care to look around at today's San Francisco—at today's America, even—and see how those who passed through the Angel Island experience, directly or indirectly, have made their mark. Clever, tough, determined, resilient—and, to judge by the poems, cultured and sensitive also—the Chinese are making an impression on America today like few others of those hundreds of racial groups that now make up the country. To the extent that this strength and resilience and doggedness can be said to have been born of their experiences following the events of April 1906, one can suggest, without too great a stretch of the imagination, that the earthquake did have its effect: It tempered the will of some of the Chinese who immigrated, and it helped to render them particularly able to succeed, in a society where success is all.

The Flight of Creation

Aside from the film *San Francisco,* which Clark Gable, Jeanette MacDonald, and Spencer Tracy made in 1936, and which has many devoted admirers but few pretensions to greatness, what lasting art was born from the tragedy? The short answer appears to be: Not much.

Although there were a number of talented and nationally known men and women working in town at the time—people like Ambrose Bierce, Jack London, Bret Harte, and Gertrude Atherton, some of whom stayed on and became intensely curious about the developing state of affairs—the creative energies of turn-of-the-century San Francisco truly lay with that colorful, highly vocal, and somewhat self-regarding movement known as the Bohemians, a movement that was in full cry at the time of the earthquake.

They had come at first as adventurers, as had many in early San Francisco, except that what distinguished them was that they carried, as someone put it, "slim volumes of Virgil" in their rucksacks for the long overland nights, and, once in town, they met up with other Virgil-lovers, got drunk with them, and promptly stuck together, isolating themselves from the more philistine companionship of the Forty-Niners. In those early days there was spontaneity and freshness to their writings, a kind of innocence of the newly arrived; by the turn of the century, however, it had been replaced by an air of contrivance, with even their name, the Bohemians, suggesting a certain preciousness. Robert Louis Stevenson was at first charmed by them; the *Overland Monthly* published them; Oscar Wilde visited them; and then for a while they sagged, grew bloated and uninteresting, and faded from the scene. (The Bohemian Club, which some of their journalistic fellow travelers inaugurated in 1872 for the city's cleverest and most intellectually alive, survived them, though, and remains today a gathering place of the West Coast's power elite.)

Very little that the Bohemians wrote from the turn of the century, whether the first fresh material or the later more jaded maunderings, has enjoyed much lasting popularity, nor did it really deserve to: So much of it was too relentlessly romantic and introspective and self-absorbed to enjoy wide currency or longevity. Moreover, many of this movement's finest writers and poets—if soi-disant—decamped to the coast after the earthquake. Many of them quite specifically went off to the hamlet of Carmel-by-the-Sea, which was fast becoming fashionable as an artists' colony, in order get away from the unrefined vulgarity of the city's demolition and reconstruction. One of the archpriests of the movement was the poet George Sterling—later to kill himself by drinking cyanide. His best-remembered phrase described San Francisco as "the cool gray city of love." But he composed the words in Carmel, a hundred miles away. Indeed, once characters like Sterling had all left town, it did seem as though the creative heart of the city had been torn out, too. There were briefly held fears, never realized, that San Francisco as a cultural epicenter might have been ruined for all time.

The painters left in droves as well, some demoralized, others made homeless, a few ruined. Collections both private and public were reduced to piles of smoking canvas—the carefully gathered selection of works at the Mark Hopkins Institute of Art, thirty-five years in the making, was utterly destroyed. A few notables stayed: William Keith, perhaps the greatest western landscape artist of the time, held on, despite losing fully a thousand works, including forty major oils of the Sierra Nevada mountains. Maynard Dixon carried on sketching, writing poems, and painting covers for the local magazines. And the photographer Arnold Genthe stayed as well, even though he lost everything other than his Chinatown pictures—which in time would make him a fortune.

But otherwise the bulk of the community took to its heels—some went down to Carmel also, others fled across the Bay to Oakland and Berkeley, still more went to New York or across the Atlantic to Paris and London, never wishing or daring to return. And so for a while the city remained gloomily silent, no one writing, no one painting, no one composing, no one drinking absinthe into the small hours and cursing the moon and the stars. The creative juices had flowed away, and there seemed at first precious little chance of bringing them back.

Not that the poetasters and hacks and the penny-a-line merchants fell totally silent. In the archives lie reams of the most execrable poetry dedicated to the city and its troubles, and scores of unreadable short stories about the earthquake. Most of the poetry is couched in the favored Bohemian high romantic style that would have made William Topaz McGonagall, the Scot whose standard set that by which all poetic dreadfulness is measured, green with envy. The greater number of those poems dedicated specifically to the future of the city appeared either in *Sunset* or the *Overland Monthly,* both of which were determined to boost the morale of the citizens by producing editions urging and cajoling everyone to work hard to re-create the new San Francisco. This, from the Irish immigrant–Bohemian Harry Cowell,* is perhaps typical, and memorably forgettable:

Now, waked from painlessness of pain bespent,
The prophet Faith foresees her doubly fair—
By fire transfigured, fire and dew and air—
All-beauteous with a beauty yet to be;
A city-soul in utmost wonderment
Renascent by the immemorial sea.

But then, out from the ruin and abandonment and the making of so much temporary dross, stepped a young married couple who would make a game attempt to change everything. They were Arthur and Lucia Mathews, painters and furniture makers, whose atelier on California Street, which they opened as soon as the ruins had cooled, became a center for what the pair hoped would be rapid revitalization of the arts in the city.

The couple are perhaps best known for their handmade chairs, tables, and mirrors, all invested with a florid and flamboyant style—a sort of Art Nouveau meets Beaux Arts meets Arts and Crafts meets Celestial Empire mélange—that came to be known as California dec-

*Father of Henry Cowell, the great American composer.

Noble, proud, beautiful, resilient, and defiant—one of the many images that graced the covers of magazines that were published in San Francisco after the earthquake, and which sought to help boost public morale and lure settlers—and their money—back to the ruined region.

orative, which has its devotees, and so its price, still today. Arthur Mathews produced an enormous number of majestic and forceful paintings, most in a decidedly Renaissance palette and many with a classical, allegorical symbolism about them; his twelve great murals, made in 1913 for the California state capitol in Sacramento, famously provide a vision of the postearthquake state that, with their exaggerated visions of a high-colored new Byzantium rising on the Bosporus-like Pacific coast, can be regarded as either noble or incorrigibly cheesy.* Mathews was not a well-liked man: He was forceful and bombastic, and not at all progressive. He would happily have chased the Bohemians out of town and kept them at arm's length down the coast in Carmel.

But the couple did have a small following, a circle of neoclassical painters and decorative craftsmen, and it was through them that Arthur and Lucia Mathews tried to pump a measure of artistic adrenaline back into the city's broken system. They did this by setting up a small club and a publishing house, the Philopolis Society and the Philopolis Press, and they started a journal: *Philopolis: A Monthly Magazine for Those Who Care.*

But it didn't last. Philopolis was out of business just ten years later, and with its collapse in 1916, and with the continuing absence of so much other talent, the city's litterateurs seemed to enter into a prolonged dry spell. No one can be sure why this was: The only explanation that seems reasonable is that San Francisco's attempt at artistic resurrection, though heroic in intent, had a kind of artifice about it, a lack of spontaneity, that may have limited its real value quite severely. It was as though the art, instead of being permitted to bubble up naturally through the cracks in the wrecked pavement, was being briefly boosted by the chamber of commerce, or the Lions Club, or the Kiwanis—with the 1915 Pan-Pacific Exposition that was staged as an official celebration of the city's rebirth all a part of this relentless boosterism. Artists generally prefer to work at their own pace, with their own instincts, gathering themselves into groups and movements and schools at their own behest. Though there are exceptions,† artists generally do not care much to create at the whim of officialdom.

---

*The legislators have evidently held mixed feelings about the murals, first displaying them in the capitol rotunda, then demoting them to storage, finally bringing them back—but this time to a smaller rotunda in the basement, where they will probably remain for good.

†Such as the Works Progress Administration and the Federal Writers Project of the Depression years. Had the artists of the time not been supported, many would likely have starved. Such was not the case in San Francisco in 1906: Artists could always push off to places less likely to be ruined, and evidently did.

One can reasonably say that it was not until the fifties, when the Beat poets began their howling, down at the City Lights bookshop and in the bars and cafés of North Beach, and when Bohemianism and the beatnik style of life came welling up from nowhere, that creative regeneration began in the city in earnest once again. The damage that was done to the intellectual and artistic spirit of the city by the events of 1906 took much longer to heal than did the physical damage. The buildings eventually came back, but not the soul entire, not for a very long while.

A Fretworked City, Pinned with Steel

They had a chance. They had an empty slate. There was an opportunity for the city's elected leaders to re-create San Francisco in the way that London had done after its great destruction in 1666, a chance to emulate Chicago's bold rebuilding after the sorrows of 1871. Moreover, a written plan already existed to accomplish all this—and, even more than that, the plan had been delivered, formally, in black and white and fresh from the printers, on the afternoon of Tuesday, April 17, just a matter of hours before the earthquake struck. The plan's author was in Paris at the time of the quake; he was telegraphed with the news of the tragedy and came scurrying back to town like an eager terrier, panting to begin building. "San Francisco of the future will be the most beautiful city of the continent," he said when he arrived, "with the possible exception of Washington."

His name was Daniel Hudson Burnham, he came from Chicago, and so far as Washington, D.C., was concerned, he was in a position to know, since he had, in essence, designed the place. The extension of the Mall, the siting and design of Union Station, the creation of the Lincoln Memorial—all these were decisions made by Daniel Burnham. Having designed, in postfire Chicago, any number of noble, classically based, and massive office buildings and houses, he had become a nationally known and revered planner of cities. He was a man of bold ambition. "Make no little plans," he once famously remarked. "They have no magic to stir men's blood and probably themselves will not be realized. Make big plans; aim high in hope and work, remembering that a noble, logical diagram once recorded will never die, but long after we are gone will be a living thing, asserting itself with ever-growing insistency."

Those San Franciscans who, at the turn of the century, had a grand vision for the future of their city—a vision of a cosmopolitan, sophisticated western capital, no longer the brawling frontier city of Gold Rush days—turned inevitably to Burnham. They first approached him informally; and then in 1902 the semiofficial Association for the Improvement and Adornment of San Francisco asked him formally for a plan in 1904, not least because they liked his designs for three of the city's best-known buildings.* He promptly came to town, set up a studio near the top of Twin Peaks, worked (without a fee) for the following year, and came up with a vision of a newborn imperial city that he felt could rival any then existing in the New World. His design, he thought, should and would be a metaphor for the urban future of the American West. It would be a thing of beauty, magnificence, and style, and it would endure—though no one uttered such a heresy in the earshot of the planners—for just as long as fickle nature allowed.

The design, when unveiled for the Board of Supervisors in September 1905, was ambitious, to say the least. The underpinning notion was triumphalism, the style baroque, the model Paris, the scale prodigious. Nine immense and die-straight boulevards radiated from the great new palace of City Hall, intersecting, as they speared across the map with *périphériques,* with huge parks, colonnades, marble subways, and castellated mansions that looked down from the city's famous hills. One park, to the south of the city, was three times the size of Golden Gate Park, which was already monstrous. And between all these grand marble confections ran water—streams, cascades, reflecting pools, and lakes set throughout the city at staggered heights, thereby providing headwaters for a score of huge fountains that could keep the city

---

*These were the *Chronicle* and the Mills buildings and the Merchants' Exchange: All three would be badly damaged in 1906.

bathed in even more mist and moisture than nature provided on its own.

The public was due to see the Burnham plan in April of the following year, and with impeccable timing the printer delivered fat bundles of the final edition to City Hall just hours before the earthquake. They were all burned in the fires, and new copies were printed and distributed to citizens a while later—but not before the newspapers had seen what had been planned and had, at first, given it their most enthusiastic blessing. The city, after all, was now a tabula rasa: Why not let Daniel Burnham loose on its wreckage, to re-create the city with a magnificence beyond all imaginings? SAN FRANCISCO WILL ARISE FROM THE ASHES / A GREAT AND MORE BEAUTIFUL CITY THAN EVER read one newspaper headline.

But it was not to be. As it turned out, the city and its commercial oligarchs had neither the time nor the patience nor the vision to see the Burnham plan through. "The crying need of San Francisco," wrote Michael De Young, the surviving founder* of the *Chronicle* newspaper, "is not more parks and boulevards; it is business." The plan would take a generation to bring to fruition, and, in the view of the newspapers and the Downtown Business Men's Association, that was too long to wait. Besides, the principal oiler-of-wheels and greaser-of-gears within the city administration, Abraham Ruef, was on trial for corruption (and indeed went to prison), and the boards that he had overseen and the city management engines he had supervised were now riven by faction and dissent, with decisions going unmade for months at a time.

So Daniel Burnham and his West Coast ambitions fell victim to both the indolence of the city and the impatience of its business community. Such building as was started after the fires continued along exactly the same lines as before, without hesitation, consideration, or the vaguest hint of a grand idea. To be sure, a few streets were widened—

---

*He and his brother, Charles, teenagers from St. Louis, founded the paper in 1865, supposedly with a borrowed $20 gold coin. The paper was a muckraking scandal sheet, and Charles was shot dead by an enraged reader; Michael was attacked by a gunman some years later, but lived.

the planners, such as they were, taking advantage of the great swaths that had been cut by the dynamiters—but none was redirected, no massive demolitions were undertaken to create tree-lined boulevards, no Greek temples were made, no subways were dug, no fountains erected, and no tracts were ever leveled and landscaped to make parks and public pleasure grounds. The proposed epicenter of it all, City Hall remained unbuilt, a ruin, for years; and even today, though immense in scale and tricked out with acanthus leaves in marble and fineries of gold leaf, it has dark alleys beside it full of unfortunates, lacks charm and grandeur, and possesses little sense of once having been central to something great, imperial, and intended to last for all time.

And without Burnham, without a settled sense of urban purpose, the city allowed itself to grow organically, with neither direction nor design. Architects today mourn the fact that no San Francisco school of architecture was ever allowed or encouraged to flourish—in the way that the Chicago school, with Burnham one of its members, did so energetically in the aftermath of that city's destruction by fire in 1871. The city center's commercial buildings were hastily put back up, with very few of them either nobly or loftily made; and the houses that were then crowded into the outer boroughs were made less lovely than they might have been, their architectural styles often merely sentimental, nostalgic, or plain faux.

The charm and loveliness of today's San Francisco derives mostly from its setting and its climate. Its architecture is modest at best, its planning less worthy than it might have been. Hence the regret over the speed with which the city was re-created after 1906, and the fact that it was re-created so very much in its own earlier image. Maybe Burnham's plan was not ideal, or without its shortcomings; but it did offer the city a chance for civic greatness, and more than a few today lament that this chance—which was offered, uniquely, by the tragedy—was so swiftly and thoughtlessly passed up.

Getting Safely Out of Dodge

Unlovely though so many of San Francisco's replacement buildings may have been, the more substantial of them were at least designed and built in ways that made them less vulnerable to the kind of ruin that was suffered by so many of their predecessors. Architects and contractors learned a lot from 1906 about the way that rigid buildings behave when the earth below goes insane. The new commercial buildings of the city's downtown—and of other cities around the world, for the implications of San Francisco were global, in all kinds of senses—were made stronger and more resilient as a result—better able to keep standing when the world's foundations were shaking deep below. Building science is a discipline that advanced hugely in the wake of the disaster: The use of reinforced concrete supports, steel skeletons, fire-protection systems all rapidly eclipsed the gimcrack building practices of before; a host of other structural reforms were implemented and new building codes were written and rigorously enforced, often with great dispatch. And the city has not suffered a disastrous fire since.

There has, however, been another earthquake, as inevitably there was bound to be. This came in 1989, when a section of a small fault that runs parallel to the San Andreas ruptured near a hitherto-unknown Santa Cruz mountain called Loma Prieta and sent powerful shock waves to San Francisco, Oakland, and beyond, as well as to all the smaller towns close to the epicenter. Officially—because modern earthquakes are always timed in Greenwich Time—it took place on October 18; in local terms it struck the day before, Tuesday, October 17, at four minutes past five in the afternoon.

It was not as strong as the 1906 event—it is believed to have had a moment magnitude* of 6.9, compared to around 7.9 (or 8.3, depending on the science of the day) of the earlier event. Nor was there nearly as much ground displacement. The twenty-one feet of earth movement that was seen in Olema in 1906 was matched nowhere in this later

*The all-too-complicated system of magnitude scales and intensity scales is explained, as well as possible, in the appendix.

quake: A displacement of only about five feet was recorded, at a spot some twenty-five miles south of the 1906 epicenter.

The damage, however, was quite phenomenal: 63 people died, nearly 4,000 were injured, and the cost of all the destroyed buildings and collapsed roadways and bridges was estimated at some $6 billion, a record for a natural disaster at the time. And yet—the improvements in building techniques did save the greater part of the cities that were shaken. Virtually none of the structures in the city center of San Francisco was badly damaged—a friend of mine staying in a hotel at the corner of Sansome and California Streets was shaken, but he was able to stay in his room watching coverage on television, and his room-service order was completed once a small fire in the kitchen had been put out, with only a trivial delay.

Generally speaking, only those structures that had not been made sufficiently strong and flexible suffered; and only those parts of the city that were on freshly made, reclaimed land saw their building subside, their gas lines break, and their water pipes rupture. San Francisco's Marina District, with its pretty houses hastily and greedily thrown up on land so unstable that it pulped thixotropically the moment it was shaken (just as the notorious "made-ground" areas of landfill liquefied in 1906), was particularly hard-hit.* The great San Francisco Bay Bridge—over which half a million commuters pass each day—was broken by the shuddering; and an entire section of double-decker roadway in Oakland pancaked down upon itself, flattening all the cars that at that moment were passing on the lower roadway.

The victims in all three of these particular tragedies died or were hurt because builders had not fully learned the lessons of 1906. Elsewhere in the city, and in Oakland and most parts of Santa Cruz—which was very badly damaged—builders and architects seemed to have absorbed the realities, and the structures they had created were

---

*Reclaimed land is always vulnerable to seismic interruption. Should a major quake hit Manhattan, the many towers built on its only significant area of landfill—Battery Park City—will all probably collapse. And if Hong Kong, a landfill capital, should ever be hit, beware.

tempered well enough to withstand the shaking. In general, they survived.

The most worrisome aspect of the 1989 earthquake goes beyond the imperfect nature of Northern California's construction industry. The most troublesome reality seems not to have sunk in fully: *The Loma Prieta Earthquake was not a result of a rupture along the San Andreas Fault.*

The physical characteristics of the 1989 rupture, which have been measured and examined by geophysicists ever since, turn out to be very different indeed from the signature patterns of the San Andreas. And so the comforting thought that the 1989 event might have been the big fault suddenly blowing off steam, suddenly relieving itself a little of the pressure that must have been building up since it was last relieved by the dramas of 1906, has proved ill founded. There is, sad to say, no such comfort available.

The San Andreas Fault is still sitting above (or is caused by, or is the superficial manifestation of) two tectonic plates that have been moving relative to each other at a fairly constant rate of (in places) as much as an inch and a half a year. If the assumption about Loma Prieta is correct, then the last time the San Andreas Fault moved in Northern California was not in 1989 but in fact a very long time ago, back in 1906—which means that, with the steady annual movement of its foundations, the two plates are now nearly 150 inches, more than 12 feet, out of kilter. This means that an unimaginably enormous amount of kinetic energy is currently stored in the rocks of the Bay Area; one day, and probably very soon, this energy will all be relieved, without warning.

The estimates made by the scientific community that works with this assumption make for deeply disturbing reading. They begin with the stark reality that there were seven major earthquakes along the entire length of the San Andreas Fault in the recorded years before 1906, and there have been only five such earthquakes since, and not one of those has been in Northern California. In 2003, working from this assumption, the U.S. Geological Survey—not a body known for making outlandish claims—issued a formal forecast: Sometime before the year 2032, along one of the seven fault systems that belong to the

San Andreas cluster and that spear their way through the Bay Area, there is now a machine-computed probability of 62 percent that an earthquake with a moment magnitude of 6.7 or greater will strike. There will be damage and casualties on an impressive scale.

It is not a question of whether a big earthquake will occur, nor even a question of precisely where it will hit. There *will* be a quake, it *will* be considerable, it *will* be somewhere in the vicinity of San Francisco, it will more than likely affect the San Andreas Fault or one of its cadet branches—and it *will* take place, most probably, before 2032. The only true unknown is the precise year, month, day, and time.

Which is why now all the communities in and around San Francisco—as well as agencies like the American Red Cross—have published big, fat, and very comprehensive manuals telling all who are authorized to know what their duties and responsibilities are in the event of the most statistically likely major disaster that is anticipated in the region—and that is a very, very large earthquake.

The city of San Francisco has drawn up detailed response plans for dealing with many possible scenarios: various earthquakes of various magnitudes, and with the involvement of various faults, all taking place at various times of day or night or season. The two faults that, if ruptured, will inevitably do damage to the city are the San Andreas—which runs from offshore at Daly City to a point two miles off the Golden Gate Bridge—and the Hayward, which courses up through the mountains to the east of Oakland and Berkeley. The response plan's best estimates for a modern rerun of the 1906 earthquake, involving the San Andreas, hold that some $15.3 billion in damage would be done to the city's buildings (nearly 30 percent of them would be wrecked) and some $24.7 billion in direct economic damage would result. Some 30,000 people would be dislocated. Bridges, tunnels, rail links, communications would be injured—though perhaps not disrupted as comprehensively as in the 1906 event.

There may well also be extensive damage to entities of kinds that did not exist in 1906—to oil terminals, to centers for advanced biological research, to laboratories that study genetic engineering and nuclear weaponry (much exotic research of this nature continues at the

Lawrence Livermore Laboratories in the hills above the East Bay, for example). There are public plans for dealing with very obvious challenges to the system, and there are secret plans for dealing with those aspects of society that already lurk in the shadows. And the various cities and counties that have jurisdiction in the region, and over the lives of many millions of people, have all agreed to work in concert if and when a major catastrophe engulfs any one of them.

THE HOLIEST OF GRAILS, sought essentially by all who perform research on the deep mysteries of the earth, is the ability to predict with some accuracy where and when an event will strike. But no one, despite claims that occasionally excite the newspapers, has yet succeeded in making such a forecast. Thus far there seems, in short, no measurable certainty that earthquake prediction is anything more than a long-sought chimera.

All manner of possible early-warning indicators are being examined—from the behavior of birds and snakes to the release of gases deep in the earth, to changes in chemical mixes of groundwaters and to the appearance on seismic records of small, machine-gun-rapid microearthquakes, looking and sounding like the rivets ripping from a ship's stressed hull. Such events, when examined in retrospect, do indeed all seem to be the precursors of major seismic events. But where will the earthquake of which these might be the auguries actually strike? And when? And how big will it be? So far such questions are utterly impossible to answer, and there is no real possibility of anyone, anywhere, being so bold as to try to warn anyone on the basis of what is and can currently be known.

Would they even dare, considering just what is at stake? Would a member of the staff at the U.S. Geological Survey, or at the Berkeley Seismological Laboratory, or at one of the numberless agencies and university departments whose business it is to analyze and decode some of the ceaseless flow of puzzling data from the earth, ever formally dare to ask the elected or appointed chieftains of any city to shut it down, evacuate, take to the hills? And then to shoulder the blame and anger and unimaginable cost for having predicted incorrectly? Earth's immeasurable capacity to surprise permits the earth, and only the earth, to order up the when and the where of its great interruptions.* Humankind has merely the ability to prepare for them and to respond to them as best it can; and then once more, when the dust has settled, to go back to the records and attempt to work out why.

And then there is, ever present, the hubris—which is surely humankind's greatest failing when it comes to its dealings with the world. The plans so carefully laid by all the administrators and all the seers can help, of course; but they can be rendered quite worthless if, for reasons born of pride or insouciance or both, the inhabitants forget, ignore or otherwise calmly assume that there is no danger, or that the risk is too small when balanced against the comforting notions of hedonism, beauty, and pleasure with which Northern California, so earthquake prone, is peculiarly endowed.

No greater monument to hubris can be found than in a pretty little town forty miles south of San Francisco, where people have lately made untold millions from their work on designing computers and the vitals that make them work. The town is called Portola Valley, and it was formally created in 1964 in a fold in the green and pleasant coast hill ranges that keep the Pacific fogs at bay. Some 4,500 people now live there, in circumstances of what appears to be unalloyed delight. The weather is warm, the fields are green, the trees are noble. There are pleasant little shops selling exquisite and costly goods; Volvos hum quietly along winding country roads; there are bicycle trails and horse

---

*Volcanoes offer up myriad subtle signs of impending eruptions, and by and large humankind nowadays gets out of their way when danger threatens. The onrush of the 2004 Bay of Bengal tsunamis could have been predicted—in places alerts could have been given two or three hours before the waves struck, had a system been in place. But earthquakes, alone of the trinity of seismic danger, are still entirely unpredictable; and huge sums are being spent on research to see if some clue, somewhere, might be found that would allow for a moderately reliable and infallible warning. So far, nothing.

paths and golden retrievers, and when runners come out in the evening cool, all of them seem good-looking and tanned with that peculiarly honey-colored skin that will seem so well suited to a Brioni evening at the quiet cocktail parties held later underneath the redwood glades and the star-filled skies.

But all is in fact nowhere near so well as it looks. The town of Portola Valley, it turns out, has been built exactly astride two of the most active traces of the San Andreas Fault. It is a deeply dangerous place, liable to be destroyed at any moment. Much of the community is quite uninsurable, and many of its houses and offices deserve to be evacuated and abandoned.

For the last twenty years geologists—fascinated by the hubris that they seem to encounter more than most, and shaking their heads at humankind's insistent folly in living in places where they shouldn't—have been arriving on Portola Valley's elegant doorsteps, making risk analyses and recommendations, some of them called for, some not, and all, in essence, telling the townsfolk they are errant blockheads for remaining where they are. One report says the town hall must be moved. Another suggests the school be shifted. The firehouse has to be somewhere else. The water lines, the electricity cables, the sewers—all are at risk from this trace or that trench or that area of expected liquefaction. But as to where each and all should go—a hundred yards this way or that, a mile away, ten miles away—no one ever agrees.

Certainly everyone purports to care. All of the various reports are read, digested, and discussed, and town meeting are staged and PowerPoint presentations are made—quite possibly by the very same scientists whose millions permitting them to live in so pleasing a place were made from inventing PowerPoint and the machines on which the presentations are shown in the first place.

But then the reports are quietly shelved, and another bottle of sauvignon blanc is uncorked, and Portola Valley's well-contented and outwardly happy residents settle down once more to watch the stars come out, and to reflect perhaps on how splendidly American is their way of life. It is a way of life quite unrivaled in its quality anywhere in the world, and certainly is in no way like the lives of those countless thousands in more obviously earthquake-prone places like Bam or Agadir or Tangshan or Banda Aceh—or even, for heaven's sake, like the lives of those who had perforce to go off to live in the great tent cities of San Francisco almost a hundred years before.

EPILOGUE

Perspective: Ice and Fire

Fell Giesar roar'd, and struggling shook the ground;
Pour'd from red nostrils, with her scalding breath,
A boiling deluge o'er the blasted heath;
And, wide in the air, in misty volumes hurl'd
Contagious atoms o'er the alarmed world.

ERASMUS DARWIN, The Botanic Garden, *1791*

IT WAS THE MIDDLE OF MAY 2004, AND I HAD FINISHED MY spring term of teaching in San Francisco. I packed up—the paltry amount of luggage I had brought with me from Massachusetts somehow having expanded alarmingly—then pointed my car northward to cross the Golden Gate Bridge. From there—no northbound toll, I was pleased to see—I drove briskly off homeward, though somewhat indirectly, since I was adding an extra 4,000 miles going by way of Alaska.

I had long cherished an ambition to drive through early-melting snowfields along the entire length of the Alaska Highway. But now there was more than a little geological resonance to making such a journey—not least because two of America's largest-ever earthquakes occurred in the state in 1964 and in 2002, with the latter taking place on a fault system that has provable links with the San Andreas. The connection between the events of San Francisco and Alaska serves as a reminder—in just the way the Gaia theory likes to suppose—that everything that happens in the natural world is connected in one enormous and living global system.

Everyone who is old enough and who was there remembers every last detail of the first, the great Good Friday Earthquake of 1964 that ruined Anchorage, the biggest city in Alaska. Not all who experienced it chose to stay. When I was on the San Andreas Fault, investigating a site south of Mendocino close to where the pygmy redwood trees grow, I went flying with an Alaskan, a pilot who told me he had lived through it. He had fled but was still disturbed, still undergoing a prolonged period of therapy, and still terrified by the sudden onset of anything unexpected.

His memory is vivid and indelible. He remembers being parked by the side of the road in his parents' Ford station wagon, maybe five miles south of the city. He recalls, as though it were yesterday, being thrown out of the car onto the road, and then holding on to the asphalt with his fingers for what seemed to him like five full minutes while the road bucked and kicked like a bronco beneath him, never giving up, roaring with the sound of a typhoon. Today, forty years later, he is terrified by any sudden noise, and he feels sick crossing bridges and standing beside tall buildings, anything that looks as though it might be vulnerable to an almighty quake. Only when he is aloft, far from the danger that the earth can bring, does he feel perfectly safe.

The Prince William Sound Earthquake of March 27, 1964, as the Anchorage quake is officially known, remains the second largest ever recorded in the world, with a moment magnitude of between 8.9 and 9.3—far more powerful than San Francisco's. It was only marginally less gigantic than the 1960 earthquake in Chile, which holds the record for all modern observed events: This had a magnitude of 9.5, killed 2,000 people, and left 2 million homeless. It was slightly stronger than the Sumatran submarine quake of 2004, which killed more than 275,000 with its terrible tsunamis.

Although far fewer people lived nearby, the 1964 Alaska quake could have been proportionately just as terrible, had it not been a bank holiday in the state. The event occurred at 5:36 P.M., and, as the pilot in California recalled with more accuracy than most, it did indeed last a very long time: Almost four minutes of continuous and highly destructive shaking occurred.

The ultimate cause of Alaskan earthquakes generally is, once again, the northward movement of the Pacific Plate—the same plate whose same northward movement in California, where it rubs up against the North American Plate, triggers the notorious events that occur along the San Andreas Fault. There is, however, a difference. Up in Alaska the curves of the plates and the topographical setting are such that the northward movement of the plate does not necessarily cause it to slide along beside the North American Plate when it encounters it. Instead it dives underneath it, dragging trillions of tons of material down as it does so and causing the North American Plate to bulge up and wrinkle.

That is the ultimate cause. The proximate cause of the event was that the northbound Pacific Plate had for some reason become stuck beneath its neighbor in one place; and, because it had been stuck like that for several centuries, enormous stresses built up around it. Suddenly on this particular black Good Friday, whatever had been holding it back suddenly released. The Pacific Plate dived down and northward, the North American Plate jumped up and southward—and a vast shaking occurred and a huge amount of damage was done in consequence.

A barely imaginable 100,000 square miles of territory was deformed by the event, all of it centered around the earthquake's focus in Prince William Sound. In some places whole tracts of landscape were thrust upward by more than thirty feet, by far the greatest vertical displacement of any modern American earthquake. Much of central Anchorage was ruined: Huge cracks opened up in the ground, scores of houses sank without trace into liquefied earth, and throughout the region tsunamis—one in particular topped with blazing oil from a Texaco tank farm it had destroyed en route—came roaring up narrow creeks, flattened villages, and carried boats and flotsam far inland. Some 135 people were killed, and damage said to be worth $300 million was done. Waves killed and injured people in Washington and Oregon and caused mayhem in Northern California. Power stations, docks, bridges, roads, and railway lines were damaged all around the central part of Alaska—a cruel blow to a state only five years old and barely much beyond the pioneer stage of its young history.

The second earthquake, which occurred on November 3, 2002, some seventy miles north of Anchorage, was by contrast a true strike-slip event, much like the 1906 San Francisco Earthquake. Moreover it took place on a fault—the Denali Fault—which looks, from a cursory glance at any good tectonic map, to be a northern extension of the San Andreas. It killed no one and did precious little damage, but it was enormous. It caused sideslips of as much as eighteen feet, having struck with a magnitude of 7.9—almost the same as the 1906 event. It was of particular concern because for 200 miles it tore along a fault trace that crosses the state's most vulnerable asset—the huge four-foot-diameter pipeline that brings crude oil down from the drilling sites beside the Arctic Ocean to the docks on the Pacific, where tankers can load it to take it to fuel the world.

ONE OF THE MAIN reasons for my taking this trip to Alaska was to see how the pipeline had fared, curiosity tourism aside. But not only that: The journey up to the Arctic would also take me along the natural northern extension of the Californian fault systems, and it would allow me to see, for one final time, the great Pacific Plate in all its northern manifestations.

And so, one by one, I would pass by—and on occasion stop beside—the monumental pieces of scenery that owe their existence to the relentless movement of this plate. Mount Shasta, an enormous snow-covered volcano in Northern California; Crater Lake in southern Oregon; St. Helens and Rainier and Olympus and the peaks of British Columbia—all were active and spectacular pieces of evidence of the movement of the confused mélange of plates that jostle for space to the north of the Mendocino Triple Junction.

DECIDUOUS TREES began to fade into the forests full of evergreens a couple of hundred miles north of San Francisco, and there was still thick snow on the rim of Crater Lake. I stopped in a comfortably eccentric country inn a few miles south of the old volcano—eccentric because the owners, great dog enthusiasts, were in the middle of organizing a boot camp for husky mushers. Dog teams, most of them preparing for that most Olympian of dogsled races, the Iditarod, were practicing their skills in the meadows around the hotel. The huskies, friendly and with ever-searching tongues, were everywhere; they rose at 5:00 A.M. with a chorus of happy howls, and so I rose, too, and pushed on northward, past other mountains, past other breathtaking manifestations of raw North American geology.

Once across the Canadian frontier, there are three possible routes north, each around 500 miles long, between Vancouver and the southern end of the Alaska Highway. I chose the most westerly, through Squamish and Whistler, up to the small town of Lillooet, where the scenery became more rugged and the feel of the country suddenly more remote. I met a farmer who grew ginseng, and a forester who worked in a place called Bella Coola, as pleasing a name as I had heard for many months.

And indeed, the names of places began now to have a pioneering sound to them: 93 Mile House, 150 Mile House, Soda Creek, Moose Heights, Summit Lake. Others were simply peculiar: Cinema, Hydraulic, Horsefly, Stoner. In even the smallest towns there were Chinese restaurants, their owners all new immigrants: I spoke more Cantonese there in the heart of the Rockies than I did when I lived in Hong Kong. And between the towns, the rivers began to rage more wildly, the mountains were needle sharp, and up on the passes there were still drifts of late-spring snow.

And there were bears. Big black bears, lumbering along the now all-but-deserted roadway, looking up briefly as I passed, hoping for something more interesting than the grubs for which they were foraging. But the highway code in these parts discourages feeding: In a bear's brain, it is a mere step from the idea that *roads mean pantry* to *child means lunch.*

I reached the Alaska Highway after a day and a half of driving: I turned onto a side road to go down the Peace River Valley and joined the highway itself a few miles up from where it officially begins, at Dawson Creek—I had a deep desire not to take a photograph of my Land Rover beside the sign marking Mile Zero, as every other driver with a bull bar and a jerrican of fuel likes to do. So, after turning left at Chetwynd (with chain-saw sculptors plying their unamusing trade on all sides), I joined it at Fort St. John, beside Milepost 47. To reach the official northern end, next to where the American government is putting up a new antimissile site at Delta Junction, Alaska, would be a further 1,343 miles. Three days' hard driving.

The Alaska Highway, work on which was started by the U.S. Army Corps of Engineers at President Roosevelt's order in the spring of 1942, took 10,000 soldiers just eight months to build. It had two avowed purposes: to connect a set of airfields that allowed aircraft to be ferried from Great Falls, Montana, up to Fairbanks and from there to Russia, as part of the Lend Lease program; and to allow troops to be sent quickly to Alaska in the event the Japanese tried to invade. (They did, however, by seizing a scattering of islands along several hundred miles of Alaska's Aleutian chain; but the logistical threat posed by the newly built Alaska Highway prevented their pressing any farther east.)

The route that the engineers took followed old Indian trails, logging routes, and rivers, and the unpaved gravel Alcan Highway, as it was first known, was formally opened to military traffic in September 1942, a scant five months after the Japanese had taken Attu and Kiska Islands. The two teams that had been constructing it from either end met at Contact Creek, beside the present-day Milepost 568. Which is close to where I spent my second and only bad night on the road.

The first night had been spent in an inn at Fort St. John; the second was at Milepost 613—and the only respite from a thirteen-hour day of driving those 566 intervening miles was (aside from my first caribou, moose, Stone sheep, mountain goat, and baby buffalo sightings) a boiling water pool at Liard Hot Springs, a park in northern British Columbia that supports small gray fish that seem to thrive in scalding water, and any number of passersby who imagine they will be refreshed by the sulfurous waters that, courtesy of the Pacific Plate's activity, are bubbling up out of the ground. There were six supersized San

ected and without a warning sign or a fragment of barbed wire, nding high off the ground to allow migrating caribou to pass be- t, was the gray immensity of the line itself.

ot out of the car, amazed. So this was it—the 800-mile-long ne of four-foot-diameter-steel Mitsubishi-made tube, the source much controversy when it was built, the source of so much con- ow that terrorism had erupted, the fount of so much of Amer- conomic prosperity and current need for energy. It carries a n barrels of oil each day down from the oil fields at Prudhoe Bay oil port at Valdez—and here I was, standing entirely alone un- ath it. I had brief thoughts about plastic explosives, of how easy uld all be.

he line is supported for almost all of its journey aboveground on a s of twelve-foot-high piers, each one shock-absorbed, thus en- g the line to move up and down if the earth chooses to misbehave. w miles north of here is the very section where the pipeline crosses Denali Fault, and where the earth does misbehave often, and spec- larly. The shock-absorbing piers on this half-mile section are in- iously protected by being set down on enormous horizontal steel ms, with Teflon-coated sliders that allow the line to move from side ide should the fault slip, as it is wont to do. The line has a series of lt-in bends here as well, so that it is too flexible to snap.

During the November 2002 earthquake all worked perfectly. The th moved eighteen feet to the right, the Teflon sliders permitted it slip beneath the line, the built-in curves absorbed the energy. The e at daybreak had an enormous kink in it—but nothing broke, not a op of oil was spilled, and not even a flicker appeared on the dials atched by the pipeline engineers, even though one of the biggest rthquakes ever recorded was tearing apart the earth directly be- eath their precious and vulnerable charge.

Now I had seen it—now that I had seen what is arguably the single ost important item of humankind's making that is potentially hreatened by the conjunction of the two tectonic plates—I could at ast turn back. It remained only to get to Anchorage itself, to rest, and efuel the car with decent gas. And so, after humming for another half

Diegans in the waters, disobligingly stripped to their underwear. I kept my clothes on, and drove ever northward.

I passed Contact Creek at 10:00 P.M.—it was still brilliantly light outside, the sun setting so late this far north—and found myself eventually foodless and uncomfortable in one of the world's nastiest towns, Watson Lake, BC. *The Milepost* guide tries to be kind: "an important service stop," "a staging post." But the truth is that the place is a complete dump, and even the presence of a "forest" of left-behind place-name boards and license plates (51,842 of them, from every imaginable place in creation) at the north end of town cannot mitigate its awfulness. Watson Lake is a place to avoid. What do the locals do? I asked the grim-faced innkeeper where I eventually found a room. "They drink and fight," she said. "Nothing else to do." Quite so.

I crossed the Continental Divide eighty miles on into my third highway day. From this point onward all waters drain into the Yukon River and thence into the Bering Sea and the Pacific; the rivers now behind me drained into the Mackenzie and thence into the Arctic Ocean. Pacific salmon migrate up the Yukon and its tributaries. A subtle set of differences, cultural, geographic, and anthropological, was thus settled on the land that lay to the west and north of this barely marked demarcation line on the road. Walker's Continental Divide Roadhouse made the point a little less eloquently but memorably: The young woman who made the rhubarb pie I had for breakfast was so beautiful, and her cooking so intoxicatingly good, that I threatened to marry her on the spot.

By now I was firmly in the Yukon, a vast tract of Canadian real estate that English speakers are told is Canada's Real North, while Francophones have to suffer its description as *Le Nord avec un grand "N."* Compared with the British Columbia now far behind, this was harsher, chillier, and more beautiful by the mile. (Except for Whitehorse, which, with 22,000 inhabitants, must be one of the smaller capital cities around, probably rivaling Pierre, South Dakota. Even though Whitehorse tries hard, with its railroad to Skagway and its wonderful paddle wheel ferry on the Yukon River, it does also have a Wal-Mart, and that, for me, is the kiss of death. The notion that the ghost of Sam Walton and the Brutes of Bentonville have come to linger anywhere at all in the Yukon sets me fretting about the state of the world even more than usual. There is worse, however: Someone suggested taking the road farther north still to Dawson City and being initiated into a drinking club that has as its signature libation a whiskey in which is marinated some unfortunate's frostbitten toe. It seemed almost as repellent an attraction as Wal-Mart.)

But then all depression lifted once I reached Haines Junction. If ever there was a pretty subarctic town, it is the small group of houses and businesses that cluster beside Milepost 985, near the great range of ice-covered peaks at the edge of Kluane National Park, one of the Canadian park system's least-visited and most beautiful possessions. I took a small plane up from here to look down on the ice field and the great glaciers that pour down from the Kluane into the not-too-distant Pacific; and on the way down I saw bull moose drinking from the streams in the low, late-evening sun, and thought I had perhaps never seen anything quite so majestic in my life. The Kluane Fault shakes every so often here; I was researching earthquakes in San Francisco and Anchorage, and the locals knew full well why I had come along their valley. This was a link, they all said. They read about it in their weekly newspapers. The same fault line, essentially.

There are Swiss and German migrants here, and one couple has for the last five years run a hotel—the Raven—which has food in its restaurant that can match that of any European café in New York or Toronto. There is a bakery, too, run by a Scots family, and there can be few more agreeable moments in northern Canada than having a latte on the terrace of the Village Bakery, a scone fresh from the oven, the Kluane peaks glistening in the morning sunshine, the air crisp and cool and full of promise.

It was after this that I got a speeding ticket, from a Mountie. "Yukon Highway Safety Week!" he informed me with a bright smile, after he had zoomed from out of nowhere with his lights flashing. The road deteriorated rapidly in the miles thereafter, as frost heaves and construction plans and long sections of gravel forced me to slow way,

way down and crawl along the lake shores up clo
tier. There was a small plaque at Soldier's Sum
cutting ceremony for the Alaska Highway was
half-century anniversary in 1992.

The American immigration and customs chec
might once have been perfunctory; these days, g
world, it was fierce, unsmiling, severe. But there v
was all but alone, and, apart from the occasional t
into Canada, and the mewing of the whirling eag
was the quietest of all frontier crossings, a lonely ou
severity. It hadn't always been American, of course:
spearing its way due north and south along a line
marked by stone pillars in the muskeg, had first be
tween the British and the Russians, the treaties
London and St. Petersburg.

I had a picnic by a lake, then sped on to Tok—sh
the army camp was once called—and spent a night
ultra-Christian bikers whose visitors' book read lik
for the evangelizing energies of Pat Robertson's 700
not to tell them about Azusa Street down in Los Ang
how their particular faith had its roots in a disaster
caused by the same geological phenomenon that had cre
in which they live, and in which Tok is sited.

It was here, at Milepost 1,314, that I was to turn left f
and so formally left the Alaska Highway. (The roadway
on north, bound for Delta Junction and the antimissile ba
tually reaches Fairbanks. It terminates at Milepost 1,422.)

And it was at a village called Glenallen that I finally rea
tended goal, the pipeline. Ordinarily you don't see it from
this is one of the few places where it passes underground—
map it seemed possible to take a short diversion and get a l
And so I found a dirt track, thick with low brush and rutte
mud, that seemed to go in the right direction. I bumped d
two miles or so, tipped over a low berm—and suddenly the

unprot
and sta
neath
I g
pipeli
of so
cern
ica's
millio
to th
dern
it wo
T
serie
ablin
A fe
the
tac
gen
bea
to
bu
ea
to
lin
dr
w
e
n
r
t

At the precise point where the trans-Alaska oil pipeline crosses the trace of the Denali Fault, engineers have built carefully calculated expansion curves into the line, and have placed Teflon-coated sliders beneath the supports so that the four-foot-diameter pipe will remain undamaged by any rupture that might occur beneath it. So far the protective measures have worked, and not a drop of oil has been spilled as a result of seismic movement.

---

day down a road known unglamorously as the Tok Cutoff, I reached the city itself—a prosperous, contented-seeming place of low skyscrapers and unexceptional suburban houses with barely anything—aside from a few parks and museums—to show for having been so desperately ruined forty years before.

That the owners protect their pipeline from earthquakes is understandable; that the city of Anchorage seems more blithely unconcerned, and at this time of year merely basks in the near-endless Arctic sunlight, is a little less easy to comprehend. Like San Francisco, it will be struck again one day; unlike San Francisco, Anchorage—at least in the brief pleasantness of its summertime—seems more content to forget the fate that lies in store.

But I was now eight days out from San Francisco, the car had come 3,850 miles, and I was weary. So I found a good hotel, got my hands on a cold bottle of champagne (this is a prosperous town—with oil, for one thing, helping to make it so), took a long bath, and had dinner while watching the sun set over the waters of Turnagain Arm at a quarter to midnight. The oysters were from Halibut Bay, the salmon was freshly caught that morning, the wine was crisp and cold, and someone was smoking Gitanes at the bar. It seemed like heaven.

Now that I had seen the Denali Fault and the pipeline there was, however, one final thing to see—and that required my driving yet another 4,000 miles back down south. So the following morning, and for the better part of the following week, I was on the road again: more caribou, more moose, and even a pair of bald eagles in the Rockies near Calgary. I crossed the American border in the windswept plains of northern Montana, zipped past the Minutemen nuclear-weapons silos that I thought had long ago been dismantled and grassed over; and, after a day spent meandering in the rain shadow of the mountains, came to the pretty university town of Bozeman and finally to my goal, by way of the imposing stone gateway into the first of America's great national parks, Yellowstone.

THE PARK IS A PLACE of almost indescribable spectacle, rightly popular and in consequence frequently, especially in the high summer, more crowded than is good for it. The wildlife, the mountains, the lakes, and the geysers are all the very obvious lures for the hundreds of thousands who each season drive in through the park's main gates. And these days there is a new reason: The widely publicized knowledge that Yellowstone Park sits on top of a potential supervolcano, whose eruption—at some unpredictable moment in the geological near term—will devastate nearly all of western America.

Most of Yellowstone is, in fact, the relic of a family of great volcanoes. There have been three periods of eruption, the first about 2 million years ago, the latest finishing around 600,000 years ago, with each spitting out, very violently, immeasurable quantities of lava and dust and ash. The last eruption left behind today's caldera with its jagged-mountain edge; this is filled at its southeastern corner with the large body of water that is Lake Yellowstone, drained northward by the spectacular river of the same name. The last eruption, in other words, left behind the very reason for the existence of the national park.

In recent years seismologists and geochemists have been poring over the park's geology and its character deep underground, and have found something that they long suspected but have not been able fully to prove until now—and it is that specific something that is drawing in the crowds. For it turns out that only a few thousand feet below the ground level of the park, the enormous body of molten magma that fed all the volcanoes of the past is not only still there—but, to use a metaphor of which the ancients would have approved, it is alive. Moreover, it is moving. It appears to be doing something that the imaginative would say is akin to *breathing.*

For recent measurements have shown that the bed of Lake Yellowstone, as well as the bottom of the river close to the point where it flows out of the lake, is rising and falling, slowly and rhythmically, year in, year out. It is as though a giant is slumbering beneath the lake, readying itself—for what? To awaken? To turn in its sleep? To snore? To stand suddenly and tear loose all that lies on top?

Such images are potent, and they have captured the American imagination in recent years—not least because even the sober scientists of the survey, who come in ever more frequently from their regional offices in Colorado and California and beyond, shake their heads in wonder at the scale of the eruption-in-waiting.

The greatest of the unknowns and the indeterminables, though, is when something might happen. Those who study the sleeping beast in the caldera offer as their safest answer only that it will happen sometime *soon,* using that word in its strictly geological sense. And that could be a quarter of a million years. After all, the first eruption began 2.1 million years ago, the second 1.3 million years ago, and the third 1.1 million years ago—and this last eruption lasted, if it is possible to imagine such a thing, for 600,000 years. It completed its work 600,000 years ago, confusingly—meaning that there has been no activity for that length of time, a period that is fully (to remind ourselves of the insignificance of mankind) five times as long as recognizable *Homo sapiens* has been bipedally extant on the face of the planet.

Yellowstone is thus, on purely statistical grounds, ready for an eruption almost any day. And when it happens, it will almost certainly be vast. Yellowstone is already one of the biggest explosive volcanic complexes on the surface of the planet, and when it goes, huge tracts of the western states will be covered with immense thicknesses of volcanic products. But as to what will trigger it—what cosmic whim might begin the ruin of the West—no one can say, or really even imagine.

Except, that is, for one small thing, and an even more alluring mystery. Researchers who were working in the park in the autumn of 2002 noticed that for some unexplained reason a number of the famous geysers that were regularly hurling steam and boiling water high into the air, to the delight of the tourists, were doing so *slightly more frequently than usual.*

There was no indication that the famous Old Faithful geyser was affected; but a number of others, smaller but equally faithful to the chronometers, did begin to display a puzzling change. The Daisy Geyser in the Upper Geyser Basin, for example, would in September and October that year invariably send its column of blue water shooting into the air every two hours, thirty minutes, and twelve seconds. But in November and December that rate changed: It shot out its water and steam every one hour, fifty-seven minutes, and forty-one seconds. It was an enormous change, and it occurred without any obvious reason.

And the other geysers nearby—the Castle, Plate and Plume Geysers in the Upper Geyser Valley, the Pink geyser in the Lower Geyser Valley, and the Lone Pine Geyser—showed much the same pattern. They suddenly speeded up—and no one had the slightest idea why. Everything seemed to change in early November. There was a flurry of small earthquakes—and the geysers all began to go mad. And for no reason—except, as someone suddenly realized, Alaska.

For it turned out that the timing of the changes all coincided, to the hour, with the arrival of the big surface waves from the Denali Earthquake of November 3—the quake that had prompted my visit to Alaska to see how the pipeline had fared. There was, it seemed, a link.

Even though the two places—the Yellowstone caldera and the Denali Fault—are separated by 1,800 miles of rock and mountain and river and lake, the occurrence of trauma in one place seems to have an effect on the other, as though the whole of western America were ringing like an immense brass bell.

Which brought me back to the premise with which this account began—the notion that this fragile planet, suspended in the blackness of space, is now something to be considered as an immense whole, with all of its elements interlinked and interconnected, the one happenstance triggering another and another and another, for as long as the world exists. The butterfly effect, written into the rocks of the American West, and into the rest of the world as well.

One day the researcher who discovered the effects of Alaskan quakes on Yellowstone geysers took me to see Daisy: Might it signal by its timing some event in the distant West? So I waited with him in the late spring sunshine, looking at my wristwatch, gazing across from behind a pinewood palisade toward the yellow patch of sulfur on the ground, surrounded by pools of blue groundwater. All was quiet.

The minute hand ticked slowly up to twenty-eight, then twenty-nine, and then thirty—and then suddenly, with a brief subterranean burp and a gurgle and a moan and a great whoosh, Daisy exploded, releasing its pure-white fountain of boiling water high up into the chilly air. A hundred children cheered, and a hundred cameras clicked the record. It was all a classic American spectacle.

And it had happened right on time. The geyser erupted just when it was supposed to. Not seconds earlier, not seconds later. My scientist guide looked a little disappointed, I thought. The world beyond Yellowstone was evidently quiet that day. Nothing had happened to change Daisy's eruptive timetable. Alaska had not had an earthquake, and nor, quite probably, had San Francisco.

But one day each of these places will have an earthquake. There is

One of the geysers at Yellowstone National Park in Wyoming; their periodic explosive bursts of waters and steam have been found to speed up each time there is an earthquake thousands of miles away in Alaska. This is the first provable indication that all the geology of the American West may be intricately linked: An event in one place may possibly trigger a sympathetic but very different seismic event somewhere else far away.

not a scintilla of doubt about it. Earthquakes, volcanoes, tsunamis are all as inevitable a part of the earth's story as sunrise and sunset are a part of the quotidian routine, the only signal difference being the rhythm and the pitiless irregularity of their occurrence. One day this place of mountains, lakes, and rivers will break and explode. New volcanoes will thrust out of the earth, produce even more rock, and lay it down on top of whatever monuments humans may have imprudently chosen to place in the way.

All that humans do, and everywhere that humans inhabit, is for the moment only—like the cherry blossoms in a Japanese springtime that are exquisite simply by virtue of their very impermanence. Geology, particularly the dramatic New Geology one sees in a place like Yellowstone, or on the Denali, or on the great San Andreas Fault, serves as an ever-present reminder of this—of the fragility of humankind, the evanescent nature of even our most impressive achievements.

It serves as a reminder that it is only by the planet's consent that places like the mountains of Montana and Wyoming exist, and only by the planet's consent that all towns and all cities—New Madrid, Charleston, Anchorage, Banda Aceh—and San Francisco—survive for as long as they do. It is a reminder, too, that this consent is a privilege, and one that may be snatched away suddenly, and without any warning at all.

APPENDIX

On Taking an Earthquake's Measure

MOST OF THE PHYSICAL PHENOMENA THAT AFFECT OR afflict the planet can be measured, and thus recorded numerically. Temperature, for example, can be measured by a thermometer of some kind, and noted in degrees Fahrenheit or Celsius. Atmospheric pressure's effects on a closed cylinder can similarly be related in millibars or the more modern units of hectopascals. Wind speed can be measured in knots or miles per hour, or, in cases where the wind affects the sea, by employing the famous scale that was first adopted by Admiral Sir Francis Beaufort* at the beginning of the nineteenth century. Tsunami waves have measurable heights and speeds. And even the

---

*Although Beaufort is inextricably linked in the public mind with the measurement of wind, he is better known among sailors as perhaps the greatest hydrographer of all time, with more than a thousand nautical charts of every corner of the maritime world to his credit. He also performed his work under some physical duress: In a battle at Malaga, before he began his work as mapmaker and gale-measurer, he was wounded no fewer than nineteen times, sixteen times by musket balls and three times by a Spanish cutlass. He was given a pension of £45 for his pains.

eruption of a volcano can be assigned an index, although it is one determined with some difficulty: The amount of material estimated to have been expelled from the crater is multiplied by the height to which it is thrown up into the air.

On Intensity

But earthquakes cannot be tidily transformed into numbers. They are so huge, unwieldy, and unpredictable that for most of human history they could only be suffered, defying classification by any kind of standard numerical arrangement. Instead, in the days before seismographs were invented, the only way of quantifying a quake was to do so subjectively, by looking at the damage it had caused, asking people exactly how they had experienced its shocks, and then assigning to it a number that described what by common consent came to be termed the quake's *intensity.* This figure would very obviously vary—the intensity would be greatest near the earthquake's epicenter, and it would diminish the farther an observer was away from it. The maximum intensity would provide some approximate measure of how big the event was when compared with another—but it would be a subjective assessment only and, moreover, one that required people who had been on hand, who had been affected and were fully aware, and who were intelligent enough to report. An earthquake that happened in an unpopulated area would pass unrecorded and unmeasured.

The first systematic investigation of intensity was probably the one conducted by Robert Mallet, the British engineer who first came up with the term *seismograph* in 1854. In late December 1857 he applied to the Royal Society for a travel grant to allow him to study in detail the great earthquake that had just occurred in Naples, and he spent two months there systematically observing what had evidently been a very major seismic event. Eventually he was able to draw maps showing the areas of the earthquake's greatest evident intensity, with lines—which he called isoseismals—connecting places that had suffered the same amount of damage, or where people or animals had experienced the same amount of shaking. From his map he was able to infer the center of the earthquake's shaking, and was also able to show how the intensity diminished with distance from this center. Once he compared his survey with the crude information that he was able to glean about other earthquakes, he was able to assign an approximate relative size to this Italian event: There was no possibility of giving the event an absolute size, but to be able to say whether it was greater or less great than another event in another part of the world did have some utility and purpose. His approach was, in consequence, seized upon by the earthquake scientists of the day as the best means available to determine a quake's size—and over the years a number of these scientists examined data in more detail and developed formal scales of intensity, some of them still in use today.

The first of these intensity scales was developed in the 1880s by the Swiss scientists M. S. de Rossi and François Forel: It assigned ten values to earthquake intensities, which were numbered between I and X (Roman numerals were the most popular convention for intensity scales, and they are still in general use today). In the Rossi-Forel scheme the number I signified a slight quake, one rather confusingly defined as having been *recorded by a single seismograph or by some seismographs of the same pattern, but not by several seismographs of different kinds; and with the shock felt also by an experienced observer.* At the other end of the scale the number "X" was used to designate a seismic happening that was marked by *great disasters, ruins, disturbance of strata, fissures in the earth's crust, rock falls from mountains.* Between I and X were all the more common earthquakes that wrought more or less damage and occasioned more or less havoc or casualties—and so a quake with intensity VIII was very bad, while an event of recorded intensity III was just this side of bearable.

Despite this vague and somewhat ambiguous calibration, the Rossi-Forel Scale became in short order enormously popular among seismologists and, perhaps more important, with the public, which found it easy to appreciate. It was in consequence used to describe most of the major seismic events that occurred in the Western world during the last quarter of the nineteenth century.

Before long, however, it was seen that the Rossi–Forel Scale had some important shortcomings. Engineers reported that the damage sustained by a city's buildings—whose collapses and other sufferings were used as prime indicators of the intensity of the earthquake that had affected them—was determined not only by the strength of an earthquake, but also by how the buildings had been constructed in the first place (which meant that Japanese earthquakes, for instance, caused very different kinds of damage to the local wood-and-paper houses—a realization that caused a whole new scale, the seven-step Omori Scale, to be invented to deal with that problem. The scale is still used today, but only in Japan).

Moreover, it also turned out that earthquake waves propagate themselves in different ways, at different rates, and with different consequences, according to the nature of the rocks through which they pass. Intensity scales were thus rapidly understood to be not simply subjective and ambiguous but also local. Any one earthquake might display a number of very different apparent intensities in locations that were just a few hundred yards apart, since buildings might react (collapsing or not as the case might be) first, according to whether they were built of brick, masonry, iron, paper, mud, felt, or wattle, and second, according to whether they had been constructed on a base made up of mud or granite, schist or shale.

So this first, very basic intensity scale had to be modified and modified as more and more information about more and more variables poured in. The Italian seismologist and vulcanologist Giuseppe Mercalli came up with what he felt was a more accurate ten-step scale in 1902; but it was soon found wanting in its own way and modified by a man named August Sieberg who expanded it into a twelve-step scale—I to XII—which was to become (under the name the Mercalli-Cancani-Sieberg Scale, or MCS) the basis of all modern measurings, particularly in southern Europe. Two Americans then translated this scale into English in the 1930s and called it the Modified Mercalli Intensity Scale, or MMI, versions of which were popular around most of the Western world for many years.

It became ever more complicated from this point on. In 1956

Charles Richter—the nudist, vegetarian, womanizing, Asperger's syndrome–afflicted seismologist from CalTech, who long before had created an entirely different and much more widely used scale of his own, as we shall see—weighed in and fine-tuned the MCS to become the "Modified Mercalli Scale of 1956," which is generally known as the MM56. And even though this scale is thought to be perfectly good and accurate (at least for most of the world beyond Japan), the process of refinement and universalization went on and on and on. A supposedly ultrasophisticated and more "powerful" scale known as the Medvedev-Sponheuer-Karnik Scale, or MSK64, became popular in the 1960s and 1970s,* and then the most recent intensity scale to be fully agreed on by the world seismological authorities was put forward at an international congress of earthquake scientists held in Zurich in 1990. It was refined at meetings in Potsdam, Munich, and Reykjavik and then finally published in 1998 under the name the European Macroseismic Intensity Scale, or the EMS. And this, it seems universally agreed, is to be the current gold standard.

For though the EMS is made to look rather complicated—because it includes the classification of seismically affected buildings into different categories of construction—it is now fully internationally agreed to and, for the time being, absolute. It represents the world consensus (aside from Japan, which sticks stubbornly to its very outdated Omori Scale) on the best way to calibrate earthquakes *on a human scale.* It succeeds all the others, including the time-honored Rossi-Forel, the elegant Mercalli, and the Mercalli's various offspring, such as the MM56, the MSK64 and the MSK81 scales.

In the table that follows, which presents in summary what is now formally known as the EMS-98 Intensity Scale, there are a number of variables. The designation (a), for example, represents the earthquake's

---

*The 1906 St. Lucia quake mentioned in chapter 1 was reinterpreted in the 1970s using the MSK64 Scale, and, as mentioned, was given an intensity rating of between VII and VIII. It was not assigned a Richter *magnitude* because of the very small number of seismograms that recorded it—seismographic information being crucial for working out magnitude, as we shall see.

effects on humans, while (b) represents the same event's effects on "objects and nature" and (c) its effects on buildings. The five categories of buildings, listed as A to F, represent, on the other hand, a scale of *construction vulnerability*—one that runs from buildings made of unreinforced masonry (which is very vulnerable) to those built using highly earthquake-resistant steel modules that by law have to be erected in earthquake-prone cities. The damage scale of 1 to 5 indicates differing degrees of effect—from slight cracking to total destruction. It is complex but in its essence is self-explanatory. The EMS-98 Scale, then, reads as follows:

I. Not felt

(a) Not felt, even under the most favorable circumstances.
(b) No effect.
(c) No damage.

II. Scarcely felt

(a) The tremor is felt only at isolated instances (<1 %) of individuals at rest and in a specially receptive position indoors.
(b) No effect.
(c) No damage.

III. Weak

(a) The earthquake is felt indoors by a few. People at rest feel a swaying or light trembling.
(b) Hanging objects swing slightly.
(c) No damage.

IV. Largely observed

(a) The earthquake is felt indoors by many and felt outdoors only by very few. A few people are awakened. The level of vibration is not frightening. The vibration is moderate. Observers feel a slight trembling or swaying of the building, room or bed, chair, etc.

(b) China, glasses, windows and doors rattle. Hanging objects swing. Light furniture shakes visibly in a few cases. Woodwork creaks in a few cases.
(c) No damage.

V. Strong

(a) The earthquake is felt indoors by most, outdoors by few. A few people are frightened and run outdoors. Many sleeping people awake. Observers feel a strong shaking or rocking of the whole building, room or furniture.
(b) Hanging objects swing considerably. China and glasses clatter together. Small, top-heavy and/or precariously supported objects may be shifted or fall down. Doors and windows swing open or shut. In a few cases window panes break. Liquids oscillate and may spill from well-filled containers. Animals indoors may become uneasy.
(c) Damage of grade 1 to a few buildings of vulnerability class A and B.

VI. Slightly damaging

(a) Felt by most indoors and by many outdoors. A few persons lose their balance. Many people are frightened and run outdoors.
(b) Small objects of ordinary stability may fall and furniture may be shifted. In few instances dishes and glassware may break. Farm animals (even outdoors) may be frightened.
(c) Damage of grade 1 is sustained by many buildings of vulnerability class A and B; a few of class A and B suffer damage of grade 2; a few of class C suffer damage of grade 1.

VII. Damaging

(a) Most people are frightened and try to run outdoors. Many find it difficult to stand, especially on upper floors.
(b) Furniture is shifted and top-heavy furniture may be overturned. Objects fall from shelves in large numbers. Water splashes from containers, tanks and pools.

(c) Many buildings of vulnerability class A suffer damage of grade 3; a few of grade 4.

Many buildings of vulnerability class B suffer damage of grade 2; a few of grade 3.

A few buildings of vulnerability class C sustain damage of grade 2.

A few buildings of vulnerability class D sustain damage of grade 1.

VIII. Heavily damaging

(a) Many people find it difficult to stand, even outdoors.

(b) Furniture may be overturned. Objects like TV sets, typewriters, etc., fall to the ground. Tombstones may occasionally be displaced, twisted or overturned. Waves may be seen on very soft ground.

(c) Many buildings of vulnerability class A suffer damage of grade 4; a few of grade 5.

Many buildings of vulnerability class B suffer damage of grade 3; a few of grade 4.

Many buildings of vulnerability class C suffer damage of grade 2; a few of grade 3.

A few buildings of vulnerability class D sustain damage of grade 2.

IX. Destructive

(a) General panic. People may be forcibly thrown to the ground.

(b) Many monuments and columns fall or are twisted. Waves are seen on soft ground.

(c) Many buildings of vulnerability class A sustain damage of grade 5.

Many buildings of vulnerability class B suffer damage of grade 4; a few of grade 5.

Many buildings of vulnerability class C suffer damage of grade 3; a few of grade 4.

Many buildings of vulnerability class D suffer damage of grade 2; a few of grade 3.

A few buildings of vulnerability class E sustain damage of grade 2.

X. Very destructive

(c) Most buildings of vulnerability class A sustain damage of grade 5.

Many buildings of vulnerability class B sustain damage of grade 5.

Many buildings of vulnerability class C suffer damage of grade 4; a few of grade 5.

Many buildings of vulnerability class D suffer damage of grade 3; a few of grade 4.

Many buildings of vulnerability class E suffer damage of grade 2; a few of grade 3.

A few buildings of vulnerability class F sustain damage of grade 2.

XI. Devastating

(c) Most buildings of vulnerability class B sustain damage of grade 5.

Most buildings of vulnerability class C suffer damage of grade 4; many of grade 5.

Many buildings of vulnerability class D suffer damage of grade 4; a few of grade 5.

Many buildings of vulnerability class E suffer damage of grade 3; a few of grade 4.

Many buildings of vulnerability class F suffer damage of grade 2; a few of grade 3.

XII. Completely devastating

(c) All buildings of vulnerability class A, B and practically all of vulnerability class C are destroyed. Most buildings of vulnerability class D, E and F are destroyed. The earthquake effects have reached the maximum conceivable effects.

There is, however, a problem. Scales that describe the *intensity* of an earthquake positively require buildings to be damaged or people to be hurt—for without being able to employ phrases like those found above—"Hanging objects swing lightly" or "All buildings . . . destroyed"—there is, quite simply, no scale. But what of earthquakes that occur where there are no people to be hurt, or where no cities exist that are vulnerable, where there are no structures that may be damaged or leveled? Might there be a way of measuring earthquakes that could be applied equally to both populated and unpopulated corners of the earth, and to all places on the planet, no matter how geologically or culturally or architecturally different?

On Magnitude

The answer to this fundamental problem was first divined in the 1930s, initially by a Japanese seismologist named Kiyoo Wadati, and then, more famously, by California's aforementioned Charles Richter. Both of these geophysicists came up with the idea that earthquakes could best be measured not by relying on the experience of observers or by reading the records of damage but by the much more dispassionate and mathematically precise means of examining very closely the traces of quakes that were written onto very accurate seismographs. A measure divined in this way would be, both Wadati and Richter decreed, an indication of something far more absolute than the earthquake's intensity, which would be called the earthquake's *magnitude.* Western chauvinism being what it is—and Richter being by far the better known of these two scientists—the measure that was the result of this new brand of rumination came generally to be known as the Richter Magnitude Scale.

The Richter scale, which is arguably among the most familiar of any of the eponymous scales or units of measurements,* is a very dif-

*The volt, watt, ohm, hertz, kelvin, farad, henry, newton, pascal, Beaufort Scale, Planck's constant, and Avogadro number are among the Richter scale's close contenders.

ferent kind of measure from those that deal with earthquake intensity. It is not at all subjective; rather, it is a mathematical indication derived from a mechanical record. And these days, thanks to Richter's inexhaustible capacity for promoting his creation, it has come to be the single measure that is almost universally used in contemporary public discussion—in the press and in common conversation—as the most reliable indicator of the strength or power of an earthquake.

We speak quite casually of an event as having been of "magnitude 6" or "magnitude 8." In the specific case of San Francisco, we speak of it not so much as having occurred in 1906—I found in writing this book that the year seems to have been forgotten by all but the most dedicated historians—but as an event that had a specific magnitude: 7.9 or 8.3, depending on how the figure is calculated. That is what seems to be remembered: *San Francisco 7.9.* Not *San Francisco 1906.* And in the case of the tsunami-triggering quake of December 2004 in northern Sumatra, we refer with awe to its size, its power—and thus its magnitude of 9.3. And that is quite probably how we will remember this event in half a century: People will probably speak of *Banda Aceh 9.3,* and they may well remember where they were when they heard the news of such an enormous event. Few seem to care especially whether it had an intensity designated by a certain Roman numeral, or whether it was rated this or that on the EMS-98 or the MKS scales. They speak, quite simply, of *how big it was,* of how big according to the scale that was propagated and publicized by that one otherwise long-forgotten man who first created it: Richter.

In essence the calculation of a Richter magnitude is simplicity itself. It depends entirely on two things: on the size of the waves recorded on a seismograph, and on the distance that this seismograph is from the focus of the earthquake that caused them. Once you have those two pieces of information—which anyone with a seismograph will have simply by reading the ink record on the paper drum—then it is perfectly easy to work out the magnitude.

The seismogram illustrated here shows the Sumatran event of December 26, 2004, as it was recorded by an observatory in Pennsylvania.

At the left end of the trace, the pressure, or P-wave—the nature of which was explained in chapter 7—is shown on the seismogram as arriving first, as it always does, about 17 minutes into the record, almost exactly 1 minute after the known time of the quake. (The horizontal scale is time, in minutes; the vertical scale, here omitted, is usually offered in millimeters.)

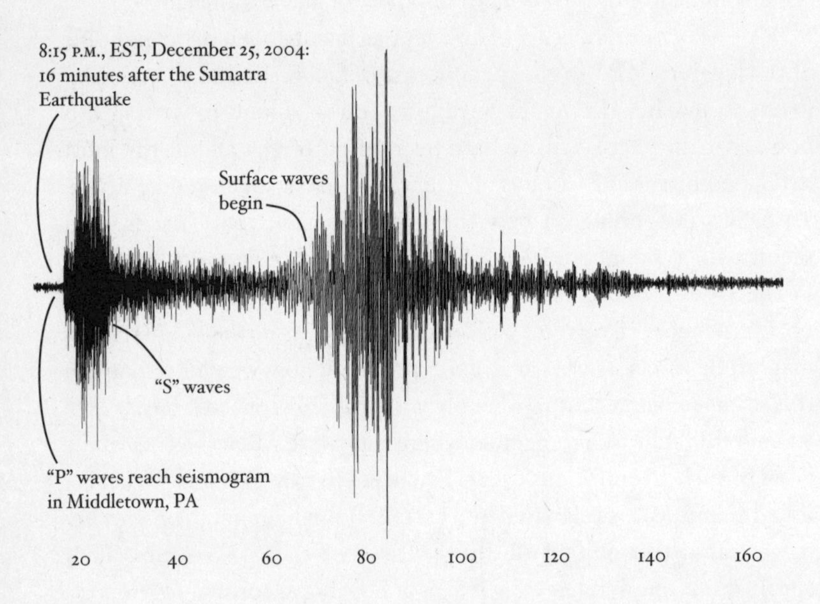

The shear, or S-wave, then begins to hit the observatory machine at some 27 minutes into the trace—a delay between the arrival of the two sets of waves of almost exactly 10 minutes, or 600 seconds. Since the average velocity of earthquake waves as they are transmitted through the earth is well known, even in this long-distance case where some of the waves pass close to the earth's core, it is fairly easy to calculate that the earthquake that caused this sudden vibration of the suspended seismograph weight occurred about 9,750 miles—15,700 kilometers—away from Pennsylvania. (By using the traces from three or more seismographic recording stations, as mentioned earlier, it is then possible to show not merely how far away the earthquake was, but exactly where on the surface of the planet it occurred. This is how earthquake epicenters are, in essence, computed. However, we are concerned here more with size than with position.)

To divine the earthquake's magnitude, it remains now simply to measure the amplitude of the greatest wave that is found on the seismogram—in this case, the first of the surface waves, which appear rather dramatically at about 68 minutes into the record. This ground movement—caused by the suspended weight bouncing as the earth quakes—was quite large, in terms that seismologists understand. It shows a ground movement of about four millimeters—almost sufficient for the Sumatran earthquake actually to have been *felt* by a few very sensitively minded people in faraway Pennsylvania.

Once the peak amplitude has been accurately determined, together with the length of time that this amplitude endures (maybe 20 seconds, give or take, although the trace would need to be very closely examined to work this out), then all that then has to happen is (for simple numerical reasons that underlie the basis of the calculations) for the amplitude of 4mm to be converted into microns, by multiplying by a 1,000, and then to be compressed by some mathematical device—usually a logarithm, to base 10—to sort out, to *damp down,* the enormous range of variations in seismogram amplitudes that are recorded by the world's machines. Finally, the resulting figure has to be recomputed to allow for the quake occurring a notional 100 kilometers (and not the actual 15,700 kilometers) away from the earthquake epicenter. And at this point, Lo!, a number appears. And in this specific case, it is the number 9.1.

And that number is the Richter *local* magnitude for this event—a number that can be refined and refined and refined, by adding or subtracting new factors and by recomputing with the insertion of otherwise unregarded additional variables, until a whole slew of other more sophisticated modern magnitudes—numbers representing such indicators as moment magnitudes, seismic magnitudes, surface-wave magnitudes, and so on, each with its specific usefulness to geophysicists—appear, as if from a magician's hat.

The basic number, though, is already there, ready to be broadcast and offered up as the public statement of the size of the event. In short: The earthquake that caused the Middletown Observatory's machine in Pennsylvania to experience such a dramatic few minutes of bouncing and shaking on that Boxing Day in 2004 was a faraway seismic event, some 9,750 miles away, and it measured 9.1 on the Richter scale. This number was one that Charles Richter had decreed quite simply would be the logarithm to base 10 of the maximum seismic-wave amplitude (in thousandths of a millimeter) recorded on a standard seismograph* at a distance of 100 kilometers from the epicenter.

The Richter figure thus calculated is of a great value to science, because it has an absolute mathematical purity to it. It does not require anyone to observe the event; nor is it necessary for any building to be damaged or for anyone to die or for any cattle or sheep to be scattered and spooked by it. Human perception is of no consequence in the making of the scale, nor in the assigning of the magnitude number to any event. One cannot publish the Richter scale in the same manner that one can publish an intensity scale, and say that a magnitude 9 event will do such and such damage while a magnitude 8 will do something quantifiably different. All one can say is that a 9 is a very great deal more significant than an 8—and that the amount of energy released by the event is proportionately very much more also.

But, having said this, Charles Richter did try to make his scale somewhat user-friendly. He worked his mathematics around the idea that the number zero should represent the smallest kind of episode that could be recorded on a seismometer; that values of around 3 would describe the smallest earthquakes that could be felt; and that values between 5 and 7 would represent the strongest ruptures that the average Californian—and all the early calculations of this very Californian man were specifically aimed at the fears and apprehensions and past experiences of the locals, his neighbors—might ever experience in a lifetime.

---

*Specifically Richter decreed that the base measurements should be those derived from traces recorded on a simple seismograph with a torsion suspension of the mass, of the type named for H. O. Wood and J. Anderson, the geophysicists who first created it.

The logarithmic nature of the scale can render it unfamiliar to most. A magnitude 5 earthquake is not twice as great as a magnitude 2.5, for example. But 900 earthquakes of magnitude 5, or thirty of magnitude 6, would be needed to release as much energy as a single quake of magnitude 7. And, while small-magnitude earthquakes can be very inefficient at releasing the stored-up energy in the earth, large-magnitude quakes can be devastatingly good: The 7.9 earthquake in San Francisco released 10^{17} joules of stored energy in the sixty seconds of ground shaking—and only a very small fraction of that went into making the ground move. Large earthquakes have the power to move mountains, quite literally—and Charles Richter's scale can show how, and offer figures to prove it.

In conclusion: Intensity requires an audience; magnitude requires only the divining powers of machines. The numbers that are offered up by each kind of measurement, and that bear no relation to the other numbers that are offered up for assigning numerical value to the world's various physical phenomena, inhabit a curious half world of their own, all invented by geophysicists and each enjoying its own utility. To tell of the intensity of a quake is to relate rather romantically something local and subjective and vaguely arbitrary; to record the magnitude, on the other hand, is to note something that is arithmetical, absolute, and able to be presented for instant comparison. So, while intensity may seem the more amusing indicator to relate, it has to be admitted that earthquake magnitude, and the Richter scale that since 1935 has recorded it, are the stuff of history, and of science, and a matter of record. And thus, in summary, they, the magnitude readings, are of a rather more lasting value. Which is why Rossi, Forel, as well as Medvedev, Sponheuer, and Karnik are known by a very few; and why Charles Richter is a famous figure, now and forever more.

Geological Time Scale

This stratigraphical diagram gives the presently recognized time span of each geological interval for the 545-million-year period, or eon, that can be readily dated and delineated using fossils, and which is known as the Phanerozoic.

The earth is, however, very much older than the life that has existed on it, having been created some 4,200 million years ago. This vastly long earlier period, not shown in the diagram, is divided by geology into two eons: the Proterozoic (which stretches from 2,500 million years to the start of the fossil-rich Cambrian, 545 million years ago) and the Archaean (dating from the estimated coalescence of the solid earth, 4,200 million years ago, up to the beginning of the Proterozoic).

Please note the different scales used in the diagram.

For further information on a time scale, which is constantly being refined, I suggest visiting the website of the International Commission on Stratigraphy at www.stratigraphy.org.

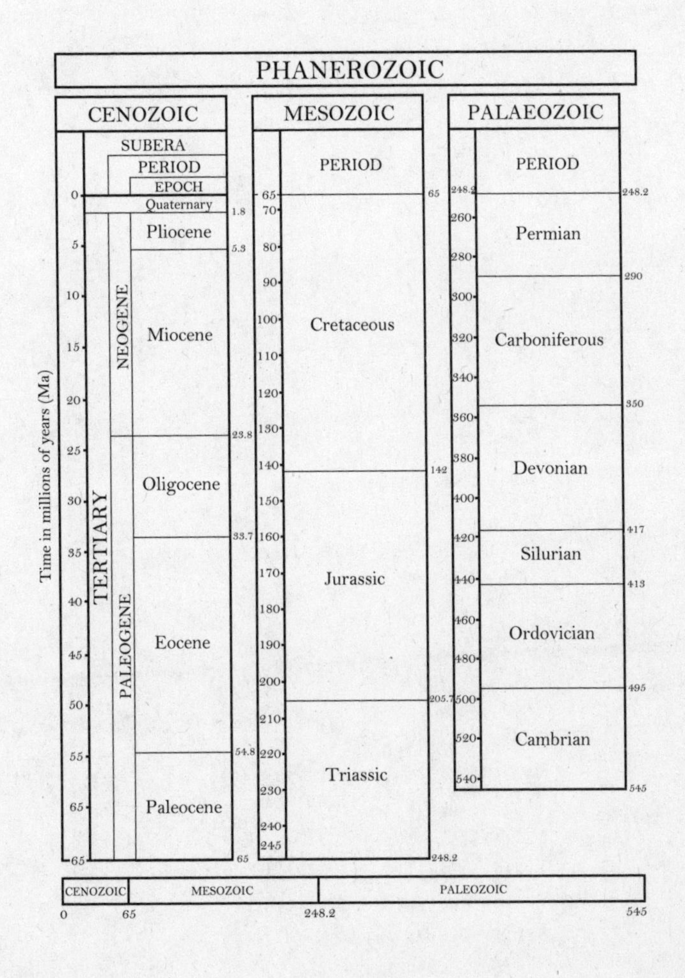
PHANEROZOIC
CENOZOIC
MESOZOIC
PALAEOZOIC
SUBERA
PERIOD
EPOCH
Quaternary
Pliocene
Miocene
Oligocene
Eocene
Paleocene
NEOGENE
PALEOGENE
TERTIARY
Time in millions of years (Ma)
PERIOD
Cretaceous
Jurassic
Triassic
PERIOD
Permian
Carboniferous
Devonian
Silurian
Ordovician
Cambrian
CENOZOIC
MESOZOIC
PALEOZOIC
0
65
248.2
545

WITH GRATITUDE

One afternoon in the early spring of 2003, as I was beginning to plan the research for this book, I remarked to my companion that to get the flavor of the city that would be its central subject I really ought to move there. We talked for a while about the details, and then we agreed: It all seemed like an eminently sensible plan.

The idea then hung in the air, but for no more than five minutes, for suddenly the telephone rang. There was an unfamiliar voice, a man's, on the other end. He introduced himself as Scott Rice, chairman of the Department of English at San Jose State University in Northern California.

He was calling, he said, to ask if the following winter I might like to move across to San Jose and teach there for a few months. It was, I thought to myself, the most uncannily propitious of synchronicities. And so it is with pleasure that I perform here my first and happiest duty—to record to whoever (or whatever) it was that prompted Professor Rice to call me that day my most earnest and sincere thanks.

For I did indeed take up the offer: I drove out west and I spent a small part of those first four months of 2004 teaching, lecturing, and generally bloviating to a most intelligent and attentive group of students in California. And then, from the pleasant flat in central San Francisco with which the university supplied me, I spent the rest of my time exploring and riffling my way through the manifold fascinations with which the cities of California, and the disaster that afflicted and otherwise affected them a century ago, are so richly endowed.

Without the university's help it would have been all so very much more difficult. I am grateful also to the friends I made at the university: Alan Soldofsky (who is academically fascinated by that most geologically minded of California poets, Robinson Jeffers), Mitch Berman, Paul Douglass, and the delightful Connie Lurie, whose generosity funded my visiting professorship to the university of which she is one of the more stellar alumni.

GIVEN THE SEISMIC PROPENSITIES of California, it is hardly surprising that so many geologists are drawn there, most hovering as moths around the ever-flickering flame of the San Andreas Fault. They were universally helpful to me, enthusiastically sharing their knowledge and research, and trying hard to keep me from committing elementary blunders—and if I have failed them then it is my failure, not theirs. Many work for or with the U.S. Geological Survey office at Menlo Park, making up what is surely the jolliest coven of bureaucrats within this or any other government—and I happily record my thanks in particular to Mary Lou Zoback and Mark Zoback, Andy Snyder, Carol Prentice, William Ellsworth, and Leslie Gordon. Susan Hough, who also works for the USGS but on the Caltech campus in Pasadena, was enormously helpful, and as a writer of popular geology books also was particularly sympathetic to the challenges of conveying complex seismic matters in an assimilable way: I owe Susan, now a good friend for life, a formidable debt.

I would also like to record my gratitude to other geologists elsewhere: Eldridge Moores and John Dewey at the University of California, Davis; Tina Niemi, who is an expert in digging trenches into the San Andreas Fault but who is nominally based far away at the University of Missouri in Kansas City; Bill Cotton, who is deeply involved with the geological bizzareries of the town of Portola Valley; Ann Crawford at the Geological Society of America in Boulder, Colorado; Peter Haeussler of the USGS office in Anchorage, Alaska; Ted Nield of the Geological Society of London; Anthony Lomax, whose recent interests from his headquarters in France include new research on the location of the 1906 earthquake's hypocenter; the authors Damien Nance and Brendan Murphy, who offered me help on the origins of the names of early supercontinents; and John Rogers, who named the first and oldest of them Ur, from his department at the University of North Carolina.

A brief and unexpected encounter in an elevator in a Dallas hotel with Dave Ruppert, an archaeologist with the National Park Service in Denver, led me to two further Park Service scientists based at Yellowstone, Hank Heasler and John Varley, who helped me understand the significance of the variations in the periodicity of the geysers there. And Fred Ganders, a botanist at the University of British Columbia, advised me on the yerba buena plant and the man who first named it.

Harold Brooks of the San Francisco office of the American Red Cross, and his colleague on the other side of the Bay, Jean Nickaloff, were both immensely helpful in showing me their plans for dealing with any future citywide disasters; as was Jim Aldrich of the City of San Francisco's Office of Emergency Services, whose latest plans are necessarily the most sophisticated of all those that have been prepared for the skein of cities that lie between Santa Rosa and San Jose, and which can be expected to suffer the consequences of any dramatic slippage on the various faults that run beneath them.

I am grateful also to Claire Perry, Curator of American Art at the Cantor Center at Stanford University; to Martin Uden, British Consul General in San Francisco, and his assistant, Julia Cottam, who uncovered original correspondence from the diplomats who were based in the city at the time of the earthquake; to Harold Hallin of Lake Tahoe,

Nevada, an insurance underwriter whose syndicate was involved in writing 1906 policies; and to my San Jose student Pat O'Laughlin of Los Gatos, who kindly introduced me to a number of his friends who had papers, images, memories, and opinions related to the seismic history of the region.

By chance I found that my apartment in San Francisco was just one floor above the flat belonging to California's most renowned living historian and author, Kevin Starr, and I am happy to record my thanks to him for his many courtesies and for his advice (and for his forbearance whenever I had heavy-footed friends over for tea).

The staff at Acorn Books was universally helpful in supplying me with long-forgotten publications that related to the 1906 events. I hope they consider that I, as customer, once helped them, too, if in a small way, on that occasion when about six of us felt obliged to jump on and subdue an extremely tall but somewhat deranged gentleman who, after lunching unwisely one day on a cocktail of unfamiliar chemicals, decided to come in and try to wreck the shop. We pinned him to the ground while waiting for the police to come, and—though I hardly recommend it as an ideal way to break the ice between bookseller and customer—we came to know one another extremely well in consequence, and discounts were offered once in a while.

And, finally, among those who gave professional help, I must give enormous thanks to Ron Henggeler, whose peerless performance as a waiter at the Big Four Restaurant on Nob Hill is matched only by his talent as an artist and his infectious enthusiasm for collecting San Francisco memorabilia. Not a single aspect of Bay Area life—from the debates over the siting of wartime antiaircraft batteries to the menu choices for lunches served at the Sutro Baths, from the mating habits of sea lions to the casualty tolls from earlier earthquakes—has ever escaped Ron's jackdawlike appetite, and his collections are immense and impeccably organized. He was kind enough to allow me to borrow from almost all and everything he possessed—film clips, books, magazines, paintings, vinyl records—and never once pressed me to return them until I was good and ready to do so. Ron is an enormously proud waiter, as he should be, and for anyone in need of some arcane fact about old San Francisco, I recommend a visit to his restaurant and a seat in his station. But do please remember the *pourboire*.

MATTHEW SCHECHMEISTER of Stanford and Amy Biegelsen of Richmond, Virginia, helped with many aspects of my researches, and did so ably and with great energy. I wish to thank my friend Renee Braden of the National Geographic Society archives, together with her admirable colleague Mark Jenkins. Other old friends, Kate Schermerhorn and Matt Clarke, recorded much of my West Coast research on film, in the probably vain hope that one day someone, somewhere, might be interested in seeing just how a book of this kind is made: That they put up with my whims so cheerfully for so long is testament to their undeniable talent as filmmakers. Richard Price of Robinson, Illinois, offered many valuable suggestions for improving the text of the paperback edition of the book; and Steven Marshank of Ashland, Oregon, made some very useful comments that I similarly incorporated into my revision. Sophie Purdy, as is her wont, helped tirelessly with this book, just as she helped with the three that immediately preceded it: My debt to her is incalculable.

MY SUGGESTION that I write this book was accepted, warmly and eagerly, by that most legendary of editors, Larry Ashmead—and though Larry retired during the writing of it, he never once abandoned his paternal interest in the project. My fondest hope is that when he sees it now, he will feel content with the way it all turned out. If he does, then that will in no small part be to the team that succeeded him—especially Henry Ferris, the wonderfully capable editor who agreed to take on the project in New York, together with Peter Hubbard, his meticulous assistant. In London my editor, Mary Mount, offered a thousand and one suggestions for improving the text: I believe I agreed with almost all of them, and am grateful to her for making them, and so tactfully and kindly, too. Donna Poppy and Sue Llewellyn remain the most faithful of copy editors, brilliant and faultless; and

Peter Matson and Bill Hamilton—helped by their respective assistants Nina Aron and Ben Mason—remain the finest agents for which an author could possibly wish. I raise a glass to each of you.

As I do, finally, to my partner, Elaine, who came to hear in a couple of years more about strike-slip faults and ophiolite sequences and the Chinese poetry of Angel Island than any person of reasonable firmness might expect to hear in a couple of lifetimes. She took it all with great good cheer and in a spirit of amused tolerance, and remained in the best of good humor all the time that I was hunched silently over my keyboard, ignoring all around me. So for allowing me the peace and for keeping content all the time: a thousand thanks, my dearest Elaine. You've been an absolute brick.

A GLOSSARY OF POSSIBLY UNFAMILIAR TERMS AND CONCEPTS

ALQUIST-PRIOLO This term, used with easy familiarity by many Californians today, refers to the 1972 Earthquake Fault Zoning Act, which was sponsored by the Democrat state senator for Santa Clara, Alfred Alquist, and the Republican state assemblyman for Ventura, Paul Priolo. The act compelled the state geologist to determine where all California's geological faults were located; imposed strict limits on the types of buildings that could be erected in seismic hazard zones that ran along them; and in time demanded that maps be prepared showing builders and real estate agents—as well as insurers—which parts of the state were relatively seismically safe and which were likely to be unsafe. Alquist-Priolo maps, as they are known, now run into the hundreds and are essential for anyone thinking of building or buying in the Golden State.

AMMONITE A slow-swimming creature of the warm and deep Mesozoic oceans, which propelled itself backward by flexing its muscular tentacles and was supported in the sea by a coiled chambered flotation cell, and which resembled in shape the horn of the Egyptian ram-headed god, Ammon. These attractively coiled fossils are found today in great abundance in limestone and shale of the Cretaceous and Jurassic.

ANDESITE An extruded volcanic rock with a very high silica content and which seems to be involved in highly explosive volcanoes. It contrasts with its much-lower-silica-content colleague-rock basalt, which tends to be found in the more gently oozing volcanoes of the world.

BORE A riverine tsunami, usually caused by unusually high flood-tides colliding with the estuarine outflow from a powerful stream. Bores can also be caused by earthquakes that strike near major rivers. The word probably has the same Old Norse root as the modern word *billow*.

BRACHIOPOD A common bivalved marine creature, found through almost all periods of the geological time scale, from the Lower Cambrian onward. One of their species—*Lingula*, which can still be seen in modern oceans—has remained evolutionarily unchanged for more than 500 million years. The creatures—often, because of their similarity in shape, known as *lamp-shells*—are fixed to the seabed by a muscular foot, and they waft extendable feeding devices through the passing seawater currents, trawling for nutrients.

CABLE As a unit of maritime measurement amounting to a tenth of a nautical mile, or about 600 feet. (A nautical mile, defined as one minute of arc of longitude measured along the equator, is 6,082 feet; a speed of one nautical mile per hour is known as one knot.)

CLINKER From the Dutch word *klinker,* from the ringing sound it makes when hit with a hammer, the incombustible residue of a volcanic eruption—a mixture of lava, bedrock, and other debris that often chokes a volcanic crater.

CONESTOGA WAGON The large covered wagons with which the earliest westbound settlers traveled across the American plains were built in Conestoga, Pennsylvania, a town named after a local Iroquois tribe. These clumsy vehicles had a postilion—a nicety not found in the smaller and swifter wagons that came to be known as *prairie schooners*—and which most emigrant expeditions favored before the coming of the railways (and which led to the demise of that most sweetly named craft of whiffletree-making).

CRATON The name given to those immense tracts of the earth's surface that have remained tectonically stable and seismically inactive for uncountable millions of years. These areas—like much of north central Canada—are composed invariably of ancient shield rock.

CWM A Welsh word for a mountain valley, still much used in mountaineering vernacular: the Western Cwm on Mount Everest, for example. Its equivalents in France, Germany, and Scotland are the words *cirque, Kar,* and *corrie;* despite the limitations of Welsh, only *cwm* appears to have found wide international acceptance among the climbing fraternity.

DIP The angle between a geological structure and the horizontal: Sedimentary rock layers are said to be steeply dipping, for example, if they are tilted close to vertical. See also *strike*.

GABBRO Basaltic magma (q.v. *andesite*) that cools slowly, in a pluton (q.v.) and that in consequence assumes a more crystalline appearance and structure than fine-grained basalts themselves

GAIA AND THE GAIA HYPOTHESIS James Lovelock, a British space scientist now regarded variously as either a visionary or a harmlessly misguided eccentric, first came up with the idea in the 1960s that the earth might be a living entity, a single, highly complex enclosed system. His belief, in essence, held then—and still holds now, after four decades of experimentation and observation—that the Gaia (the word formed is from the Greek word for "mother earth") is a complex entity that involves the earth's biosphere, atmosphere, oceans, and soil, "the totality constituting a feedback or cybernetic system which seeks an optimal physical and chemical environment for life on this planet." Through Gaia, says Lovelock, the earth sustains a kind of homeostasis, the maintenance of the relatively constant conditions that exist upon its surfaces.

GNEISS A coarse-grained metamorphic rock, marked by foliations of alternately dark- and light-colored mineral bands. Even more foliated rocks that readily split into slatelike slabs are generally called *schists:* the two types of rock are frequently found together.

GRABEN When two faults—not strike-strike-slip faults, like the San Andreas, but the more common vertical or near-vertical faults—occur close by and approximately parallel with each other, and the land between them drops, then the resulting valley is known as a *graben*. If it rises, pinched upward between two fault lines, the resulting uplifted block is called a *horst*.

GRANITE Found essentially only in continental bodies, granite is a slow-crystallizing volcanic rock, invariably cooled in a pluton, and with a substantial percentage of light-colored quartz, as well as feldspar and other minerals in large or relatively large crystalline forms.

GRAYWACKE A hard, dark, silty sandstone that has been mildly metamorphosed by heat and pressure and is thus rather harder and more dense than other sedimentary rocks of similar constitution.

INTRUSION The rock mass that results from the invasion of one rock by a molten version of another. Dolerites—dark, fine-grained igneous rocks—invariably form intrusions, known either as *dikes,* if they invaded vertically and cut across the bedding planes of sedimentary rocks, or *sills,* if they inserted themselves in parallel, like the igneous ham in a sedimentary sandwich. The Palisades Sill, on the New Jersey shore of the Lower Hudson River opposite Manhattan, is a classic dolerite sill.

K-T BOUNDARY EVENT The fossil record at the end of the Cretaceous and just before the beginning of the Tertiary, 65 million years ago, becomes suddenly and inexplicably interrupted, and is marked by the extinctions of nearly three-quarters of all the then-current lifeforms. There was no ready explanation for what might have happened—until the 1970s, when researchers in Italy detected an anomalous spike in the amount of the rare metal iridium found in a thin clay layer that marks the boundary between the two eras. This—together with their discoveries of soot and pressure-strained minerals in the clay—led them to deduce that one or more meteors (which produce iridium as a by-product) might have struck the earth 65 million years ago, causing a major catastrophe for all living things. Eventually a meteor crater—the smoking gun—was discovered off the coast of the Yucatán, at Chicxulub, and others were detected off western India, in the North Sea, and in the Ukraine, all from the same period. It now looks very much to believers as though a shower of enormous meteors struck the planet at the end of the Cretaceous—bringing the era to an end and causing infinitely more destruction worldwide than was occasioned by any earthquake, San Francisco's included.

NUNATAK This Inuit word denotes a mountain peak rising through or between glaciers or land ice in an arctic environment. In Greenland, where most of the world's nunataks are to be found, the base rock is invariably basalt, extruded from the mid–Atlantic ridge volcanoes in and around Iceland.

OOLITH An egg-shaped calcareous particle precipitated in the warm seas of the (usually) Jurassic period and which, once consolidated into rock, forms an attractive and decorative limestone.

OROGENY A period of mountain formation in which the earth's crust is deformed by one or more of a variety of processes that include vulcanism, subduction, plate collision, folding, plate divergence, and seafloor spreading or faulting.

PLATE TECTONICS The fundamental theory of the New Geology, first adduced in the mid-1960s by a variety of geologists and geophysicists working independently around the world. The Canadian J. Tuzo Wilson is popularly credited with first using the word *plates* in a paper in *Nature* in 1965. The theory holds that the earth's lithosphere consists of large, rigid plates that move horizontally in response to the flow of the asthenosphere beneath them, and that interactions among the plates at their borders cause most major geologic activity, including the creation of oceans, continents, mountains, volcanoes, and earthquakes.

PLUTON Another name for a large intruded body of igneous rock, inevitably younger than the country rock into which it has been intruded.

RECUMBENT FOLD When the folding of a body of rock becomes so severe that the apex of the fold passes the vertical and the fold bends over upon itself (as in the folding seen from California Route 14 and illustrated on page 192), the structure is known as recumbent. It illustrates the intensity of the orogeny that was involved in its creation.

STRIKE The line, or the compass direction of this line, that marks where an inclined geological feature intersects with the earth's horizontal surface.

TERRANE One of the key components of the New Geology, a terrane is a massive body of rock or rocks that manages to preserve its original identity more or less intact, despite having been moved a long way from where the rocks were first formed. In California and many American states to the west of the Rocky Mountains, a number of readily recognizable terranes have become accreted onto the original continent by the tectonic processes involving the Pacific and North American Plates. Thus, in the Basin and Range region, in states like Arizona and Nevada, immense blocks of very foreign rocks—each a terrane—have been carried thousands of miles east to form the spectacular mountains that so define the region.

THIXOTROPIC The property of certain solid or nearly solid substances to become fluid when agitated, and then to resume their solid or semisolid state when the agitation ends. Thixotropic reactions in areas of landfill—the marina in San Francisco, Battery Park City in Manhattan—can lead to disastrous consequences in the event of earthquakes: The land turns to liquid under the influence of the shaking, and buildings collapse in great numbers.

UMBO This knob on the very tip of a bivalved shell, such as a brachiopod (q.v.), represents the oldest, originating part of the shell, and its size and shape is much used in determining the precise type, species, and, thus, the relative age of fossils.

SUGGESTIONS FOR FURTHER READING, WITH CAVEATS

IN A BOOK CALLED *SIXPENCE HOUSE,* A WRY AND PLEASANTLY amusing recent account by an American named Paul Collins of his move to Britain's book-dealing village of Hay-on-Wye, the author tells of finding a long-faded copy of a volume titled *The San Francisco Calamity by Earthquake and Fire,* written by an equally long-faded writer named Charles Morris. I found a copy in my own collection—it is one of those books with a title page that tries to tell all, breathlessly. (It begins by describing the book as "A Complete and Accurate Account of the Fearful Disaster which Visited the Great City and the Pacific Coast, the Reign of Panic and Lawlessness, the Plight of 300,000 Homeless People and the World-wide Rush to Rescue Told by Eye Witnesses including Graphic and Reliable Accounts . . ." and so on and so on.)

Paul Collins notes that the book he bought had evidently been sent as a gift (to relatives in Winnipeg) in October 1906—meaning that the 446-page book had been written, published, bought, wrapped, and mailed no more than six months after the event that it sought to describe. It becomes amply clear to any reader, Mr. Collins among them, that in part as a consequence of this rush to publication, the book is almost entirely fiction.

The cover illustration sets the tone, dominated as it is by a lurid engraving of a skyscraper busily collapsing itself into a sea of licking flames. The building is easily recognizable as the beehive-domed Claus Spreckels Building (popularly known as the Call Building, because it housed the city's preeminant newspaper of the day). And true, it did burn, spectacularly. But far from ever collapsing, as the book's cover suggests, it survived. Moreover it still stands today, remodeled but proudly wearing its well-known beehive roof, on San Francisco's Market Street.

The jacket's fanciful design hints at the myriad other inventions within. Dubious accounts turn out to be legion, all collected and repeated by a writer whom Collins describe as one of the many local hacks all too eager "to make quick bucks off cataclysms." One passage, which he admits he did find "viscerally affecting," concerned a man who apparently threw himself onto the body of a woman who he claimed was his dead mother. He was supposedly smothering her face with grief-stricken kisses when a soldier noticed that what in fact he was doing was *chewing off her ears,* to get to hold of her diamond earrings. The soldier shot him—or, as Morris naturally wrote, "he put a bullet through the ghoul."

And all, so far as one can check, total nonsense.

There are gentler and more affecting reported scenes in the book, too—a refugee walking his huge Newfoundland dog and carrying a kitten, talking to the kitten all the while. A lone woman pushing an upright piano along the road, a few inches at a time. But then there is the account of a group of soldiers watching three men who were standing on top of the roof of the Hotel Windsor, two blocks from the Call Building, which was blazing furiously. "Rather than see the crazed men fall in with the roof and be roasted alive," Morris writes, "the military officer directed his men to shoot them, which they did in the presence of 5,000 people."

Is there any truth to these stories—to the tale of the earrings, or of the sad lady with her pianoforte, or the assassinated men plunging from the top of a burning hotel? Almost certainly not, say today's more clinically detached historians. Official records note that a great number of dreadful and memorable things happened in San Francisco—sufficiently numerous, indeed, for one to have little need of invention. Yet many writers of the day, and some subsequently, seemingly found in the San Francisco story such a seductive combination of glitter, sin, and violence that it seemed entirely meet and proper to them to add a seasoning of fiction, to make the spicy even spicier.

Hence one of the most vexing problems when examining, soberly, reports of the more lurid aspects of the disaster—and in particular those that relate specifically to the fire, which seem to be the most lurid of all. What is to be believed today, and what is more prudent to disregard?

Charles Morris's book certainly belongs to the latter category, as do all too many others that were written at the time. Official reports, though they are more staid and not necessarily entirely correct themselves, seem generally to be the better bet. And modern accounts have generally kept well away from the suspect material, such as the ear gnawing, unless there is written proof—a police report, a note in a coroner's record—that it did in fact occur.

The bibliography that follows includes many books that consider very specific aspects of the story: the insurance claims, say, or the mechanics behind strike-slip faulting. These are all invariably entirely reliable. But for anyone wishing to read unimpeachably sound accounts of the event itself, books in which the earthquake stands front and center, then the choice, particularly among the older volumes, is necessarily limited. Readers would thus be well advised to choose rather carefully, and to bear in mind that, in the light of the peculiarly flexible and liberal flamboyance that has long been associated with literary San Francisco and its most notable tragedy, one piece of advice rules: *caveat lector.*

Aidala, Thomas. *The Great Houses of San Francisco.* New York: Arch Cape Press, 1987.

Allen, Terence Beckington. *San Francisco Coroner's Office: A History 1850–1980.* San Francisco: Redactors' Press, 1999.

Asbury, Herbert. *The Barbary Coast.* New York: Alfred A. Knopf, 1933.

Bailey, Janet. *The Great San Francisco Trivia and Fact Book.* Nashville: Cumberland House, 1999.

Bain, David Howard. *Empire Express: Building the First Transcontinental Railroad.* New York: Viking Penguin, 1999.

Bally, Albert W., and Allison R. Palmer, eds. *The Geology of North America: An Overview.* Boulder, CO: Geological Society of America, 1989.

Bancroft, Hubert Howe. *Some Cities and San Francisco.* New York: Bancroft, 1907.

Barker, Malcolm E. *Three Fearful Days.* San Francisco: Londonborn Publications, 1998.

Bartleman, Frank. *Azusa Street.* New Kensington, PA: Whitaker House, 1982.

Begich, Nick, and Jeane Manning. *Angels Don't Play This HAARP: Advances in Tesla Technology.* 2nd ed. Anchorage, AK: Earthpulse Press, 2004.

Behr, H. H. *The Hoot of the Owl.* San Francisco: A. M. Robertson, 1904.

Benet, James. *A Guide to San Francisco and the Bay Region.* New York: Random House, 1963.

Bogardus, John A. *Spreading the Risks: Insuring the American Experience.* Chevy Chase, MD: Prosperity Press, 2003.

Bolt, Bruce A. *Earthquakes.* 5th ed. New York: W. H. Freeman, 2003.

Bolton, Herbert Eugene. *Outpost of Empire: The Story of the Founding of San Francisco.* New York: Alfred A. Knopf, 1939.

Brechin, Gray. *Imperial San Francisco: Urban Power, Earthly Ruin.* Berkeley: University of California Press, 1999.

Bronson, William. *The Earth Shook, the Sky Burned.* San Francisco: Chronicle Books, 1986.

Brook, James, Chris Carlsson, and Nancy J. Peters, eds. *Reclaiming San Francisco: History, Politics, Culture.* San Francisco: City Lights Books, 1998.

Butler, Jon, Grant Wacker, and Randall Balmer. *Religion in American Life: A Short History.* New York: Oxford University Press, 2002.

Caen, Herb. *The San Francisco Book.* Boston: Houghton Mifflin, 1948.

——, and Dong Kingman. *San Francisco: City on Golden Hills.* New York: Doubleday, 1967.

Caughey, John Walton. *California.* 2nd ed. Englewood Cliffs, NJ: Prentice-Hall, 1953.

Chang, Iris. *The Chinese in America.* New York: Viking Penguin, 2003.

Chase, Marilyn. *The Barbary Plague: The Black Death in Victorian San Francisco.* New York: Random House, 2003.

Chen, Young. *Chinese San Francisco 1850–1943.* Palo Alto, CA: Stanford University Press, 2000.

Clarke, Thurston. *California Fault: Searching for the Spirit of a State along the San Andreas.* New York: Ballantine Books, 1996.

Cleland, Robert Glass. *California in Our Time.* New York: Alfred A. Knopf, 1947.

——. *From Wilderness to Empire.* New York: Alfred A. Knopf, 1944.

Cohen, David, Doug Menuez, and Ron Grant Tussy, eds. *Fifteen Seconds: The Great California Earthquake of 1989.* San Francisco: The Tides Foundation, 1989.

Cohen, Stan. *8.6: The Great Alaska Earthquake.* Missoula, MT: Pictorial Histories Publishing, 1995.

Collier, Michael. *A Land in Motion: California's San Andreas Fault.* Berkeley: University of California Press, 1999.

Conrad, Barnaby. *San Francisco: A Profile with Pictures.* New York: Bramhall House, 1959.

Dalessandro, James. *1906.* San Francisco: Chronicle Books, 2004.

Dana, Richard Henry. *Two Years Before the Mast.* Penguin Classics. New York: Penguin Books, 1986.

Davis, Mike. *Ecology of Fear: Los Angeles and the Imagination of Disaster.* London: Picador, 2000.

Decker, Robert, and Barbara Decker. *Volcanoes.* 3rd ed. New York: W. H. Freeman, 1997.

Delehanty, Randolph. *The Ultimate Guide to San Francisco.* San Francisco: Chronicle Books, 1989.

Dickson, Samuel. *San Francisco Is Your Home.* Palo Alto, CA: Stanford University Press, 1947.

Dillon, Richard H.. *Embarcadero.* New York: Coward-McCann, 1959.

——. *Shanghaiing Days.* New York: Coward-McCann, 1961.

——. *The Hatchet Men.* New York: Coward-McCann, 1962.

Downs, Tom. *San Francisco.* Victoria, Australia: Lonely Planet Publications, 2002.

Eames, David B. *San Francisco Street Secrets: The Stories Behind San Francisco Street Names.* N.p.: Gem Guide Books, 1995.

Elliott, James Welsh. *Taylor & Taylor: A Reminiscence.* San Francisco: The Kemble Collections, 1985.

Erickson, Jon. *Quakes, Eruptions, and Other Geologic Cataclysms.* New York: Checkmark Books, 2001.

Fallows, Samuel. *Complete Story of the San Francisco Horror.* [Chicago?]: Hubert O. Russell, 1906.

Feldman, Jay. *When the Mississippi Ran Backwards: Empire, Intrigue, Murder and the New Madrid Earthquakes.* New York: Free Press, 2005.

Field, Maria Antonia. *Chimes of Mission Bells.* San Francisco: The Philopolis Press, 1914.

Fortey, Richard. *The Earth: An Intimate History.* London: HarperCollins, 2004.

Fradkin, Philip L. *Magnitude 8.* Berkeley: University of California Press, 1998.

Fritz, William J. *Roadside Geology of the Yellowstone Country.* Missoula, MT: Mountain Press, 1985.

Fuller, Myron L. *The New Madrid Earthquake: A Scientific Factual Field Account.* 4th ed. Marble Hill, MO: Gutenberg-Richter Publications, 1995.

Gagey, Edmond M. *The San Francisco Stage: A History.* New York: Columbia University Press, 1950.

Gentry, Curt. *The Madams of San Francisco.* Garden City, NY: Doubleday, 1964.

Gere, James M., and Haresh C. Shah. *Terra Non Firma.* Stanford: Stanford Alumni Association, 1984.

Gilliam, Harold. *The San Francisco Experience: The Romantic Lore behind the Fabulous Façade of the Bay Area.* New York: Doubleday, 1972.

Greer, Andrew Sean. *The Confessions of Max Tivoli.* New York: Farrar, Straus and Giroux, 2004.

Gudde, Erwin G. *1,000 California Place Names: The Story Behind the Naming of Mountains, Rivers, Lakes, Capes, Bays, Counties and Cities.* Berkeley: University of California Press, 1959.

Hansen, Gladys. *San Francisco Almanac.* San Rafael, CA: Presidio Press, 1980.

——, and Emmet Condon. *Denial of Disaster: The Untold Story and Photographs of the San Francisco Earthquake and Fire of 1906.* San Francisco: Cameron, 1989.

Harris, Stephen L. *Agents of Chaos.* Missoula, MT: Mountain Press, 1990.

Hartman, Chester. *City for Sale: The Transformation of San Francisco.* Berkeley: University of California Press, 2002.

Heizer, Robert F., and Alan F. Almquist. *The Other Californians: Prejudice and Discrimination under Spain, Mexico and the United States to 1920.* Berkeley: University of California Press, 1971.

Heppenheimer, T. A. *The Coming Quake.* New York: Times Books, 1990.

Herron, Don. *The Literary World of San Francisco and Its Environs.* San Francisco: City Lights Books, 1985.

Hicks, Geoff, and Hamish Campbell. *Awesome Forces: The Natural Hazards That Threaten New Zealand.* Wellington: Te Papa Press, 1999.

Hillenbrand, Laura. *Seabiscuit: An American Legend.* New York: Random House, 2001.

Holbrook, Stewart H. *The Story of American Railroads.* New York: Crown Publishers, 1947.

Hough, Susan Elizabeth. *Earthshaking Science: What We Know (and Don't Know) about Earthquakes.* Princeton: Princeton University Press, 2002.

———. *Finding Fault in California: An Earthquake Tourist's Guide.* Missoula, MT: Mountain Press, 2004.

Howard, Arthur D. *Geologic History of Middle California.* Berkeley: University of California Press, 1979.

Jeffers, H. Paul. *Disaster by the Bay.* Guilford, CT: Lyons Press, 2003.

Jenkins, Olaf P., ed. *Geologic Guidebook of the San Francisco Bay Counties.* San Francisco: Department of Natural Resources Division of Mines, 1951.

Kahn, Edgar M. *Cable Car Days in San Francisco.* Oakland: Friends of the San Francisco Public Library, 1976.

Kemble, John Haskell. *San Francisco Bay: A Pictorial Maritime History.* New York: Cornell Maritime Press, 1957.

Knox, Ray, and David Stewart. *The New Madrid Fault Finders Guide.* Marble Hill, MO: Gutenberg-Richter Publications, 1995.

Konigsmark, Ted. *Geologic Trips: San Francisco and the Bay Area.* Gualala, CA: GeoPress, 1998.

Kovach, Robert L. *Early Earthquakes of the Americas.* Cambridge: Cambridge University Press, 2004.

Kurzman, Dan. *Disaster! The Great San Francisco Earthquake and Fire of 1906.* New York: HarperCollins, 2001.

Leach, Frank A. *Recollections of a Mint Director.* Wolfeboro, NH: Bowers and Marina Galleries, 1987.

Lockwood, Charles. *Suddenly San Francisco: The Early Years of an Instant City.* San Francisco: San Francisco Examiner Division of the Hearst Corporation, 1978.

Longstreet, Stephen. *The Wilder Shore.* New York: Doubleday, 1968.

McDowell, Jack, ed.. *San Francisco.* Menlo Park, CA: Lane Publishing, 1977.

McGroarty, John S. *California: Its History and Romance.* Los Angeles: Grafton Publishing, 1911.

McLeod, Alexander. *Pigtails and Gold Dust.* Caldwell, ID: The Caxton Printers, 1947.

McPhee, John. *Assembling California.* New York: Farrar, Straus and Giroux, 1993.

——. *Annals of the Former World.* New York: Farrar, Straus and Giroux, 1998.

Mader, George G., et al. *Geology and Planning: The Portola Valley Experience.* Portola Valley, CA: William Spangle, 1988.

Martin, Don, and Kay Martin. *Hiking Marin.* San Anselmo, CA: Martin Press, 1999.

Marty, Martin E., and R. Scott Appleby. *The Glory and the Power: The Fundamentalist Challenge to the Modern World.* Boston: Beacon Press, 1992.

Morris, Charles, ed. *The San Francisco Calamity by Earthquake and Fire.* Philadelphia: Geographical Society of Philadelphia, 1906.

Morris, Jan. *The World: Travels 1950–2000.* New York: W.W. Norton, 2003.

Neville, Amelia Ransome. *The Fantastic City: Memoirs of the Social and Romantic Life of Old San Francisco.* Boston: Houghton Mifflin, 1932.

Nuttli, Otto W. *The Effects of Earthquakes in the Central United States.* 3rd ed. Marble Hill, MO: Gutenberg-Richter Publications, 1995.

Oreskes, Naomi, ed. *Plate Tectonics: An Insider's History of the Modern Theory of the Earth.* Boulder, CO: Westview Press, 2001.

Orme, Antony R., ed. *The Physical Geography of North America.* New York: Oxford University Press, 2002.

Page, Jake, and Charles Officer. *The Big One.* New York: Houghton Mifflin, 2004.

Penick, James Lal. *The New Madrid Earthquakes.* Columbia: University of Missouri Press, 1981.

Pollock, Christopher. *San Francisco's Golden Gate Park: A Thousand and Seventeen Acres of Stories.* Portland, OR: West Winds Press, 2001.

Reisner, Marc. *A Dangerous Place: California's Unsettling Fate.* New York: Pantheon Books, 2003.

Richards, Rand. *Historic San Francisco: A Concise History and Guide.* San Francisco: Heritage House Publishers, 2003.

Robinson, Andrew. *Earth Shock.* London: Thames and Hudson, 2002.

Rogers, John J., and M. Santosh. *Continents and Supercontinents.* New York: Oxford University Press, 2004.

Saul, Eric, and Don Denevi. *The Great San Francisco Earthquake and Fire 1906.* Millbrae, CA: Celestial Arts, 1981.

Searight, Frank Thompson. *The Doomed City: A Thrilling Tale.* Chicago: Laird and Lee, 1906.

Shearer, Frederic E., ed. *The Pacific Tourist.* New York: Adams and Bishop, 1886.

Shearer, Peter M. *Introduction to Seismology.* Cambridge: Cambridge University Press, 1999.

Solnit, Rebecca. *River of Shadows: Eadweard Muybridge and the Technological Wild West.* New York: Viking Penguin, 2003.

Soule, Frank, John H. Gihon, and James Nisbet. *The Annals of San Francisco.* New York: D. Appleton, 1854.

Starr, Kevin. *Americans and the California Dream 1850–1915.* New York: Oxford University Press, 1973.

——. *The Dream Endures: California Enters the 1940s.* New York: Oxford University Press, 1997.

——. *Embattled Dreams; California in War and Peace 1940–1950.* New York: Oxford University Press, 2002.

——. *Endangered Dreams: The Great Depression in California.* New York: Oxford University Press, 1996.

——. *Inventing the Dream: California through the Progressive Era.* New York: Oxford University Press, 1985.

——. *Material Dreams: Southern California through the 1920s.* New York: Oxford University Press, 1990.

Steele, Rufus. *The City That Is: The Story of the Rebuilding of San Francisco in Three Years.* San Francisco: A. M. Robertson, 1909.

Stegner, Wallace. *Angle of Repose.* New York: Penguin Books, 1992.

Stoffer, Philip, and Gordon, Leslie, eds. *Geology and Natural History of the San Francisco Bay Area.* Menlo Park, CA: US Geological Survey, 2001.

Synan, Vinson. *The Century of the Holy Spirit: 100 Years of Pentecostal and Charismatic Renewal 1901–2001.* Nashville: Thomas Nelson, 2001.

Taper, Bernard, ed. *Mark Twain's San Francisco.* Berkeley: Heyday Books, 1963.

Thomas, Gordon, and Max Morgan-Witts. *Earthquake: The Destruction of San Francisco.* London: Souvenir Press, 1971.

Thomas, Lately. *A Debonair Scoundrel.* New York: Holt, Rinehart, Winston, 1962.

Tompkins, Stuart R. *Life in America: Alaska.* Grand Rapids, MI: Fideler, 1960.

Trefil, James. *Cassell's Laws of Nature.* London: Cassell, 2002.

Turner, Patricia, ed. *1906 Remembered: First-hand Accounts of the San Francisco Disaster.* San Francisco: Friends of the San Francisco Public Library, 1981.

Watkins, T. H., and R. R. Olmsted. *Mirror of the Dream: An Illustrated History of San Francisco.* San Francisco: Scrimshaw Press, 1976.

Weber, Francis J. *A Bicentennial Compendium of Maynard J. Geiger's "The Life and Times of Fray Junipero Serra."* San Luis Obispo, CA: EZ Nature Books, 1988.

Wiley, Peter Booth. *National Trust Guide: San Francisco.* New York: John Wiley, 2000.

Wilson, James Russel ("The Well-known Author"). *San Francisco's Horror of Earthquake and Fire, Terrible Devastation and Heart-Rending Scenes, Immense Loss of Life and Hundreds of Millions of Property Destroyed. The Most Appalling Disaster of Modern Times, Containing a Vivid Description of This Overwhelming Calamity—Suddenness of the Blow—Great Number of Victims—Fall of Great Buildings—Thousands Driven from Their Homes. This Unparalleled Catastrophe Leaves San Francisco a Heap of Smouldering Ruins—Fierce Flames Sweep the Doomed City, Beautiful Buildings in Ashes, to which Is Added Graphic Accounts of the Eruptions of Vesuvius and Many Other Volcanoes, Explaining the Causes of Volcanic Eruptions and Earthquakes Embellished with a Great Number of Superb Photographic Views Taken before and after the Terrible Calamity.* San Francisco: Memorial Publishing, 1906.

Wilson, Neill C., and Frank J. Taylor. *Southern Pacific: The Roaring Story of a Fighting Railroad.* New York: McGraw-Hill, 1952.

Wood, Charles E. *Charlie's Charts North to Alaska (Victoria, B.C., to Glacier Bay, Alaska).* Canada: Charlie's Charts Cruising Guides, 2001.

Woods, Samuel D. *Lights and Shadows of Life on the Pacific Coast.* New York: Funk and Wagnalls, 1910.

Worster, Donald. *A River Running West: The Life of John Wesley Powell.* New York: Oxford University Press, 2001.

Wuerthner, George. *California's Wilderness Areas: The Complete Guide.* Englewood, CA: Westcliffe Publishers, 1997.

Wurman, Richard Saul. *Access: San Francisco.* New York: HarperCollins, 2003.

Wyatt, David. *Five Fires: Race, Catastrophe, and the Shaping of California.* Reading, MA: Addison-Wesley, 1997.

Zeigler, Wilbur Gleason. *Story of the Earthquake and Fire.* San Francisco: Leon C. Osteyee, 1906.

Zollinger, James Peter. *Sutter: The Man and His Empire.* New York: Oxford University Press, 1939.

Reports

Bailey, Edgar H., William P. Irwin, and David L. Jones. *Bulletin 183: Franciscan and Related Rocks, and Their Significance in the Geology of Western California.* Sacramento: California Division of Mines and Geology, 1964.

Behan, John E. (Clerk, City and County of San Francisco Board of Supervisors). *San Francisco Municipal Reports for the Fiscal Year 1905/6, Ending June 30, 1906, and Fiscal Year 1906/7, Ending June 30, 1907.* San Francisco: Neal Publishing, 1908.

——. *General Ordinances of the Board of Supervisors of the City and County of San Francisco in Effect December 1, 1907.* San Francisco: Carlisle, 1907.

Borchardt, Glenn, ed. *Proceedings of the Second Conference on Earthquake Hazards in the Eastern San Francisco Bay Area.* Special Publication 113. N.p.: California Department of Conservation, 1992.

Britton, Joseph (President, Board of Freeholders). *Charter for the City and County of San Francisco Prepared and Proposed by the Board of Freeholders.* San Francisco: Geo. Spaulding, 1895.

Burchfiel, B. C., P. W. Lipman, and M. L. Zoback, eds. *The Cordilleran Orogen: Conterminous U.S.* Volume G-3 of *The Geology of North America.* Boulder, CO: Geological Society of America, 1992.

Cable Railway Company. *System of Traction Railways for Cities and Towns.* San Francisco: Cable Railway Company, 1881.

Christiansen, Robert L. *Geology of Yellowstone National Park: The Quaternary and Pliocene Yellowstone Plateau Volcanic Field of Wyoming, Idaho and Montana.* Professional Paper 729-G. Menlo Park, CA: U. S. Geological Survey, 2000.

Cork, P. C. (Acting Governor). *St. Lucia: Report for 1907.* Colonial Reports Number 579. London: Darling & Son, 1908.

Crowell, John C., and Arthur G. Sylvester. *Tectonics of the Juncture between the San Andreas Fault System and the Salton Trough, Southeastern California: A Guidebook.* Santa Barbara: Department of Geological Sciences, University of California, 1979.

Davis, James F. *Recommended Criteria for Delineating Seismic Hazard Zones in California.* Sacramento: California Division of Mines and Geology, 1992.

Dutton, Captain Clarence Edward, U.S. Ordnance Corps. *The Charleston Earthquake of August 31, 1886.* N.p.: U.S. Geological Survey, 1888.

Gilbert, Grove Karl, et al. *The San Francisco Earthquake and Fire of April 18,*

1906, and Their Effects on Structures and Structural Materials. Washington, D.C.: U.S. Geological Survey, 1907.

Hansen, Roger, and Rodney Combellick. *Planning Scenario Earthquakes for Southeast Alaska.* Fairbanks: Alaska Division of Geological and Geophysical Surveys, 1998.

Holmdahl, DeWayne, et al. *Guidelines for Evaluating and Mitigating Seismic Hazards in California.* Special Publication 117. Sacramento: California Department of Conservation, 1997.

Kirkland, W. G. *Earthquakes: Agadir, Morocco—February 29, 1960; Skopje, Yugoslavia—July 26, 1963; Anchorage, Alaska—March 27, 1964; Caracas, Venezuela—July 29, 1967.* Washington, DC: American Iron and Steel Institute, 1975.

Lawson, Andrew C., et al. *Atlas of Maps and Seismographs Accompanying the Report of the State Earthquake Investigation Commission upon the California Earthquake of April 18, 1906.* Publication Number 87. Washington, D.C.: Carnegie Institution of Washington, 1908.

——. *The California Earthquake of April 18, 1906: Report of the State Earthquake Investigation Commission.* Vol. 1. Publication Number 87. Washington, D.C.: Carnegie Institution of Washington, 1908.

McKeown, F. A., and L. C. Pakiser, eds. *Investigations of the New Madrid, Missouri, Earthquake Region.* U.S. Geological Survey Professional Paper 1,236. Washington, DC: U.S. Geological Survey, 1982.

Moores, Eldridge M., Doris Sloan, and Dorothy L. Stout, eds. *Classic Concepts in Cordilleran Geology: A View from California.* Special Paper 338. Boulder, CO: Geological Society of America, 1999.

Plafker, George, and John Galloway, eds. *Lessons Learned from the Loma Prieta, CA, Earthquake of October 17, 1989.* U.S. Geological Survey Circular 1,045. Washington, DC: U.S. Geological Survey, 1989.

Rabbitt, Mary C. *Minerals, Lands and Geology for the Common Defense and General Welfare. Volume 1: Before 1879.* Washington, D.C.: U.S. Geological Survey, 1982.

Rankin, Douglas W., ed. *Studies Related to the Charleston, South Carolina, Earthquake of 1886—A Preliminary Report.* U.S. Geological Survey Professional Paper 1,028. Washington, D.C.: U.S. Geological Survey, 1977.

Rickard, T. A., ed. *After Earthquake and Fire: A Reprint of the Articles and Editorial Comment Appearing in the Mining and Scientific Press Immediately After the Disaster at San Francisco, April 18, 1906.* San Francisco: Mining and Scientific Press, 1906.

Simpson, David, et al. *EarthScope: Acquisition, Construction, Integration and Facility Management. A Collaborative Proposal to the National Science Foundation Major Research Equipment and Facilities Construction Account.* N.p.: EarthScope, 2003.

Sloan, Doris, and David L. Wagner. *Geologic Excursions in Northern California: San Francisco to the Sierra Nevada.* Special Publication 109. Sacramento: California Department of Conservation, 1991.

Steinbrugge, Karl V., et al. *Earthquake Planning Scenario for a Magnitude 7.5 Earthquake on the Hayward Fault in the San Francisco Bay Area.* Special Publication 78. Sacramento: California Department of Conservation, 1987.

Wagner, David L., and Stephan A. Graham, eds. *Geologic Field Trips in Northern California.* Special Publication 119. N.p.: California Department of Conservation, 1999.

Wallace, Robert E., ed. *The San Andreas Fault System, California.* U.S. Geological Survey Professional Paper 1,515. Washington, D.C.: U.S. Geological Survey, 1990.

Webb, U. S. *Biennial Report of the Attorney-General of the State of California 1904–1906.* Sacramento: W. W. Shannon, Supt. State Printing, 1906.

Woods, Mary C., and W. Ray Seiple, eds. *The Northridge, California, Earthquake of January 17, 1994.* Special Publication 116. Sacramento: California Department of Conservation, 1995.

Papers

Elders, Wilfred, et al. "Crustal Spreading in Southern California." *Science* 178 (October 1972): 15–24.

Husen, S., et al. "Changes in Geyser Eruption Behavior and Remotely Triggered Seismicity in Yellowstone National Park Produced by the 2002 M7.9 Denali Fault Earthquake, Alaska." *Geology* 32, no. 6 (June 2004): 537–40.

Moores, Eldridge. "Ultramafics and Orogeny, with Models of the U.S. Cordillera and the Tethys." *Nature* 228 (1970): 837–42.

Murphy, J. Brendan, and R. Damian Nance. "How Do Supercontinents Assemble?" *American Scientist* 92 (August 2004): 324–33.

Prentice, Carol, et al. "Northern San Andreas Fault near Shelter Cove, CA." *GSA Bulletin* 111 (April 1999): 512–23.

Prentice, Carol, and Daniel Ponti. "Coseismic Deformation of the Wrights

Tunnel during the 1906 San Francisco Earthquake: A Key to Understanding 1906 Fault Slip and 1989 Surface Ruptures in the Southern Santa Cruz Mountains, CA." *Journal of Geophysical Research* 102 (January 1997): 635–48.

——, and David Schwartz. "Re-evaluation of 1906 Surface Faulting, Geomorphic Expression and Seismic Hazard along the San Andreas Fault in the Southern Santa Cruz Mountains." *Bulletin of the Seismological Society of America* 81, no. 5 (October 1991): 1,424–79.

Sankaran, A. V. "The Supercontinent Medley: Recent Views." *Current Science* 85 (October 2003): 1,121–23.

Sieh, Kerry, and Richard H. Jahns. "Holocene Activity of the San Andreas Fault at Wallace Creek, California." *GSA Bulletin* 95 (1984): 883–96.

——, and Robert E. Wallace. "The San Andreas Fault at Wallace Creek, San Luis Obispo County, California." Cordilleran Section, *GSA Centennial Field Guide* 1987: 85–90.

Stern, Robert J. "Subduction Zones." *Reviews of Geophysics* 40 (December 2002): 1–24.

Sturz, Anne, et al. "Mud Volcanoes and Mud Pots, Salton Sea Geothermal Area, Imperial Valley, California." University of San Diego Marine and Environmental Studies Program, 1987.

Udías, Agustin, and William Stauder. "The Jesuit Contribution to Seismology." *Seismological Research Letters* 67 (May 1996): 10–19.

Zoback, Mark D., et al. "New Evidence on the State of Stress of the San Andreas Fault System." *Science* 238 (November 1987): 1,105–11.

Miscellaneous

British Consulate General, San Francisco. Correspondence file for 1906.

City of Oakland's Standardized Emergency Management System (SEMS) Emergency Plan. Oakland: City of Oakland, 2002.

Disaster Mitigation Plan for the City of Berkeley. Berkeley: City of Berkeley, 2004.

Emergency Operations Plan: City and County of San Francisco Operational Area. San Francisco: Mayor's Office of Emergency Services, 1996.

Great Earthquake April 18, 1906: Views of Its Calamitous Results in San Francisco and Vicinity. Mountain View, CA: Pacific Press, 1906.

Johnson, N. E. *Views of San Jose and Santa Clara after the Earthquake and Fire—April 18, 1906.* San Jose: N. E. Johnson, Photographer and Publisher, 1906.

Major Disaster Response Plan. Version 2.0. San Francisco: American Red Cross, Bay Area, 1999.

Moores, Eldridge. Interview by John McPhee, 2003. Transcript. Davis, CA: University of California, Pacific Regional Humanities Center.

Pierce, C.C. and Co. [photographs by]. *The Picture Story of the San Francisco Earthquake—Wednesday, April 18, 1906.* Los Angeles: Geo. Rice [n.d.]

San Francisco Directory for the Year Ending 1907. San Francisco: H. S. Crocker, 1907.

The San Francisco Disaster Photographed: Fifty Glimpses of Havoc by Earthquake and Fire. New York: C. S. Hammond, 1906.

Sonoma County Emergency Operations Plan. Santa Rosa, CA: Sonoma County Department of Emergency Services, 2000.

William Lettis and Associates. *Seismic Hazard Evaluation: Proposed Portola Valley Town Center, 765 Portola Road, Portola Valley, CA.* Walnut Creek: William Lettis, 2003.

Works Progress Administration. *The WPA Guide to California: The Federal Writers' Project Guide to 1930s California.* New York: Pantheon Books, 1984.

INDEX

Grateful acknowledgment is made to the following for the use of the illustrations and data that appear in this book: Page 31: Cleet Carlton/www.goldengate photo.com. Page 55: EERC, University of California, Berkeley. Page 60: Plate boundary data from Peter Bird, *An updated digital model of plate boundaries,* Geochemistry Geophysics Geosystems, 4(3), 1027. Volcanoes selected from the Smithsonian Institution, Global Volcanism Program. Earthquake locations from the U.S. Geological Survey, Earthquake Hazards Program. Page 67: Wolfgang Kaehler/CORBIS. Page 75: Continent coastlines and locations provided by Ronald C. Blakey, Northern Arizona University. Page 80: Continent coastlines and locations provided by John J. W. Rogers, University of North Carolina. Page 92: South Caroliniana Library, University of South Carolina, Columbia. Page 100: State Historical Society of Missouri. Page 107: Cynthia Yow. Page 135: Charles & Josette Lenars/CORBIS. Page 152: Courtesy of Eldridge Moores; photograph by Bud Turner. Page 160: Ben A. van der Pluijm. Pages 168, 183, and 190: Fault lines from California Geologic Survey, *Digital Database of Faults from the Fault Activity Map of California and Adjacent Areas*. Page 170: Tom Bean/CORBIS. Page 173: Bancroft Library, University of California, Berkeley. Page 192: Susan Hough, from her book *Finding Fault in California: An Earthquake Tourist's Guide*. Pages 202 and 262: illustrations by Laura Hartman Maestro. Page 251: Originally appeared in *Theatre Magazine*, July 1906. Page 260: Shake intensity contours derived from data provided in U.S. Geological Survey, Open-File Report 2005-1135, Version 1.0, "Modified Mercalli Intensity Maps for the 1906 San Francisco Earthquake Plotted in ShakeMap Format," by John Boatwright and Howard Bundock. Page 268: Hoen & Co., Baltimore, MD. Page 278: U.S. Geological Survey. Page 289: Ansel Adams Publishing Rights Trust/CORBIS. Page 327: Heath Lambert Group. Page 340: Flower Pentecostal Heritage Center. Page 354: *Sunset* magazine. Page 379: Bill Perkins/Shannon & Wilson, Inc. Page 398: Pennsylvania Geological Survey. Page 402: From F. M. Gradstein and J. Ogg, "A Phanerozoic Time Scale," *Episodes* 19, nos. 1 and 2 (1996). Pages 205, 221, 226–227, 234, 300, 306, 309, 349, 384: CORBIS. Pages 123, 145, 146, 281, 301: Bettmann/CORBIS.

Insights,
Interviews
& More . . .

About the author

About the book

Read on

About the author

Meet Simon Winchester

Bettina Strauss

AUTHOR, JOURNALIST, AND BROADCASTER Simon Winchester has worked as a foreign correspondent for most of his career. Before joining his first newspaper in 1967, however, he graduated from Oxford with a degree in geology and spent a year working as a geologist in the Ruwenzori Mountains in western Uganda and on oil rigs in the North Sea.

His journalistic work, mainly for the *Guardian* and the *Sunday Times*, has seen him based in Belfast; Washington, D.C.; New Delhi; New York; London; and Hong Kong, where he covered such stories as the Ulster crisis, the creation of Bangladesh, the fall of President Marcos, the Watergate affair, the Jonestown Massacre, the assassination of Egypt's President Sadat, the death and cremation of Pol Pot, and the 1982 Falklands War. During the Falklands conflict he was arrested and spent three months in prison in Ushuaia, Tierra del Fuego, on spying charges. Winchester has been a freelance writer since 1987.

“During the Falklands War, Winchester was arrested and spent three months in prison in Ushuaia, Tierra del Fuego, on spying charges.”

He now works principally as an author, although he contributes to a number of

American and British magazines and journals, including *Harper's, Smithsonian, National Geographic,* the *Spectator, Granta,* the *New York Times,* and the *Atlantic Monthly.* He was appointed Asia-Pacific editor of *Condé Nast Traveler* at its inception in 1987, and later became editor-at-large. His writing has won him several awards, including British Journalist of the Year.

He writes and presents television films on a variety of historical topics—including a series on the final years of colonial Hong Kong—and is a frequent contributor to the BBC radio program *From Our Own Correspondent.* Winchester also lectures widely—most recently before London's Royal Geographical Society (of which he is a Fellow) and to audiences aboard the cruise liners *QE2* and *Seabourn Pride.*

His books cover a wide range of subjects: the remnants of the British Empire; the colonial architecture of India; aristocracy; the American Midwest; his months in an Argentine prison on spying charges; his description of a six-month walk through the Korean Peninsula; and the Pacific Ocean and the future of China. More recently he has written *The River at the Center of the World,* about China's Yangtze River; the best-selling *The Professor and the Madman,* which is to be made into a major motion picture by distinguished French director Luc Besson; *The Fracture Zone: My Return to the Balkans,* which recounts his journey from Austria to Turkey during the 1999 Kosovo crisis; and the bestselling *The Map That Changed the World,* about the nineteenth-century geologist William Smith. His latest books, *Krakatoa: The Day the World Exploded: August 27, 1883* (April 2003) and *A Crack in the Edge of the World: America and the Great California Earthquake of 1906* (October 2005), have both been *New York Times* bestsellers and appeared on numerous best of and notable lists. ▶

Meet Simon Winchester ***(continued)***

Simon Winchester lives in New York City and has a small farm in the Berkshires in Massachusetts. Mr. Winchester was made an Officer of the Order of the British Empire (OBE) by Her Majesty The Queen in 2006. He received the honor in a ceremony at Buckingham Palace.

"Mr. Winchester was made an Officer of the Order of the British Empire (OBE) by Her Majesty The Queen in 2006."

A Conversation with Simon Winchester

The following interview was conducted in July 2005.

Your last book was about the explosion of the Krakatoa volcano in the late nineteenth century. Before that you wrote about William Smith and the geological "map that changed the world." Should we see a connection between these books and your new one?

Well, I really suppose so. I studied geology in Oxford back in the 1960s; I wouldn't say that I'd forgotten it all but I'd certainly not been a geologist. I did work in that field one year in Uganda (1967), and then essentially abandoned it and went off and worked as a foreign correspondent and did all sorts of other things. But after I wrote a book called *The Professor and the Madman* about seven or eight years ago, we were casting around wondering what to do next. My then editor, Larry Ashmead, said, "Was there anyone fascinating from the world of geology?"; oddly enough, Larry had been a geologist in his youth before he began working as a publisher. I remembered this chap William Smith, who was the creator of the first-ever geological map, and it did seem that his life was interesting. Not only had he created this map, but it had been plagiarized; he lost a lot of money, went bankrupt, and was sent to debtors' prison. His wife went mad. She turned into a raging nymphomaniac. The story had all the sort of ingredients that made *The Professor and the Madman* so fascinating.

It was agreed that I'd write that book, and it whetted my appetite for the geology I'd abandoned all those . . . I mean, four decades before. *The Map That Changed the World,* the story of William Smith, did rather well. I ▸

“I studied geology in Oxford back in the 1960s; I wouldn't say that I'd forgotten it all but I'd certainly not been a geologist.”

A Conversation with Simon Winchester

(continued)

“Not only is it a great story illustrating all the new geology (and how much it's changed since I studied it), but April 2006 is of course the one hundredth anniversary of the earthquake.”

thought, My word, geology is not, so far as the reading public is concerned, as dull as I had anticipated. So Larry and I had a discussion. I said, “Well you know, if geology is setting people on fire, then there's a story I'd love to tell.” This was the story of the Krakatoa eruption. *Krakatoa* followed *The Map That Changed the World* as the second geological book and also, by good fortune, did quite well. Then I thought, Well my gosh, let's go for broke, because there's another fantastically good story—the 1906 San Francisco Earthquake—which has all the elements that underpin the geology making Krakatoa so interesting. And not only is it a great story illustrating all the new geology (and how much it's changed since I studied it), but April 2006 is of course the one hundredth anniversary of the earthquake. All the curves coincided to make it seem a pretty decent idea to write a book about San Francisco.

So, yes, the three books are interconnected. I suppose they reflect my real enthusiasm and excitement and reconnection with a field of study that I had so long ago abandoned. And the one person who's particularly pleased by all of this is my old tutor Harold Reading at Oxford, because he knew I'd become a journalist and an author, and I think he was ever so slightly disappointed. He thought he had taught me well enough and that I would remain enthusiastic. When he heard I was writing about William Smith he was thrilled, and helped me so enormously that I dedicated the book to him. He liked *Krakatoa* and I very much hope that he is going to like the new book on San Francisco.

courtesy of the author

Simon Winchester at Meteor Crater, Arizona, on research for *A Crack in the Edge of the World.*

I'm sure he'll love it. One of the most fascinating things about your book is how

you place the earthquake in historical context. Tell us a little bit about how this particular earthquake, the 1906 earthquake, affected the future of San Francisco, the state of California, and (for that matter) the rest of the twentieth century in America?

A very complicated series of things seems to have happened. For instance, it drove all the artists out of San Francisco. This city had been famous for Mark Twain, Jack London, Ambrose Bierce, and all these luminaries and Bohemians. All these people suddenly scuttled off to the hills, and for a good twenty, thirty, or forty years San Francisco was a cultural wilderness. But that, of course, isn't one of the truly important things that happened. More importantly, people started to take earthquakes seriously and engineer their buildings in a stronger fashion. So the earthquake had an effect on the way the city was rebuilt and on preparations for the future, so that should another disaster happen they would be a little bit better able to cope with it. It had interesting sociological consequences too. One of the things that fascinated me, not least because it also seems to have happened in the aftermath of the Krakatoa eruption, was the effect it had on religion. After Krakatoa, you had a lot of Muslims saying this is clearly a sign from Allah. This volcano is a sign that he's angry. We must rise up and kill our rulers, the Dutch, and drive them out. And essentially they did. One might argue that Krakatoa triggered the first militant Islamic fundamentalist uprising in the world—a long, long time before the creation of Israel in 1948, and all those things. A similar thing happened in California, not with Muslims but with fundamentalist Christians. There was a church down in Los Angeles in a place called ▸

“One might argue that Krakatoa triggered the first militant Islamic fundamentalist uprising in the world. A similar thing happened in California, not with Muslims but with fundamentalist Christians.”

A Conversation with Simon Winchester
(continued)

> "The power of the Christian right and particularly the Pentecostal brand of Evangelicals has had a crucially important effect on contemporary American politics. This movement was triggered in large part by what was perceived as a sign from God on April 18, 1906."

Azusa Street. This was a fledgling church of people who called themselves Pentecostalists. They spoke in tongues, they waved their arms around, and did all sorts of crazy things—things that would appear to others as crazy. And that sort of direction came about because of what they saw as manifestations from God. He would send them signs. Miracles would be proclaimed. People would, as I mentioned, talk in tongues. On the week before the San Francisco earthquake this little church had a modest-sized meeting, a couple of people spoke in tongues, and it was all going along quite nicely. But the pastor stood up and said, "We are expecting a sign from the Lord."

Three days later San Francisco, arguably the most sinful of all American cities and given over to drinking, whoring, gambling, and all those fun things that prevailed in the aftermath of the Gold Rush days, was destroyed by an earthquake. So the Pentecostalist pastor, not unjustifiably, said "Well, there's no doubt about it, this is the sign from God we've been waiting for." And suddenly this little church was overrun with people. I mean tens of thousands of people came, they had to have overspill locations. It became like the Crystal Cathedral that you see in Los Angeles today; the comparison is not actually an unreasonable one to make, because out of the Pentecostalist Church that essentially began in 1906 came all the great American Evangelical figures, from Aimee Semple McPherson right through to Pat Robertson and Tammy Faye and Jim Bakker. One might argue—and I don't want to make too much of this—that the power of the Christian right and particularly the Pentecostal brand of Evangelicals has had a crucially important

Bancroft Library

The ruins of St. Francis Church, 1906. It was later restored and still stands.

effect on contemporary American politics. This movement was triggered in large part by what was perceived as a sign from God on April 18, 1906. So the downstream effects of the San Francisco Earthquake are Pentecostalism, conservative Christianity, and certain political ramifications that are today being felt around the world.

It does underpin the overarching theory of this book, the Gaia theory, which holds that the entire world is an interconnected system where everything leads to everything else. To think that an earthquake in San Francisco had an effect on global politics today may be stretching credulity, but to a geologist it's not completely fanciful.

> "The year 2004 was a very bad seismic year. Lots of people died, lots of places were destroyed. And 1906 was exactly the same—uncannily the same."

Sounds reasonable. This past year we've seen a whole string of geological events, earthquakes in particular, from the tsunami in South Asia to the earthquake in Iran. What do you think we can learn from the year 1906 to help us understand these recent occurrences?

You know, that's one of the most interesting things. Here you had on Boxing Day—well, I'm an Englishman, December 26—2003 the devastating earthquake in Bam in Iran; on December 29, 2004, almost to the hour a year later, you had this devastating tsunami-causing earthquake off the north coast of Sumatra. Between those two bookends of calamity you had a year full of seismic happenings. Mount St. Helens was erupting, you had earthquakes in Peru, Chile, Taiwan, China, and Japan. I mean it was a very bad seismic year. Lots of people died, lots of places were destroyed. Such years happen infrequently, but they do happen.

And 1906 was exactly the same— ▸

A Conversation with Simon Winchester
(continued)

uncannily the same. We tend to think of it as merely being the year in which the San Francisco event occurred, but it began with a terrible earthquake off the coasts of Colombia and Ecuador in which thousands died. There was also an earthquake in Saint Lucia and an earthquake in Taiwan that was very big—eight or nine thousand people died in that one. There was an earthquake in the Caucasus. Vesuvius was erupting. Then came the San Francisco Earthquake. As if that wasn't enough, the port of Valparaiso in Chile was leveled by an earthquake in August with twenty thousand people killed.

Why are certain years so seismically active? It's beginning to seem (ever since the theory of plate tectonics was put forward) that it's not unreasonable to suppose a sort of butterfly effect happens. You get a terrible and devastating event on a plate up in the northern hemisphere, let's say in Alaska; the event triggers a sort of cascade of events all around the world. Something can happen in Sumatra, and immediately after the Sumatran event 144 earthquakes occur on exactly the opposite side of the world in northern Alaska. The world is sort of like a big brass bell; if you hit it *really* hard the whole thing vibrates, and on the far side of the world other devastating, seemingly unconnected events can happen. It seems to have occurred in 2004. It certainly seems to have occurred in 1906, and scientists are now fascinated with the idea of this seismic butterfly effect.

Before the Flood

When Hurricane Katrina destroyed New Orleans in late August 2005, much of the debate in the aftermath centered on the inadequate public response by local and federal authorities. In a New York Times *op-ed column a week after the hurricane struck, Simon Winchester remarked on the comparative response in San Francisco in April 1906.*

THE LAST TIME a great American city was destroyed by a violent caprice of nature, the response was shockingly different from what we have seen in New Orleans. In tone and tempo, residents, government institutions, and the nation as a whole responded to the earthquake that brought San Francisco to its knees a century ago in a manner that was well-nigh impeccable, something from which the country was long able to derive a considerable measure of pride.

This was all the more remarkable for taking place at a time when civilized existence was a far more grueling business, an age bereft of cell phones and Black Hawks and conditioned air, with no Federal Emergency Management Agency to give us a false sense of security and no Weather Channel to tell us what to expect.

Nobody in the "cool gray city of love," as the poet George Sterling called it, had the faintest inkling that anything might go wrong on the early morning of April 18, 1906. Enrico Caruso and John Barrymore—who both happened to be in town—and 400,000 others slumbered on, with only a slight lightening of eggshell blue in the skies over Oakland and the clank of the first cable cars suggesting the beginning of another ordinary day.

Then, at 5:12 A.M., a giant granite hand ▸

"In tone and tempo, residents, government institutions, and the nation as a whole responded to the earthquake that brought San Francisco to its knees a century ago in a manner that was well-nigh impeccable."

“People picked themselves up, dusted themselves off, took stock, and took charge.”

rose from the California earth and tore through the city. Palaces of brick held up no better than Gold Rush shanties of pine and redwood siding; hot chimneys, electric wires, and gas pipes toppled, setting a series of fires that, with the water mains broken and the hydrants dry, proceeded over the next three dreadful days and nights to destroy what remained of the imperial city. In the end, at least three thousand were dead and two hundred twenty-five thousand were left homeless.

Everyone who survived remembered: there was at first a shocked silence; then the screams of the injured; and then, in a score of ways and at a speed that matched the ferocity of the wind-whipped fires, people picked themselves up, dusted themselves off, took stock, and took charge.

Huntington Library

Golden Gate Park, 1906.

A stentorian army general named Frederick Funston realized he was on his own—his superior officer was at a daughter's wedding in Chicago—and sent orders to the Presidio military base. Within two hours, scores of soldiers were marching into the city, platoons wheeling around the fires, each man with bayonet fixed and twenty rounds of ball issued; they presented themselves to Mayor Eugene Schmitz by 7:45 A.M.—just 153 minutes after the shaking began.

The mayor, a former violinist who had previously been little more than a puppet of the city's political machine, ordered the troops to shoot any looters, demanded military dynamite and sappers to clear firebreaks, and requisitioned boats to be sent to the Oakland telegraph office to put the word out over the wires: "San Francisco is in ruins," the cables read. "Our city needs help."

America read those wires and dropped everything. The first relief train, from Los Angeles, steamed into the Berkeley marshaling yards by eleven o'clock that night. The navy and the Revenue Cutter Service, like the army not waiting for orders from back East, ran fire boats and rescue ferries. The powder companies worked overtime to make explosives to blast wreckage.

Washington learned of the calamity in the raw and unscripted form of Morse code messages, with no need for the interpolations of anchormen or pollsters. Congress met in emergency session and quickly passed legislation to pay all imaginable bills. By 4:00 A.M. on April 19, William Taft, President Theodore Roosevelt's secretary of war, ordered rescue trains to begin pounding toward the Rockies; one of them, originating in Virginia, was the longest hospital train ever assembled.

Millions of rations were sped to the city from Oregon and the Dakotas; within a week virtually every military tent in the army quartermaster general's stock was pitched in San Francisco; within three weeks some ten percent of America's standing army was on hand to help police and firefighters (whose chief had been killed early in the disaster) bring the city back to its feet.

To the great institutions go the kudos of history, and rightly so. But I delight in the lesser gestures, like that of largely forgotten San Francisco postal official Arthur Fisk, who issued an order on his personal ▸

Before the Flood *(continued)*

recognizance that no letter posted without a stamp and clearly coming from the hand of a victim would go undelivered for want of fee. Thus did hundreds of the homeless of San Francisco let their loved ones know of their condition—a courtesy of a time in which efficiency, resourcefulness, and simple human kindness were prized in a manner we'd do well to emulate today.

Originally published in the New York Times *on September 8, 2005.*

> "William Taft, President Theodore Roosevelt's secretary of war, ordered rescue trains to begin pounding toward the Rockies; one of them, originating in Virginia, was the longest hospital train ever assembled."

Have You Read?

More by Simon Winchester

KRAKATOA: THE DAY THE WORLD EXPLODED: AUGUST 27, 1883

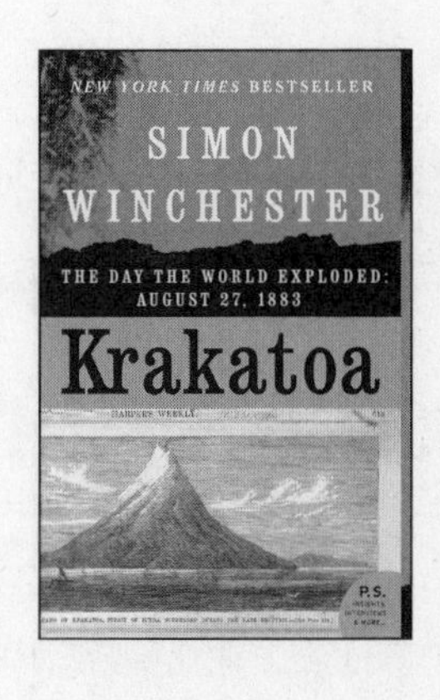

Simon Winchester details the legendary annihilation of the volcano Krakatoa in 1883 and its lasting world-changing effects, including the creation of an immense tsunami, the release of dust that swirled around the world for years, and the triggering of a wave of murderous anti-Western militancy by fundamentalist Muslims in Java.

"One of the best books ever written about the history and significance of a natural disaster."
—*New York Times Book Review*

"A real-life story bigger than any Hollywood blockbuster." —*Entertainment Weekly*

THE MAP THAT CHANGED THE WORLD: WILLIAM SMITH AND THE BIRTH OF MODERN GEOLOGY

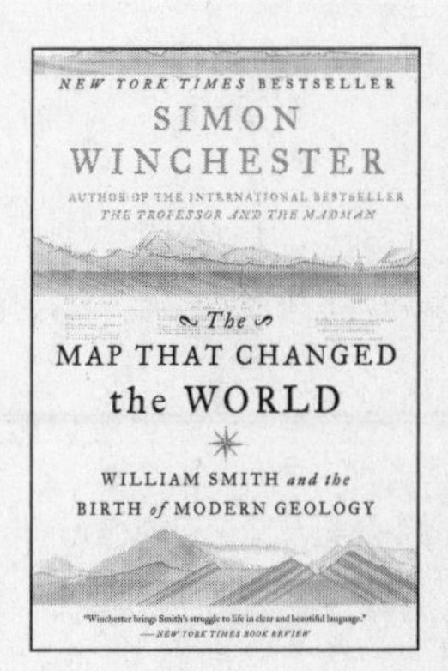

William Smith, the orphaned son of an English country blacksmith, dug canals for a living. From one particular dig emerged his obsession to create the world's first geological map. Known today as the father of modern geology, Smith spent twenty-two years researching and creating his epochal and remarkably beautiful hand-painted map. Instead of receiving accolades and honors, however, he ended up the victim of plagiarism; Smith landed in debtors' prison and was virtually homeless for ten years ▸

more. In 1831 this quiet genius finally received the Geological Society of London's highest award and was offered a lifetime pension by King William IV.

"Winchester brings Smith's struggle to life in clear and beautiful language."
—*New York Times Book Review*

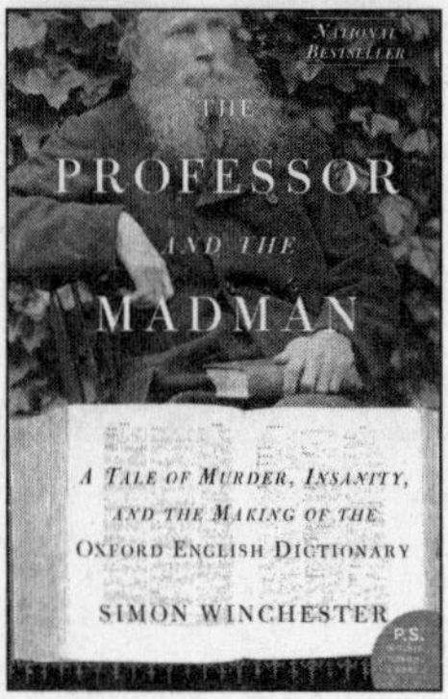

THE PROFESSOR AND THE MADMAN: A TALE OF MURDER, INSANITY, AND THE MAKING OF THE *OXFORD ENGLISH DICTIONARY*

The best-selling tale of madness, genius, and the incredible obsessions of two remarkable men that led to the making of the *Oxford English Dictionary*—one of the most ambitious projects ever undertaken. As word definitions were collected, the overseeing committee led by Professor James Murray discovered that one man, Dr. W. C. Minor, had submitted more than ten thousand. When the committee insisted on honoring him, a shocking truth came to light: Dr. Minor, an American Civil War veteran, was also an inmate at an asylum for the criminally insane.

"An extraordinary tale and Simon Winchester could not have told it better. . . . A splendid book."
—*Economist*

"The linguistic detective story of the decade."
—William Safire, *New York Times Magazine*

OUTPOSTS: JOURNEYS TO THE SURVIVING RELICS OF THE BRITISH EMPIRE

Simon Winchester, struck by a sudden need to discover exactly what was left of the British Empire, set out across the globe to visit the far-flung islands that are all that remain of what once made Britain great. He traveled thousands of miles to capture a last glint of imperial glory.

"A brilliant and delightful addition to the long and distinguished shelf of British literary odysseys." —Christopher Buckley, *Washington Post Book World*

"Winchester traveled one hundred thousand miles back and forth from Antarctica to the Caribbean, from the Mediterranean to the Far East, and has come up with a fascinating and important book." —*Times* (London)

THE FRACTURE ZONE: MY RETURN TO THE BALKANS

A true portrait of one of the world's most chaotic and beautiful regions that explains why violence has always occurred there—and why it may continue to occur there for years to come.

"A vivid, informative history of the Balkans." —*Chicago Tribune*

"Scholarly and moving . . . combines historical significance with dramatic insight." —*Independent*

Have You Read? *(continued)*

KOREA: A WALK THROUGH THE LAND OF MIRACLES

Fascinating for its vivid presentation of historical and geographic detail, *Korea* is that rare book that actually defines a land and its people, while providing Winchester's gift for writing about engaging characters in true, compelling stories.

"Immensely readable. . . . Winchester made his journey of over three hundred miles on foot, a remarkable achievement in itself and one that afforded him a unique opportunity to experience both the country and its people at a grassroots level." —*Guardian*